Programming Microcontrollers in C

Second Edition

Ted Van Sickle

A Volume in the EMBEDDED TECHNOLOGY™ Series

Newnes

Amsterdam   Boston   London   New York   Oxford   Paris
San Diego   San Francisco   Singapore   Sydney   Tokyo

Library of Congress Cataloging-in-Publication Data

ISBN: **1-878707-57-4**

British Library Cataloguing-in-Publication Data
A catalogue record for this book is available from the British Library.

The publisher offers special discounts on bulk orders of this book.
For information, please contact:

Manager of Special Sales
Elsevier Science
200 Wheeler Road
Burlington, MA 01803
Tel: 781-313-4700
Fax: 781-313-4880

For information on all Newnes publications available, contact our World Wide Web home page at: http://www.newnespress.com

10 9 8 7 6 5 4 3 2

Printed in the United States of America

Contents

Introduction   *vii*

What's on the CD-ROM?   *xv*

CHAPTER ONE: Introduction to C 1

  Some Simple Programs 1
  Names 8
  Types and Type Declarations 9
  Storage Classes, Linkage, and Scope 12
  Character Constants 15
  Arrays 18
  Other types 20
  Operators and Expressions 24
  Increment and Decrement Operators 30
  Precedence and Associativity 34
  Program Flow and Control 36
  Functions 51
  Recursion 61
  Summary 63

CHAPTER TWO: Advanced C Topics 65

  Pointers 65
  Multidimensional Arrays 80
  Structures 87
  More Structures 107
  Input and Output 110
  Memory Management 114
  Miscellaneous Functions 116
  Summary 121

CHAPTER THREE: What Are Microcontrollers? 123

  Microcontroller Memory 127
  Input/Output 129

Programming Microcontrollers 134
Coding Tips for Microcontrollers 137

CHAPTER FOUR: Small 8-Bit Systems 149

Microcontroller Memory 153
Timers 166
Analog-to-Digital Converter Operation 195
Pulse Width Modulator System 201
Other Program Items 207
Summary 209

CHAPTER FIVE: Programming Large 8-Bit Systems 211

Header File 211
Sorting Programs 230
Data Compression 237
Timer Operations 245
Summary 285

CHAPTER SIX: Large Microcontrollers 287

The MC68HC16 288
System Integration Module (SIM) 296
A Pulse Width Modulation Program 299
Cosmic MC68HC16 Compiler 305
Table Look-Up 319
Digital Signal Processor Operations 326
Other MC68HC16 Considerations 345

CHAPTER SEVEN: Advanced Topics in
Programming Embedded Systems (M68HC12) 347

Numeric Encoding 352
Numeric Decoding 354
Coding the alpha data 356
The Monitor Program 370
The SAVEIT() Routine 376
The printout() and the printafter() Functions 378
Reset 381

Input/Output Functions 382
Putting It All Together 386
Summary 391

CHAPTER EIGHT: MCORE, a RISC Machine *393*
Delay Routine 395
Delays Revisited 401
Serial Input/Output 404
Handling Interrupts 413
A Clock Program 419
Keyboard 432
Integrating Keyboard and Clock 440
Adding a Display 442
Summary 446

Index 447

Introduction to Second Edition

Today, even more than when the first edition of this book was written, the use of microcontrollers has expanded to an almost unbelievable level. A typical car has 15 microcontrollers. A modern home can have more than 50 microcontrollers controlling everything from the thermostat, to the furnace, to the microwave. Microcontrollers are everywhere! In the meantime, many new chips have been placed on the market as well.

Also, there have been significant modifications to our programming languages. The standard C language is now called C99 rather than C89. There have been several changes in the language, but most of these changes will not be available to us for some time. Many of the modifications to the language will be of little use to programs for embedded systems. For example, complex arithmetic has been added to the language. It is rare that we use even floating-point arithmetic, and I have never seen an application for an embedded system where complex arithmetic was needed. However, other additions allow improved optimization processes, such as the restrict keyword and the static keyword used to modify the index of an array. Other changes have less impact on the generation of code, such as the // opening to a single line comment. Also, today you will have no implicit int return from a function. All in all, expect the new versions of C compilers to be significant improvements over the older versions. Also, expect that the new compilers will not break older code. The features of the new standard should begin showing up in any new version of the compilers that you use.

The C++ standard committee has completed its work on the first language standard for C++. There is much clamor about the use of C++ for embedded systems. C++ as it stands is an excellent language, but it is aimed primarily at large system programs, not the small programs that we will be developing into the future. C still remains the grand champion at giving us embedded programmers the detailed control over the computer that we need and that other computer languages seem to overlook.

The first six chapters of the book have been revised and any errors that were found were corrected. Every chapter has been altered, but not so much that you would not recognize it. Chapter 7 has been added. In that chapter, a relatively complex program is developed to run on the M68HC912B32. The development system was based on this chip and it had no significant RAM to hold the code during development. Therefore, all of the code was completely designed, coded, and tested on a DOS-based system. Extensive tests were completed to make certain that there were no hidden bugs. The modules were small and easy to test. Each module was tested with a program written to exercise all parts of the module. When the several modules were integrated into a single program, the program worked in the DOS-based system. All changes needed to convert this program were implemented under the control of conditional compiler commands. When the program was converted to the M68HC12 version and compiled, it loaded correctly and ran.

Chapter 8 introduces a new chip for Motorola, the MMC2001. This chip is a RISC chip. Many of the good things to be said of RISC configurations are absolutely true. This chip is very fast. Each of its instructions requires only one word, 32 bits, of memory. Almost all instructions execute in a single clock cycle. The chip that I used here ran at 32 mHz, and you could not feel any temperature rise on the chip. It is from a great family of chips that should become a future standard.

The first edition of this book had several appendices. These were needed to show general background material that the reader should not be expected to know. Also, quite a few specialized header files used to interconnect the program to the peripheral components on the microcontroller were included. Also, with the first edition, there was a card with which the reader could order two diskettes that contained all of the source code in the book, demonstration compilers that would compile the source code, and other useful information. All of these things have been included on the CD-ROM that comes with this edition. Additionally, you will find PDF versions of all appropriate Motorola data manuals and reference manuals for all of the chips discussed in the book. Also included are copies of all header files used with the programs, and some more that will probably be useful to you.

Introduction to First Edition

Early detractors of the C language often said that C was little more than an over-grown assembler. Those early disparaging remarks were to some extent true and also prophetic. C is indeed a high level language and retains much of the contact with the underlying computer hardware that is usually lost with a high level language. It is this computer relevance that makes people say that C is a transform of an assembler, but this computer relevance also makes C the ideal high level language vehicle to deal with microcontrollers. With C we have all of the advantages of an easily understood language, a widely standardized language, a language where programmers are readily available, a language where any trained programmer can understand the work of another, and a language that is very productive.

The main purpose of this book is to explore the use of C as a programming tool for microcontrollers. We assume that you are familiar with the basic concepts of programming. A background in C is not necessary, but some experience with a programming language is required. I have been teaching C programming for microcontrollers for several years, and have found that my students are usually excellent programmers with many years of experience programming microcontrollers in assembly language. Most have little need or interest in learning a new language. I have never had a class yet where I was able to jump into programming microcontrollers without providing substantial background in the C language. In many instances, students believe that a high-level language like C and microcontrollers are incompatible. This forces me, unfortunately, to turn part of my class into a sales presentation to convince some students that microcontrollers and C have a future together. I am usually able to show that the benefits gained from using C far outweigh the costs attributed to its use. The first two chapters are included for those who are unfamiliar with C. If you are already familiar with C, feel free to skip ahead to Chapter 3.

C is a very powerful high level language that allows the programmer access to the inner workings of the computer. Access to computer details, memory maps, register bits, and so forth, are not usually available with high level languages. These features are hidden deliberately from the programmer to make the languages universal and portable between machines. The authors of C decided that it is desirable to have access to the heart of the machine because it was intended to use C to write operating systems. An operating system must be master of all aspects of the machine it is controlling. Therefore, no aspect of the machine could be hidden from the programmer. Features like bit manipulation, bit field manipulation, direct memory addressing, and the ability to manipulate function addresses as pointers have been included in C. All of these features are used in programming microcontrollers. In fact, C is probably the only popular high level language that can be conveniently used for a microcontroller.

Every effort has been made to present the C aspects of programming these machines clearly. Example programs and listings along with their compiled results are presented whenever needed. If there are problems hidden in the C code, these problems are explored and alternate methods of writing the code are shown. General rules that will result in more compact code or quicker execution of the code are developed. Example programs that demonstrate the basis for these rules will be shown.

C is a rich and powerful language. Beyond the normal high level language capability, C makes extensive use of pointers and address indirection that is usually available only with assembly language. C also provides you with a complete set of bit operations, including bit manipulations and bit fields in addition to logical bit operations. In C, the programmer knows much about the memory map which is often under programmer control. A C programmer can readily write a byte to a control register of a peripheral component to the computer. These assembly language-like features of the C language serve to make C the high level language of choice for the microcontroller programmer.

As a language, C has suffered many well-intended upgrades and changes. It was written early in the 1970s by Dennis Ritchie of Bell Laboratories. As originally written, C was a "free wheeling" language with few constraints on the programmer. It was assumed that any programmer using the language would be competent, so there was little need for the controls and hand-holding done by popular compilers of the day. Therefore, C was a typed language but it was not strongly typed. All function returns were assumed to be integer unless otherwise specified. Function arguments were typed, but these types were never checked for validity when the functions were called. The programmer could specify an integer argument and then pass a floating point number as the argument. These kinds of errors are made easily by the best programmer, and they are usually very difficult to find when debugging the program.

Another set of problems with the language was the library functions that always accompanied a compiler. No standard library was specified. C does not have built-in input/output capability. Therefore, the basic C standard contained the specifications for a set of functions needed to provide sensible input/output to the language. A few other features such as a math library, a string handling library, and so forth started out with the language. But these and other features were included along with other enhancements in a helter-skelter manner in different compilers as new compiler versions were created.

In 1983, an ANSI Committee (The X3J11 ANSI C Standards Committee) was convened to standardize the C language. The key results of the work of this committee has been to create a strongly typed language with a clear standard library. One of the constraints that the ANSI committee placed upon itself was that the existing base of C code must compile error free with an ANSI C compiler. Therefore, all of the ANSI dictated typing requirements are optional under an ANSI C compiler. In this text, it is always assumed that an ANSI compliant compiler will be used, and the ANSI C form will be used throughout.

C compilers for microcontrollers—especially the small devices— must compromise some of the features of a compiler for a large computer. The small machines have limited resources that are required to implement some of the code generated by a compiler for a large computer. When the computer is large, the compiler writer need not worry about such problems as limited stack space or a small register set. But when the computer is small, these limitations will often force the compiler writer to take extraordinary steps just to be able to have a compiler. In this book, we will discuss the C programming language, not an abbreviated version that you might expect to use with some of the smaller microcontrollers. In the range of all microcontrollers, you will find components with limited register sets, memory, and other computer necessary peripherals. You will also find computers with many megabytes of memory space, and all of the other important computer features usually found only on a large computer. Therefore, we will discuss the language C for the large computer, and when language features must be abbreviated because of computer limitations, these points will be brought out.

All of the programs found in this book have been compiled and tested. Usually source code that has been compiled has been copied directly from computer disks into the text so that there should be few errors caused by hand copying of the programs into the text. The compilers used to test these programs are available from Byte Craft Ltd. of Hamilton, Ontario, Canada (for the MC68HC05) and Intermetrics of Cambridge, Massachusetts (for the MC68HC11 and MC68HC16). If you wish to develop serious programs for any of these microcontrollers, you should purchase the appropriate compiler from the supplier.

How does one partition a book on C programming for microcontrollers? First, the text must contain a good background on the C language. Second, it is necessary to include a rather extensive background on some microcontrollers. Finally, C must be used to demonstrate the creation of code for the specified microcontrollers. This approach is used here. The C background is complete. The background on the chosen microcontrollers is presented briefly, as this book is not intended to be a text on microcontrollers. Therefore, the chapters that cover specific microcontrollers are to the point. The references found in each chapter contain texts and data books that will cover the various microcontrollers discussed. This book grew out of my teaching activities, so chapters include several exercises suitable for classroom as well as individual use. The only way to learn programming is to program, and the exercises are designed to let you put the material in each chapter to use in typical microcontroller programming situations.

Chapters 1 and 2 contain a background on ANSI C. Data in these chapters is basic to all C programs. There is no specific coverage for microcontroller programming. Chapter 3 contains a brief background on microcontrollers, and it also contains general programming guidelines that should be used when writing code for microcontrollers.

Chapter 4 is devoted to writing programs for the MC68HC05 family. In this chapter, the use of microcontroller specific header files is introduced. These header files are written for a specific part, and must be included in any program for the part.

In Chapter 5 you will find techniques for programming the MC68HC11 family of parts. Several of the peripherals on these parts are examined, and code to access these peripherals is written.

More complex microcontrollers are found in the MC68HC16 and the MC68300 families. Programming the MC68HC16 is discussed in Chapter 6. This part contains an internal bus with several peripherals placed on this bus. Access to these peripherals is through memory mapped registers and how these peripherals are accessed will be found in Chapter 6.

There are several appendices. Appendix A contains several header files that are useful in programming MC68HC05 programs. Appendix B contains some code that demonstrates the power of the types defined by structures, and how these types can be made into very convenient new types by the typedef keyword.

One of the advantages of a high level language is that it isolates the programmer from the details of the computer being programmed. There are both plusses and minuses to this idea. First, as a programmer, you do not need to know details of the register map and the programmers model of the computer being programmed because the language takes care of these details for you. On the other hand, microcontrollers all have peripherals and other components that must be accessed by the program. The programmer must be able to write C code that will set and reset bits and flags in control registers for these parts. It would be desirable to write this book with no detailed discussion of the insides of the microcontrollers you will be programming; however, I could not do it. I needed a careful discussion of the ways peripheral components are used. Appendix C and Appendix E contain detailed descriptions of the MC68HC11 and the MC68HC16 family parts respectively. I am particularly indebted to Motorola Semiconductor Products, Inc. for the contents of Appendix E. This Appendix is a very slightly modified version of the Appendix D found in the MC68HC16Z1 users manual.

Appendix C contains a header file for the MC68HC11Ex series, and Appendix F contains several header files needed to program the MC68HC16 components.

This book has taken entirely too much time to write. As the author, it is my fault, and I have been a burden to those around me while I have labored on this task. The basis for the text comes from about three years of teaching classes on programming microcontrollers in C. This class has been taught as a three or four day course, mainly to Motorola customers. I am amazed that it is possible to learn from every class that I teach. During the time I have been writing, I have learned object oriented programming and the C++ language, and I have also taught classes on this subject. It is difficult to move from one language to another, especially languages with similar roots like C and C++, and not get them mixed up. I am comfortable that this book is on C without C++ spilling into the material.

I have received much help in writing this book. My dear wife, who understands nothing about computers, has read most of the book and made comments about the contents. If this text is more readable than usual, it is her contribution. Any problems that you find are my responsibility entirely.

Motorola has provided me much time and support that I appreciate. Most of the photographs found in the book are from Motorola files. My manager, Neil Krohn, has encouraged me at every phase in the preparation of this manuscript. Neil and Motorola deserve my heartfelt thanks.

What's on the CD-ROM?

Programs

The programs on this CD-ROM will help you learn how to program small embedded control systems. The directory named *Programs* contains all of the programs from the book. Programs from each chapter are grouped together in directories named Chapter1, Chapter2, etc., where the number corresponds to the book chapter in which the code is found. The subdirectory *Header~1*, or *Header Files*, contains a series of directories that contain the specific header files needed to connect your compiled code to the peripherals found on the indicated chips. These header files have been used extensively, but you will probably still find an occasional bug in them. If you do find a bug, please notify me at the email address below.

There are demonstration compilers for the M6805, the M68HC11, and the M68HC16 families of chips. The Byte Craft Limited compiler is placed in directory *C6805*. Instructions for use of this compiler can be obtained by merely typing \c6805\c6805 with no arguments and the instruction sheet will appear.

The two Intermetrics demonstration compilers are placed in *HC11DEMO* and *HC16DEMO* respectively. When using one of these compilers, the directory name should be placed in the system path. Only one of the demo directories should be in the path at a time because the two compilers both use the same function names. Confusion will reign if both directories are in the path at the same time. In the *Software* directory, you will find files named HC16BOOK.TXT and HC11BOOK.TXT. These files are transcriptions of the books normally shipped with the Demo Kit packages from Intermetrics. There is no convenient means to copy the several figures found in these books into these ASCII files. Therefore, the files are complete with the exception of the figures. The text describes the contents of the figures. I am sorry for any inconvenience caused by these necessary omissions. Also, the contents of these books contain discussions of how you should install the various programs contained in the Demo Kits. These compilers are already installed on the CD-ROM, but the basic programs from which they are installed are found in the directories *HC16* and *HC11*. You can reinstall these demonstration compilers from the programs in these directories if you wish.

Intermetrics no longer supports the compilers found on the CD-ROM. If you wish continued support with these compilers, you should contact COSMIC Software at

Cosmic Software
400 W. Cummings Park STE6000
Woburn, MA 01801-6512
781 932 2556 x 15

Motorola Reference Manuals and Data Manuals

The CD-ROM contains full copies of several Motorola M68HC11 reference manuals and data manuals, along with similar information for the M68HC05, M68HC08, M68HC12, M68HC16, and M683XX family of chips, and the MCORE family. These reference materials have been provided with the permission of Motorola and are there for your use.

eBook

Also included on the CD-ROM is a full, searchable eBook version of the text in Adobe pdf format. In addition, there are sample chapters of other electronics engineering references available in both eBook and print versions from LLH Technology Publishing.

Good luck on your venture into C.

Ted Van Sickle
e-mail: tvansickle@a-sync.com
http://www.a-sync.com/

Chapter 1

Introduction to C

Programming is a contact sport. Programming theory is interesting, but you must sit at a keyboard and write code to become a programmer. The aim of this introductory section is to give you a brief glimpse of C so that you can quickly write simple programs. Later sections will revisit many of the concepts outlined here and provide a more in-depth look at what you are doing. For now, let's start writing code.

Some Simple Programs

C is a function based language. You will see that C uses far more functions than other procedural languages. In fact, any C program itself is merely a function. This function has a name declared by the language. In C, parameters are passed to functions as arguments. A function consists of a name followed by parentheses enclosing arguments, or perhaps an empty pair of parentheses if the function requires no arguments. If there are several arguments to be passed, the arguments are separated by commas.

The mandatory program function name in C is main. Every C program must have a function named main, and this function is the one executed when the program is run. Examine the following program:

```
#include <stdio.h>

int main(void)
{
   printf("Microcontrollers run the world!\n");
   return 0;
}
```

This program contains all of the elements of a C program. Note first that C is a "free form" language. Spaces, carriage returns, tabs, and so forth are for the programmer's convenience and are ignored by the compiler. The first line of the program

```
#include <stdio.h>
```

is called a *preprocessor command*. Preprocessor commands are identified by the # at the beginning of the line. In this case, #include tells the preprocessor to open the file stdio.h and read it into the program to be compiled with the remainder of the program. The file name is surrounded by angle brackets < >. These delimiters tell the compiler to search for the file in a region designated by the operating system as SET INCLUDE. Had the file name been delimited by double quotes, " ", the operating system would have searched only the default directory for the file. The default directory is, of course, the directory from which you are operating.

The next line of the program is a definition for a function named main. In ANSI C, as opposed to classic C, each function definition must inform the compiler of the return type from the function, and the type of the function's arguments. In this case, the function main has to return an integer and it expects no arguments. The type int preceding the function name indicates that it returns an integer and that no arguments to the function are expected.

The line following the function definition contains an opening brace {. This brace designates the beginning of a block or a compound statement. The next line of the program contains a function call to the function printf(). This function is made available to the program by the inclusion of the header file stdio.h, and it is a function that writes a message to the computer terminal screen. In this case, the message to be sent to the screen is

```
Microcontrollers run the world!
```

The *escape character* '\n' at the end of the message informs the program to insert a new line at that point. The complete message including the new line escape character is enclosed in double quotes. These double quotes identify a *string*, and the string is the argument to the function printf(). Note that the statement beginning with printf is closed with a semicolon. In C, every statement is terminated with a semicolon.

After the message is sent to the screen, there is nothing more for the program to do, so the program is terminated by executing the statement `return 0;`. This statement returns the value 0 back to the calling program, which is the operating system. Also, execution of the return statement will cause all open files to be closed. If there were no return statement at the end of the program, the normal processing at the end of the program would close open files, but there would be no value returned to the calling program.

This is an area where there is much discussion and many dissenting viewpoints. Early C did not require that `main` return a value to the calling program. When the C89 standard was written, it required that `main` return an `int`. Unfortunately, many people, set in their ways, have refused to adhere to the standard nomenclature in this case and they often use `void main(void)` instead of the form above. Most compilers will ignore this form and allow the `void main(void)` function call. For some reason, this form angers many code reviewers, so you should use the correct form shown above.

The program is closed by the inclusion of a closing brace, }, at the end. There could be many statements within the block following `main()` creating a program of any complexity. The closing brace is the terminator of a compound statement. The compound statement is the only case in C where a complete statement closure does not require a semicolon.

Another program example is as follows:

```c
#include <stdio.h>

int main (void)
{
   int a,b,c,d;

   a=10;
   b=5;
   c=2;

   d=a*b*c;
   printf("a * b * c = %d\n", d);
```

```
    d=a*b+c;
    printf("a * b + c = %d\n", d);
    d=a+b*c;
    printf("a + b * c = %d\n", d);
    return 0;
}
```

Before discussing this bit of code, we need to talk about the numbers used in it. Like most high-level languages, C provides for different classes of numbers. These classes can each be variable *types*. One class is the *integer* type and a second is the *floating point* type. We will examine these number classes in more detail later, but for now let us concentrate on the integer types. Integer numbers usually have a numeric range of about $\pm 2^{(n-1)}$, where n is the number of bits that contains the integer type. Integers are also called integral types. Integral types do not "understand" or permit fractions. Any fraction that results from a division operation will be truncated and disappear from the calculation. All variables must be declared or defined to be a specific type prior to their use in a program.

The first line of code in `main`

```
int a,b,c,d;
```

declares the variables a, b, c, and d to be integer types. This particular statement is both a declaration and a definition statement. A definition statement causes memory to be allocated for each variable, and a label name to be assigned each location. A declaration statement does not cause memory allocation, but rather it merely provides information as to the nature of the variable to the compiler. We will see more of definition and declaration statements later.

The three assignment statements

```
a=10;
b=5;
c=2;
```

assign initial values to the variables a, b, and c. The equal sign signifies assignment. The value 10 is placed in the memory location designated as a, etc. The next statement

```
d=a*b*c;
```

notifies the compiler to generate code that will cause the integer stored in location a to be multiplied by the integer in b and the result of that product to be multiplied by the integer found in c. Usually, the name a, b, or c is used to designate the content of the memory location assigned to the label name. This integer result will be stored in the location identified by d.

The print statement

```
printf("a * b * c = %d\n", d);
```

is similar to the same statement in the first example. In this case, however, the data string

```
"a * b * c = %d\n"
```

contains a printer command character %d. This character notifies the printf function that it is to take the first argument following the data string, convert it to a decimal value, and print it out to the screen. The result of this line of code will be

```
a * b * c = 100
```

printed on the screen.

The line of code

```
d=a*b+c;
```

demonstrates another characteristic of the language. Each operator is assigned a precedence that determines the order in which an expression is evaluated. The parenthesis operators are of the highest precedence. The precedence of the * operator is higher than that of the + operator, so this expression will be evaluated as

```
d=(a*b)+c;
```

In other words, the product indicated by * will be executed prior to the addition indicated by the +. The expression that follows later in the code

```
d=a+b*c;
```

will be evaluated as

```
d=a+(b*c);
```

causing the result of the third calculation to differ from that of the second.

The result obtained when running this program is as follows.

```
a * b * c = 100
a * b + c = 52
a + b * c = 20
```

Here is another example that demonstrates a primitive looping construct:

```
#include <stdio.h>
int main(void)
{
   int i;

   i=1;
   printf("\ti\ti\ti\n");
   printf("\t\t Squared Cubed\n\n");
   while(i<11)
   {
      printf("\t%d\t%d\t%d\n", i, i*i, i*i*i);
      i=i+1;
   }
   return 0;
}
```

This example was designed to produce a simple table of the values of the first ten integers, these values squared, and these values cubed. The lines

```
printf("\ti\ti\ti\n");
printf("\t\t Squared Cubed\n\n");
```

combine to produce a header that identifies the contents of the three columns generated by the program. The escape character \t is a tab character that causes the screen cursor to skip to the next tab position. The default tab value in C is eight spaces.

The command

```
while(i<11)
```

. . . . .

causes the argument of the while to be evaluated immediately, and if the argument is TRUE, the statement following the while will be executed. The argument should be read "i is less than 11." The initially assigned value for i was 1, so the argument is TRUE. The compound statement

```
{
  printf("\t%d\t%d\t%d\n", i, i*i,i*i*i);
  i=i+1;
}
```

will start execution with the value of i being equal to 1. Once this statement is evaluated, control is passed back to the while and its argument is evaluated. If the argument is TRUE, the statement following will be evaluated again. This sequence will repeat until the argument evaluates as FALSE.

In this expression, the string argument of the printf function contains three %d commands. Each %d command causes the corresponding argument following the string to be printed to the screen. There are tab characters, \t, to separate the various printed values on the screen. The first %d will cause the value of i to be printed on the screen. The second %d will cause the value i*i, or i^2, to be printed to the screen. The third %d will print the value of i*i*i, or i^3 to be printed. When C executes the function call, the values of the arguments are calculated prior to the call, so arguments like i*i are evaluated by the calling program and passed by value to the function.

The statement

```
i=i+1;
```

is an example of the use of both *precedence and association*—the direction in which expressions are evaluated—in C. The equal sign here is an operator just like the + symbol. The + operator is evaluated from left to right, and the = operator is evaluated from right to left. Also, the + operator has higher precedence than the = operator. Therefore, the above statement will add one to the value stored in i and then assign this new value to the variable i. This expression simply increments the variable i.

The above statement is the terminating statement of the compound statement following the while. Since i had an initial value of 1, control will be returned to the while with a value of 2 for i. 2, of course, is less than 11, so the statement following the `while` will be executed again and new values will be printed to the screen. This sequence will be repeated until the incremented value for `i` equals 11, at which time `i<11` will be FALSE. At that point in the program, the statement following the `while` will be skipped, and the program will have reached its end. The result of executing the above program is shown in the following table:

i	i squared	i cubed
1	1	1
2	4	8
3	9	27
4	16	64
5	25	125
6	36	216
7	49	343
8	64	512
9	81	729
10	100	1000

EXERCISES

1. Write, compile, and execute each of the example programs shown in this section.

2. Write a program to calculate the Fahrenheit temperature for the Celsius values between 0° degrees and 100° in steps of 10° each. The conversion formula is F=9*C/5+32. Use integer variables, and examine the result when you use F=C*(9/5) + 32. What went wrong?

Names

Variables, constants and functions in C are named, and the program controls operations on these named variables and constants. Variables and constants are called operands. Names can be as many as 31 characters long. The characters that make up the name can be the upper and the lower case letters, the digits 0 through 9, and the underscore character '_'. There are several defined constants and functions that are used by the compiler. All of these names begin with an underscore. Because of this convention, you should avoid the use of an underscore as the first character for either function or variable names in your code. This approach will completely avoid name conflict with these hidden or unexpected names. Compilers usually allow the names to be unique in the first 31 characters. Unfortunately, some linkers used to link various program modules require that the names be unique in the first six or eight characters, depending on the linker.

C has a collection of keywords that cannot be used for names. These keywords are listed below:

KEYWORDS

auto	double	int	struct
break	else	long	switch
case	enum	register	typedef
char	extern	return	union
const	float	short	unsigned
continue	for	signed	void
default	goto	sizeof	volatile
do	if	static	while

Types and Type Declarations

C has only a few built-in types. Here they are:

char—is usually eight bits. The character is the smallest storage unit.

int—an integer is usually the size of the basic unit of storage for the machine. An int must be at least 16 bits wide.

float—a single precision floating-point number.

double—a double precision floating-point number.

Additional qualifiers are used to modify the basic types. These qualifiers include:

short—modifies an int, and is a variable whose width is no greater than that of the int. For example, with a compiler with a 32 bit int a short int could be 16 bits. You will find examples where short and int are the same size.

long—modifies an int, and is a variable size whose width is no less than that of an int. For example, on a 16-bit machine, an int might be 16 bits, and a long int could be 32 bits. long can also modify a double to specify an extended precision floating-point number. You will find examples where a long and an int are the same size.

signed—modifies all integral numbers and produces a range of numbers that contains both positive and negative numbers. For example, if the type char is 8 bits, a signed char can contain the range of numbers –128 to +127. Default for char and int is signed when they are declared.

unsigned—modifies all integral numbers and produces a range of numbers that are positive only. For example, if the type char is 8 bits, an unsigned char can contain the range of numbers 0 to +255. It is not necessary to include the type int with the qualifiers short or long. Thus, the following statements are the same:

```
long int a,c;
short int d;
```

and

```
long a,c;
short d;
```

When a variable is defined, space is allocated in memory for its storage. The basic variable size is implementation dependent, and especially for microcontrollers, you will find that this variability will show up when you change from one microcomputer to another.

Each variable must be defined prior to being used. A variable may be defined at the beginning of any code block, and the variable's scope is the block in which it is defined. When the block in which the variable is defined is exited, the variable goes out of existence. There is no problem with defining variables with the same name in different blocks. The compiler will make certain that these variables do not get mixed up in the execution of the code.

An additional qualifier is const. When const is used as a qualifier on the declaration of any variable, an initialization value must be declared. This value cannot be changed by the program. Therefore the declaration

```
const double PI = 3.14159265;
```

will create the value for the mathematical constant pi and store it in the location provided for `PI`. Any attempt to change the value of `PI` by the program will cause compiler error.

Conventions for writing constants are straightforward. A simple number with no decimal point is an `int`. To make a number long, you must suffix it with an `l` or an `L`. For example, `6047` is an int and `6047L` is a long. The `u` or `U` suffix on a number will cause creation of a proper unsigned number.

A floating-point number must contain a decimal point or an exponent or both. The numbers 1.114 and 17.3e-5 are examples of floating point numbers. All floating point numbers are of the type `double` unless a suffix is appended to the number. Any number suffixed with an `f` or an `F` is a single precision floating-point number, and a suffix of `l` or `L` on a floating-point number will generate a type `long  double`. Octal (base 8) and hexadecimal (base 16) numbers can be created. Any number that is prefixed with a 0—a leading zero—is taken to be an octal number. Hexadecimal numbers are prefixed with a `0x` or a `0X`. The rules above for L and U also apply to octal and hexadecimal numbers.

The final type qualifier is `volatile`. The qualifier `volatile` instructs the compiler to NOT optimize any code involving the variable. In execution of an expression, a side effect refers to the fact that the expression alters something. The side effect of the following statement

```
a=b+c;
```

is that the stored value of `a` is changed. A sequence point is a point in the code where all side effects of previous evaluations are completed and no side effects from subsequent evaluations will have taken place. An important consideration of the optimization is that if an expression has no side effects, it can be eliminated by the compiler. Therefore, if a statement involves no sequence point, or alters no memory, it is subject to being discarded by the compiler. This operation is not particularly bad when writing normal code, but when working with microcontrollers where events can occur as a result of hardware operations, not the program, this optimization can utterly destroy a program. For example, whenever the hardware can alter a stored value, the compiler should be able to discard accesses to that value because the program never alters the value. In such a circumstance, if you had an analog-to-digital converter peripheral in your system, the program would never be required to read its return value more than once. "The program did not change the value stored in the input location subsequent to the first read, therefore its value has not changed and it is not necessary to read the location again." This will always produce wrong results. The key word `volatile` indicates to the program that a variable must not be optimized. Therefore, if the input location is identified as a `volatile` variable, it will not be optimized and the problem will go away. As a point of interest, accessing a `volatile` object, modifying an object, modifying a file, or calling a function that does any of those operations are all defined as side effects by the standard.

Storage Classes, Linkage, and Scope

Additional modifiers are called storage classes and designate where a variable is to be stored and how it is initialized. These storage classes are `auto` (for automatic), `static`, and `malloced`. The first two storage classes are described in the following sections. The storage class malloc provides dynamic memory allocation and is discussed in detail in Chapter 2.

Automatic variables

For local variables defined within a function, the default storage class is `auto`. An automatic variable has the scope of the block in which it is defined, and it is uninitialized when it is created. Automatic variables are usually stored on the program stack, so space for the variable is created when the function is entered. When the stack is cleaned up prior to the return at the end of the function, all variables stored on the stack are deleted.

As we saw in our first program example, variables can be initialized at the time of declaration by assigning the variable an initial value:

```
int rupt=17;
```

An automatic variable will be assigned its initial value each time the block in which it is declared is entered. If the variable is not initialized at declaration, it will contain the contents of uninitialized memory, which can be any value.

Another class of variable is `register`. A register class variable is automatic, i.e., it comes into being at the beginning of the block in which it is defined and it goes out of scope at the end of the block. If a `register` is available in the computer, a `register` variable will be stored in a register. To define a register variable, you should use the form

```
register int roger=10;
```

These variables can be `long`, `short`, `int`, or `char`.

When a register is not available, a `register` variable will be stored just like any other automatic variable. A programmer might consider the use of `register` variables in code that contains "tight loops" to save the time of memory accesses while executing the loop. A bit of advice. Compilers have improved continuously over the past years. With today's compilers, the optimizers are so efficient that the compiler can probably do a better job of assigning `register` variables than the programmer. Therefore, it makes little sense to specify a lot of `register` variables just to improve the efficiency of your code.

Static variables

Sometimes you might want to assign a value to a variable and have it retain that value for later function calls. Such a variable can be created by calling it `static` at its definition. There are two groups of `static` variables: Local `static` variables which have a scope of the function in which they are defined, and global or external `static` class variables. Unless otherwise declared, all `static` class variables are initialized to 0 when they are created.

There are two groups of external `static` variables. Any external variable is a `static` class variable. It is automatically initialized to the value 0 when the program is loaded unless the value is otherwise declared in the definition statement. An external variable that is declared as `static` in its definition statement like

```
static int redoubt;
```

will have file scope. Remember normal external variables can be accessed from any module in the entire program. A `static` external variable can be accessed only from within the file in which it is defined. Note that `static` variables are not stored on the stack, but rather stored in a `static` data memory area.

Inside of a function, the following declaration is made:

```
static int keep = 1;
```

When the program is loaded and executed, the value 1 is assigned to keep. Thereafter, each time the function is entered, keep will not be initialized but will retain the value assigned to it the last time the function was executed.

Global variables can be designated as static. A global variable that is static is similar to a conventional global variable with the exception that it can be accessed only from the file in which it is declared.

If there is an external variable that is declared in one file that is to be accessed by a function defined in another file, the function must notify the compiler that the variable is external with the use of the keyword extern. The following is an example of such an access.

In file 1:

```
int able;

int main(void)
{
   long quickstart(void);
   long r;

   .

   .

   .

   able=17;
   l=quickstart();

   .

   .

}
```

In file 2:

```
long quickstart(void)
{
   extern int able;

   .

   /* do something with able */

   .
```

```
    return result;
}
```

When the file 1 is compiled, the variable `able` is marked as external, and memory is allocated for its storage. When the file 2 is compiled, the variable `able` is recognized to be external because of the `extern` keyword, and no memory is allocated for the variable. When the link phase of the compilation is completed, all address references to `able` in file 2 will be assigned the address of `able` that was defined in file 1. The example above in which the declaration

```
extern int able;
```

allowed access to `able` from the file 2 will not work if `able` had been declared as follows in file 1:

```
static int able;
```

Character Constants

Character constants or escape sequences are data that can be stored in memory locations designated as `char`. A character constant is identified by a backslash preceding the character. We have seen the use of the character constants ` '\n' ` and ` '\t' ` in previous examples. Several of these escape sequences shown in the following table have predefined meanings.

Escape Sequence	*Meaning*
\a	bell character
\b	backspace
\f	form feed
\n	new line
\r	carriage return
\v	vertical tab
\t	horizontal tab
\?	question mark
\\	back slash
\'	single quote
\"	double quote
\ooo	octal number
\xxx	hexadecimal number

If these constants are used within a program, they must be identified by quotes. In the earlier example, the new line character was a part of a string. Therefore, it effectively was contained in quotes. If a single character constant is to be generated, the constant must be included in single quotes. For example, a test might include a statement like

```
if(c!='\t')
    . . . .
```

This statement causes the variable c to be compared with the constant `'\t'`, and the statement following the `if` will be executed if they are not the same. Another preprocessor command is `#define`. With the `#define` command, you can define a character sequence that will be placed in your code sequence whenever it is encountered. If you have character constants that you wish to use in your code, these constants can be identified as

```
#define CR '\x0d'
#define LF '\x0a'
#define BELL '\x07'
#define NULL '\x00'
```

and so forth.

We'll discuss the `#define` preprocessor command further later.

The following program shows use of an escape character.

```
/* Count lines of text in an input */

#include <stdio.h>

int main(void)
{
    int c,nl=0; /* the number of lines is in nl */
    while((c=getchar())!=EOF)
        if(c=='\n')
            nl++;

    printf("The number of lines is %d\n",nl);
    return 0;
}
```

Often you will want to leave "clues" as to what the program or line of code is supposed to do. Comments within the code provide this documentation. A C comment is delimited by

```
/*  .  .  .  .  .  .  .  */
```

and the comment can contain anything except another comment. In other words, comments may NOT be nested. The first line of code in the above program is a comment, and the sixth line contains both code and a comment. The compiler ignores all information inside the comment delimiters.

This program uses two integer variables c and nl. The variable c is the temporary storage location in which input data are stored, and nl is where the number of input lines are counted.

The while statement contains a rather complicated argument. At any point in a C program when a value is calculated, it can be stored in a specified location. For example, in the while expression

```
while((c=getchar()) != EOF)
```

the inner expression

```
c=getchar()
```

causes the function getchar() to be executed. The return from getchar() is a character from the input stream. This character is assigned to the variable c. After this operation is completed, the result returned from getchar() is compared with the constant EOF. EOF means end-of-file, and it is the value returned by getchar() when a program tries to read beyond the end of the data stream. It is defined in the file stdio.h. The symbol != is read "is not equal to." Therefore, the argument of the while will be TRUE so long as getchar() does not return an EOF and the statement following the while will be continually executed until an EOF is returned.

Operators in an expression that have the higher precedence will be executed before the lower precedence operators. In the expression

```
c= getchar() != EOF
```

the operator != has a higher precedence than that of the = operator. Therefore, when this expression is evaluated, the logical portion of the expression will be evaluated first, and the result of the logical evaluation—either TRUE or FALSE—will be assigned to the variable c. This result is of course incorrect. To avoid this problem, use

```
(c = getchar()) != EOF
```

as the while argument. In this case, the parentheses group the c=getchar() expression and it will be completed prior to execution of the comparison. The variable c will have the correct value as returned from the input stream. If the above expression is logically true, then the value that was returned from the input stream is tested to determine if it is a new line character. If a new line character is found, the counter nl is incremented. Otherwise, the next character is read in and the sequence repeated until an EOF is returned from the getchar(). Whenever an assignment is executed inside of another expression, always enclose the complete assignment expression in parentheses.

The final statement in the program

```
printf("The number of lines is %d\n",nl);
```

prints out the number of new line characters detected in reading the input file.

Arrays

An *array* is a collection of like types of data that are stored in consecutive memory locations. An array is designated at declaration time by appending a pair of square brackets to the array name. If the size of the array is to be determined at the declaration, the square brackets can contain the number of elements in the array. Following are proper array declarations.

```
extern int a[];
long rd[100];
float temperatures[1000];
char st[]={"Make a character array"};
float pressure[]={ 1.1, 2.3, 3.9, 3.7, 2.5, 1.5,
0.4};
```

As you can see, the size of an array must be designated in some manner before you can use empty square brackets in the designation. In the first case above, the array a [ ] is defined in global memory, so all that is necessary for the compiler to know is that a [ ] is an array.

The argument of an array is sometimes called its *index*. It is a number that selects a specific entry into an array. Array arguments start with zero always. Therefore, when an array of 100 elements is created, these elements are accessed by using the arguments 0 to 99. The standard requires that the first element beyond the end of the array be accessible as an array entry. Attempts to access elements beyond that will give undefined results.

Arrays can be initialized at declaration. The initialization values must be enclosed in braces, and if there are several individual numerical values, these values must be separated by commas. In the case of a string initialization, it is necessary to include the string in quotes and also enclose the string along with its quotation marks within the braces. In both of these cases, the size of the array is calculated at compile time, and it is unnecessary for the programmer to figure the size of the array.

A *string* is a special case of an array. Whenever a string is generated in C, an array of characters is created. The length of the array is one greater than the length of the string. The individual characters from the string are placed in the array entries. To be a proper C string, the array's last character must be a zero or a null. All strings in C are null terminated. If you as a programmer create a string in your program, you must append a null on the end of the character array to be guaranteed that C will treat the array as a string.

If the programmer should specify an array size and then initialize a portion of the array like

```
int time[6]={1,5,3,4};
```

the compiler will initialize the first four members of the array with the specified values and initialize the remainder of the array with zero values. This approach allows you to initialize any array with all zero values by

```
long ziggy[100]={0};
```

which will fill all of the elements of the array `ziggy[]` with zeros.

C provides you with no array boundary checking. It is the programmer's responsibility to guarantee that array arguments do not violate the boundaries of the array.

Other types

There are mechanisms for creating other types in C. The three other types are enum, union, and struct. It is often quite convenient to make use of the data types to accomplish things that are difficult with the normal types available. We will see how to use these types in this section.

The enum

The name enum is used in C in a manner similar to the #define preprocessor command. The enum statement

```
enum state { OUT, IN};
```

produces the same result as

```
#define OUT 0
#define IN 1
```

Here, the name state is called the tag name. In this case OUT will be given a value of 0 and IN a value 1. In the enum{} form, unless specifically assigned, the members will be given successively increasing values and the first will be given a value 0. Values can be assigned by an enum{};

```
enum months {Jan =1,Feb, Mar, April, May, June,
July, Aug, Sept, Oct, Nov, Dec};
```

will cause Jan to be 1, Feb 2, and so forth up to Dec which will be 12. Each member can be assigned a different value, but whenever the programmer assignments stop, the values assigned to the variables following will be successively increased. These values are, by default, of the int type. The name months in the above expression is called a tag name. An enum creates a new type and you might have several enums in your code that you would wish to create as instances. The key word enum with its tag name identifies the specific enum when it is used as a type identifier in a definition statement.

Another example

```
enum (FALSE,TRUE,Sun=1,Mon,
Tues,Wed,Thur,Fri,Sat);
```

will result in FALSE being 0, TRUE 1, Sun 1, Mon 2, and so forth to Sat 7. Note that it is not necessary to assign a tag name to an enum.

An enum is typed at declaration time. Therefore, the values created by an enum are indeed numerical values. This differs from the #define because the statement

```
#define FALSE 0
```

will cause the character '0' to be inserted into the source code whenever the label FALSE is encountered. As such, the #define construct is a character substitution technique or a macro expansion. The result of an enum is a numerical substitution. The #define construct, being a simple character substitution, has no typing attached to its arguments. Constants created by an enum are typed, and therefore, will avoid many of the potential hazards of dealing with untyped variables.

Let us examine how one might use a type created with an enum construct. The following enum defines two constants

```
enum direction {LEFT,RIGHT};
```

In a program, a definition statement

```
enum direction d;
```

will cause a variable d to be created. The acceptable values for d are the names LEFT and RIGHT. We know, of course, that the numerical value for LEFT is 0 and the value for RIGHT. Within your program, you can assign and test the value of d. For example,

```
if(d==LEFT)
   do something
```

or

```
if(d==RIGHT)
        do something else
```

or

```
d = RIGHT;
```

As stated earlier, the acceptable values for d are LEFT and RIGHT. There is no checking within the program to see if the programmer has indeed kept the trust. Therefore, it is possible to assign any integer value to d, and the program will compile. It probably will not work correctly, however.

The Union

The union was invented when memory was very dear. The main purpose of the union was to allow the storing of several variables at a single memory location. A union has a tag name much the same as the enum above.

```
union several
{
    long biggie;
    int middle_size;
    char little,another_char;
    short little_bigger;
};
```

The union several contains several members. These members are not necessarily of the same type and there can be multiple instances of the same type. To create an instance of such a union, you need a definition statement. This statement can be external or immediately following the opening of a code block, and hence local. Such a statement might be

```
union several these;
```

This definition causes a union several named these to be created with memory allocated. To access the members of the union, you can use the dot operator as

```
these.biggie = something;
```

or

```
another = these.another_char;
```

An interesting feature of a union. If you should check the size of a union, you would find that it is the size of the largest of its members. Whenever you access, either read or write, a union, the proper size data is written or read, and it overwrites any other data that might be found in the memory location. Therefore, you can use a union for storage of only one of its members at a time and writing anything to the `union` destroys any data previously stored to the `union`.

The struct

Yet another type is the `struct`. The `struct` is a collection of things much like the array. In the case of the `struct`, there are two major differences. A `struct` can contain different types, and the `struct` itself is a first class type. An array must be a collection of like types, and an array is NOT a type, so the type-like things you can do with a `struct` are not available for an array.

You create a `struct` in much the same form as was seen with a `union`. You may use a tag name.

```
struct able
{
   char a,b;
   int c,d;
};
```

This `struct` is made up of two characters and two integers. If you wish to define an instance of the `struct`, you should use

```
struct able here:
```

Access the members of the `struct` with the dot operator like

```
here.a = 'a';
here.b = 16;
here.c = 32000;
here.d = -16500;
```

We will see more of `struct` in Chapter 2 where you will learn how to make use of the new types created by `struct`.

EXERCISES

1. Write a program that reads all of the characters from an input file and prints the characters on the screen. Use the `getchar()` function used earlier to read the inputs and the `putchar(c)` to print the results to the screen.

2. Modify the above program to count the number of characters in an input stream.

3. Write a program that reads the characters from an input file and counts in an array the occurrences of each letter. Make the program "case insensitive" by treating all upper case letters as lower case.

Operators and Expressions

The variables and constants discussed in the previous section are classed as operands. They are values or objects that are operated upon by a program. The operations that take place are specified by operators. This section contains a discussion of the several operators.

Operators abound in C. All of the symbols involved in the language are operators. Each has a precedence and an associativity. This section is concerned with how operators and operands are put together to interact in a manner desired by the programmer.

Arithmetic Operators

The arithmetic operators are those used to perform arithmetic operations. These operators are:

```
+
-
*
/
%
```

These operators are called binary operators because they are always used with two operands. These operands are placed on either side of the operator. The symbol + designates arithmetic addition, and the – symbol designates subtraction. The symbols * and / designate multiplication and division, respectively. These operators are clearly different for different variable types. The compiler understands these differences and creates correct code for the operand types involved. The modulus operator % returns the remainder after an integer division. The modulus operator works only on integer types—-int, char, and long. It cannot be applied to types float, double or long double.

Two unary operators are + and −. These operators are of higher precedence than the normal arithmetic operators. They operate on only the operand written to the right of the operator and are therefore called unary. The unary minus sign causes the negative value of the operand to be calculated, and the unary positive sign causes no calculation to take place.

Among the binary operators, `*`, `/`, and `%` have equal precedence, which is higher that of + and − . The unary operators + and − have a higher precedence than `*`, `/`, or `%`. The arithmetic operators will work with any of the arithmetic types. Because the operations needed for an integer operation differ from those needed for the corresponding double operation, the compiler will place the proper arithmetic routines in the code to perform the specified operation.

The concept of a fraction is almost unknown to an integer type. If a division of two integers is executed, the result is rounded toward zero. Therefore, the result of 1/2 is 0 as is 9999/10000. This characteristic is often used in programming.

The only way that you can handle fractions with integer operations is to make use of the modulus operation. The result of `a %b` is the remainder that is left over after `a` is divided by `b`. The modulus operation can provide insight into the fractional value of what is left over after an integer divide.

EXERCISES

1. Write a program that evaluates

   $f(x) = X^2 - 3X + 2$

   for values of X in $0 \leq X \leq 3$ in steps of 0.1.

2. The roots of a quadratic equation can be evaluated by the equation

   $x = (-b + \text{sqrt}(b^2 - 4ac))/2a$

   and

   $x = (-b - \text{sqrt}(b^2 - 4ac))/2a$

   where the quadratic equation is `ax`^{2}` + bx + c = 0`. Write a program that will evaluate the roots of such an equation. Note that the term `sqrt (b`^{2}` - 4ac)` is called the discriminant. If its argument is not positive, the square root of a negative number is imaginary and the equation has complex roots. Handle both real and complex roots in your program.

Relational or Logical Operators

The relational operators are all *binary* operators. When contained in an expression, the program will evaluate the left operand and then the right operand. These operands will be compared, and if the comparison shows that the meaning of the operator is correct, the program will return 1. Otherwise, the program will return a 0. In the vocabulary of C, FALSE is always zero. If calculated by a logical expression, TRUE will always be one. However, if the argument of a conditional expression is anything but zero, it will respond as if the argument is TRUE. In other words, FALSE is always zero and TRUE is anything else. The relational operators are:

    <   (less than)
    <=   (less than or equal to)
    >   (greater than)
    >=   (greater than or equal to)

These operators all have the same precedence, which is slightly higher than the following equality operators:

    ==   (is equal to)
    !=   (is not equal to)

The *logical* operators are && and ||. The first operator indicates a logical AND and the second a logical OR. A logical AND will return TRUE if both of its operands are TRUE, and a logical OR will return TRUE if either of its operands is TRUE. The logical OR has lower precedence than the logical AND. The precedence of the logical AND is lower than the precedence of the relational operators and the equality operators.

In the evaluation of long logical expressions, the program starts on the left side of the expression and evaluates the expression until it knows whether the whole expression is true or false, and it then exits the evaluation and returns a proper value. For example, suppose there is a character c, and it is necessary to determine if this character is a letter. In such a case, the following logical expression might be used:

```
if( c >= 'A' && c <= 'Z' || c >= 'a' && c <= 'z')
```

The logical and operator && has lower precedence than any of the relational operators, so the relational expressions will each be evaluated prior to the && operations. If upon entering this expression, c is equal to the character ' 5 ', which is arithmetically smaller than any of the letters, the first term c >= 'A' will be FALSE. Therefore, the result of the first logical and expression is known to be FALSE without evaluating the term c <= 'Z'. The evaluation will then skip to the third term c >= 'a', and the term c <= 'Z' will not be evaluated. In this case, the character '5' will be smaller than the character 'a' so that the second and expression will also be FALSE. Therefore, the logical value will be known after evaluation of only two of the logical terms of the argument rather than having to evaluate all four of the terms.

EXERCISES

1. Write a function that converts a character that is a letter to lower case.

2. Leap years occur every four years unless the year happens to be divisible by 100. Any year divisible by 400 is a leap year, however. Write a logical expression that will return TRUE if the given year is a leap year and FALSE if it is not.

Type Conversions Within Expressions

Implied in our earlier discussions on variable types, different data types not only occupy different width in memory, some may be completely incompatible when attempting to execute operations involving mixed data types. In earlier languages, it was up to the programmer to guarantee that the data types involved with an operation were the same. C resolves this problem, and the compiler will select the proper data type to complete operations on mixed data types.

Each data type has an implied width. When an operation is to be executed on mixed data types, the widths of the two types are evaluated, and the lesser width operand is promoted to the type of the greater width operand prior to execution of the operation. Thus, if the program called for d = a * b, where d is of type long, a is type int, and b is type long, a will be converted to the type long prior to the multiplication.

This logic carries over to mixing of float and double types as well. If for example a program called for the division a/b where a is of the type int and b is of the type double, the program would convert a to the type double before execution of the divide.

There might be times when the programmer will want to change the type of a variable. C provides a *cast* operator which forces the program to convert the type of a variable to a different type. This unary operator has the form.

```
(type name) expression
```

where the results of the evaluation of the expression will be converted to the named type contained within the parentheses preceding the expression.

Bitwise Operators

Operators that work on the individual bits within a variable are called bitwise operators. Following is a table of all of these operators:

&	bitwise AND	>>	right shift
|	bitwise Inclusive OR	<<	left shift
^	bitwise Exclusive OR	~	one's complement

The first three bitwise operators are traditional binary operators. These binary operators operate in integer type (char, int, long, etc.) operands, and the two operands must be of the same type.

If a bitwise AND is executed, those locations in the result where both operands have bit values of 1 will have a value of 1. All other locations will be 0. For a bitwise inclusive OR, each bit in the result will be 1 when either or both operand bits are 1. All locations where both operand bits are 0 will be 0. The exclusive OR is similar to an addition with no carry. Whenever the bits in the operands are different, the result bit will be 1. If both operand bits are the same, either both bits 1 or both bits 0, the result will be 0.

The right shift operator and the left shift operator are also binary operators. Here the types of the operands need not be the same. The expression

```
x >> 3
```

causes the variable x to be shifted to the right by three bits prior to its use. Likewise,

```
y << 5
```

will cause y to be shifted to the left by five bits. In all number systems, a left shift by one digit corresponds to a multiplication by the number base. Similarly, a shift to the right by one digit causes a division by the number base. We are using the binary system in this case, so a shift left by one bit causes the number to be multiplied by two. Unlike most number systems, the binary system (or two's complement system) allows the sign of the number to be contained in the binary representation of the number itself. These considerations lead to two different types of shifts for a system of binary numbers. A shift in which bits vacated by the shift are replaced by zeros is called a *logical shift*. All left shifts are logical shifts. As the shift progresses toward the left, bits that fill the number from the right will all be zero. Bits that shift out of the number on the left side are lost. A right shift can be either a logical or an *arithmetic shift*. If the type being shifted is signed, the sign bit—which is the leftmost bit—will propagate, retaining a number of the same sign. This is an arithmetic sign. If the number being shifted is unsigned, zeros are filled into the number from the left as the shift proceeds. In all cases, bits shifted out of a number by a shift operation will be lost. The one's complement operator ~ is a unary operator that causes the bits in a variable to be reversed. Every 1 is replaced by a 0, and every 0 is replaced by a 1. The bitwise AND and OR operations are used to turn bits on and off. Suppose that we have a character variable r, and we wish to turn the least significant three bits off. Try

```
r = r & ~7;
```

In this case, the number 7 has each of the least significant bits turned on or 1. Therefore, the term ~7 has all of the bits in the number but the least significant turned on and these three bits are turned off or 0.

When this mask is ANDed with r, all of the bits of r, with the exception of the least significant three bits, will be ANDed with a 1, and these bit values will remain unchanged. The least significant three bits will be ANDed with 0 and the result in these three bits will be 0. The bitwise OR will turn bits on. Suppose you wanted to turn bits 2 and 3 of r above on. Here you would use

```
r = r | 0x0c;
```

The hexadecimal number 0x0c is a number that has bits 2 and 3 turned on and all other bits turned off. This OR operation will leave bits 2 and 3 on and all other bits will remain unchanged. Suppose that you want to complement a bit in a variable. For example, bit 0 of the memory location PORTA must be toggled each time a certain routine is entered. The expression

```
PORTA = PORTA ^ 1;
```

will perform this operation. All of the bits except for bit 1 of PORTA will remain unchanged because the exclusive OR of any bit with a 0 will not change the bit value. However, if bit 1 is 1 in PORTA the exclusive OR will force this bit to 0. If this bit is 0, the exclusive OR will force this bit to a 1. Therefore, the above expression will complement bit 0 of PORTA each time it is executed.

The bitwise operators &, |, and ^ are of lower precedence than the equality operators, and higher precedence than the logical AND operator. The bit shift operators are of the same precedence, of lower precedence than the arithmetic operators + and − , and of higher precedence than the relational operators.

Increment and Decrement Operators

When the C language was written, every effort was made to write a language that is concise and yet unambiguous. Several powerful short-hand operators were included in the language that will shorten the program. The increment and decrement operators are examples of such short-hand operators. In the examples earlier there were instances of expressions such as

```
i = i + 1;
```

Here the `i` value stored in memory is replaced by one more than the value found there at the beginning of execution of the expression. The C expression

```
++i;
```

will do exactly the same thing. The increment operator `++` causes 1 to be added to the value in the memory location `i`. The decrement operator `--` causes 1 to be subtracted from the value in the memory location. The increment and decrement operators can be either prefix or postfix operators. If, like above, the `++` operator precedes the variable, it is called a prefix operator. If the variable is used in an expression, it will be incremented prior to its use. For example, suppose `i` = 5. Then the expression

```
j = 2 * ++i;
```

will leave a 12 for the value `j` and 6 for `i`. On the other hand, if `i` again is 5, the expression

```
j = 2 * i--;
```

will leave a value of 10 for `j` and 4 for `I`.

An easy way to see how the preincrement and the postincrement works is as follows: Suppose that you have a pair of statements

```
j=j+1;
<statement with j>
```

These statements can be replaced with

```
<statement with ++j>
```

The preincrement means that you should replace j with j+1 before you evaluate the expression. Likewise the statements

```
<statement with j>
j=j+1;
```

can be replaced with

```
<statement with j++>
```

with the post increment, you should evaluate the expression and then replace j with j+1.

Often somebody will wonder what will happen if you have multiple increments, either pre or post, of a variable within a single expression. There is an easy answer for that question. Do not do it. The standard provides that between sequence points, an object shall have its value modified at most once and the prior value of the object shall be accessed only to determine its value. Interpretations of the above requirements disallow statements such as

```
j = j++;
```

or

```
a[j] = j++;
```

or

```
m = j++ + ++j;
```

Assignment Operators

Another shorthand that was included in C is called the assignment operator. When you are programming, you will find that expressions such as

```
i = i+2;
```

or

```
x = x<<1;
```

are used often. Almost any binary operator can be found on the right side of the expression. A special set of operators was created in C to simplify these expressions. The first expression can be written

```
i += 2;
```

and the second

```
x <<= 1;
```

These expressions use what is defined as an assignment operator. The operators that can be used in assignment operators are

```
+      >>
-      <<
*      &
```

```
/         ^
%         |
```

If you have two expressions `e1` and `e2`, and let the operand `$` represent any binary C operator, then

```
e1 $= e2;
```

is equivalent to

```
e1 = (e1) $ (e2);
```

The precedence of all of the operator assignments are the same and less than the precedence of the conditional operator discussed in the next section. These operators assignments and the = operator are associated from right to left.

The Conditional Expression

Another code sequence found frequently is

```
if(exp1)
    exp2 ;
else
    exp3 ;
```

The logical expression `exp1` is evaluated. If that expression is TRUE, `exp2` is executed. Otherwise, `exp3` is executed. In the compact notation of C, the above code sequence can be written

```
exp1 ? exp2 : exp3;
```

This expression is read if `exp1` is TRUE, execute `exp2`. Otherwise, execute `exp3`. Another way of stating this is that if `exp1` is TRUE, the value of the expression is `exp2`; otherwise the value of the expression is `exp3`.

The conditional expression is found often in macro definitions, which we'll discuss later.

EXERCISES

1. Write a program to determine if a number is even or odd.

2. Write a function that determines the number of bits in an integer on your machine.

3. Write a program that will rotate the bits in the number 0x5aa5 to the left by n bits. A rotate differs from a shift in that the most significant bit will be shifted into the least significant bit during the rotation. A shift merely shifts zeros into the least significant bit.

4. An arithmetic right shift propagates the most significant bit to the right when the number is shifted right. If zeros are shifted into the most significant bit, the shift is called a logical right shift. Write a program that determines whether your compiler implements a logical or arithmetic right shift with the operator >> with both signed and unsigned arithmetic.

5. Write a function upper(c) that returns the upper case letter if the character c is a lower case letter. Otherwise it shall return the character c.

6. If you used the if() else construct in problem 4, rewrite the function to use the conditional expression.

Precedence and Associativity

Here is a summary of the rules of both precedence and association of all C operators. The higher an operator falls in the table, the higher its precedence. Operators that fall on the same line are all of the same precedence. All symbols used in C are operators. Therefore, the operator () refers to the parentheses enclosing the arguments to a function call. The operator [] refers to the brackets enclosing the argument of an array. The period operator . and the comma operator , will both be discussed when introduced. Likewise, the -> and the sizeof operators will be introduced later.

Operator	Associativity
() [] -> .	left to right
! ~ ++ — + - * & (type) sizeof	right to left
* / %	left to right
+ -	left to right
<< >>	left to right

`< <= > =>`	left to right
`== !=`	left to right
`&`	left to right
`^`	left to right
<code>|</code>	left to right
`&&`	left to right
<code>||</code>	left to right
`?:`	right to left
`= += -= *= /= %= &= ^= |= <<= >>=`	right to left
`,`	left to right

Note the very high precedence of the parentheses and the square brackets. It is the high precedence of these operators that allows the programmer to force operations that are not in line with the normal precedence of the language. The second highest precedence is the list of unary operators. These operators are all associated from right to left.

EXERCISES

1. Which of the following words are valid names for use in a C program?

```
able                    toots
What_day_is_it          WindowBar
_calloc                 8arnold
Hurting?                value
constant                Constant
sizeof                  continue
```

2. Write a program to evaluate the constant

$$(1.0377 \times 10^7 + 3.1822 \times 10^3) / (7.221 \times 10^4 + 22.1 \times 10^6)$$

The answer will be 0.468162.

3. Write a function that raises the integer x to the power n. Name the function `x_to_the_n`, and write a program that evaluates `x_to_the_n` for several different values of both x and n.

4. Write a program that will examine a specified year and determine if it is a leap year.

5. Write a program that will count the number of digits in an input file. Record and print out the number of occurrences of each digit.

6. In C the term "white space" refers to the occurrence of a space, a tab character, or a new line character. Write a program that will evaluate the number of white space characters in an input file.

Program Flow and Control

Program flow and control comprise several different means to control the execution of a program. Looping constructs, for example, control the repeated execution of a program segment while adjusting parameters used in the execution at either the beginning or the end of the loop. Two way branches are created by if/else statements, and the choice of one of many operations can be accomplished with the else if or switch/case statements. The following paragraphs will provide a quick look at each of these program flow and control methods.

The While Statement

There are three looping constructs available to the C programmer: the while() statement, the for(;;) statement and the do/while() statement. The following program demonstrates the use of the while looping construct along with some other concepts. We have seen the while statement earlier, but the following program will provide a new look at its use.

```
#include <stdio.h>

int main(void)
{
   int guess,i;

   i=1;
   guess = 5;
   while(guess != i)
   {
```

```
      i = guess;
      guess = (i + (10000/i))/2;
   }
   printf("The square root of 10000 is
%d\n",guess);
   return 0;
}
```

As in the first example, the #include statement is used to bring standard input/output features into the program, and the program starts with the function definition main(). Inside of the main program, the first statement is

```
int guess,i;
```

This statement defines the variables guess and i as integers. No value is assigned to i at this time, but a space in memory is allocated to guess and i and the space is sufficient to store an integer. The first executable statement in the program is

```
i=1;
```

This statement is called an assignment statement. The equal sign here is a misnomer. The statement is read "replace the contents of the memory location assigned to i with a 1." The next statement

```
guess = 5;
```

assigns a value 5 to the variable guess. The statement

```
while(guess != i)
```

invokes a looping operation. The while operation will cause the statement following to execute repeatedly. At the beginning of each loop execution, the while argument guess!=i is checked. This argument is read "guess is not equal to i." So long as this argument is TRUE, the statement following the while will be executed. When guess becomes equal to i, the statement following the while will be skipped.

The while is followed by a compound statement that contains two statements:

```
{
   i=guess;
```

```
    guess = (i + (10000/i))/2;
}
```

This calculation is known as a Newton loop. It states that if `i` is a guess at the square root of 10000, then `(i+(10000/i))/2` is a better guess. The loop will continue to execute until `i` is exactly equal to `guess`. At this time the compound statement will be skipped.

When the statement following the `while` is skipped, program control is passed to the statement

```
printf("The square root of 10000 is %d\n",guess);
```

This statement prints out the value of the last guess, which will be the square root of 10000.

The For Loop

Many times, a sequence of code like

```
statement1;
while(statement2)
{
    .

    .

    .

    statement3;
}
```

will be found. This exact sequence was seen in the above example. There is a shorthand version of this sequence that can be used. It is as follows:

```
for(statement1;statement2;statement3)
```

The `for` construct takes three arguments, each separated by semi-colons. In operation, the `for` construct is compiled exactly the same as the above sequence. In other words, `statement1` is executed followed by a standard `while` with `statement2` as its argument. The compound statement that follows will have `statement3` placed at its end, so that `statement3` is executed just prior to completion of the statement following the `while` construct. The `for` construct can be used to write the above program in the following manner:

```
#include <stdio.h>
```

```
int main(void)
{
   int guess,i;

   for(i=1,guess=5;i!=guess;)
   {
      i=guess;
      guess=(i+(10000/i))/2;
   }
   printf("The square root of 10000 =
%d\n",guess);
   return 0;
}
```

Recall that the `for` allows three arguments. Not all arguments are necessary for proper execution of the `for`. In this case, only two arguments are included. The first argument is really two initialization arguments separated by a comma operator. When the comma operator is used, the statements separated by commas are each evaluated until the semicolon is found. At this time, the initialization is terminated. By the way, the comma operator can be used in normal code sequences so that you can string several statements in a row without separating them with semicolons. The second argument of the `for` construct is `i != guess`. The `for` loop will execute so long as this expression is TRUE. Note that there is no third statement in the `for` invocation.

This argument is where you would normally place the change in `i` that is to take place at the end of each loop. In this case, the operation on `i` is `i=guess`. If this expression were used for the third argument, at the end of the first loop, the second argument would be FALSE, and execution of the calculation would be prematurely terminated.

The Do/While Construct

Another looping structure is the `do/while` loop. Recall that the argument of a `while` statement is tested prior to executing the statement following. If the argument of the `while` is FALSE to begin with, the statement following will never be executed. Sometimes, it is desired to execute the statement at least once whether the argument is TRUE or not. In such a case, the argument should be tested at the end of the loop rather than at the beginning as with the `while`. The `do/while` construct accomplishes this operation. The construction of a do-while loop is as follows

```
    .
    .
    .
do
{
    .
    .
    .

} while (expression);
    .
```

The program will enter the `do` construct and execute the code that follows up to the `while` statement. At that time, the expression is evaluated. If it is TRUE, program control is returned to the statement following the `do`. Otherwise, if the expression evaluates to FALSE, control will pass to the statement following the `while`. Notice that there is a semicolon following the `while(expression)`. This semicolon is necessary for correct operation of the `do-while` loop.

The following function converts the integer number n into the corresponding ASCII string. The function has two parts: the first part converts the number into an ASCII string, but the result is backward in the array; the second part reverses the data in the array so that the result is correct.

```
/* convert a positive integer to an ASCII string;
valid for positive numbers only */

void itoa(unsigned int n, char s[])
{
    int i=0,j=0,temp;

    /* convert the number to ASCII */

    do
    {
```

```
   s[i++] = '0' + n % 10;
   n /=10;
} while ( n != 0);
s[i]=0; /* make the array a string */

/* but it is backwards in the array — reverse
t*/

i--; /* don't swap the NULL */
while( i > j)
{
   temp = s[j];
   s[j++] = s[i];
   s[i--] = temp;
}
}
```

The function uses three integer variables. The variables i and j are both initialized to zero, and the variable $temp$ does not need to be initialized. The first portion of the program contains a do-while loop. Within this loop, the number is converted into a string. The statement

```
s[i++] = '0' + n % 10;
```

first calculates the value of the integer modulo 10. This value is the number of 1s in the number. Adding that value to the character '0' will create the character that corresponds to the number of 1s. This value is stored in the location s[i] with i=0 and then i is incremented.

The second statement in the loop replaces n with n divided by 10. This code removes any 1s that were in the number originally, and now the original 10s are in the 1s position. Since this division is an integer division, if the result is between 0 and 1 it will be rounded to 0. Therefore, the test in the while argument allows the above two statements to repeat until the original number n is exhausted by repeated divisions by 10.

When the do-while loop is completed, s[i] will be the character immediately following the string of characters. A string is created by placing a 0 or a null in this location of the array.

To reverse the data, the program starts by decrementing i so that s[i], the last entry in the array, is the most significant character in the number, and it must be placed in the first array location s[0]:. Likewise, the character in s[0] must be placed in s[i]: and so forth. The while loop that follows accomplishes this requirement.

EXERCISES

1. Write a program atoi(s[]) that starts with a character string and converts this string to an int. Assume that the string contains no sign.

2. Write a program that reads a text file one character at a time and counts the number of words in the file.

The If/Else Statement

The if/else statement has the general form

```
if(expression)
    statement1;
else
    statement2;
```

If the logical evaluation of the expression that is the argument of the if is TRUE, statement1 will be executed. After statement1 is executed, program control will pass to the statement following statement2, and statement2 will not be executed. If the evaluation of statement is FALSE, statement2 will be executed, and statement1 will be skipped. The else statement is not necessary. If there is no else statement, the expression is evaluated. If it is TRUE, statement1 will be executed. Otherwise, statement1 will be skipped. The following program demonstrates the use of the if/else flow control method.

```
/* count number of digits and other characters in
input */

#include <stdio.h>

int main(void)
```

```
{
   int c,nn,no;
   no=0;
   nn=0;
   while((c=getchar())!=EOF)
      if(c>='0'&&c<='9')
         nn++;
   else
      no++;
   printf("Digits=%d and other characters=%d\n",nn,no);
   return 0;
}
```

The statement

```
int c,nn,no;
```

declares the three variables `c`, `nn`, and `no` to be integers. You may declare as many variables as you wish with a single declaration statement. The next statements

```
no=0;
nn=0;
```

initialize the values of `no` and `nn` to 0. Variables declared with the above sequence of instructions are automatic variables. These variables are not initialized by the compiler, and the programmer must initialize them to a required value. Otherwise the variables will contain garbage.

The code sequence

```
while((c = getchar()) !=EOF)
   if(c>='0' && c<='9')
      nn++;
else
      no++;
```

comprise the `while` and its following statement. The `if` portion of the statement tests the value of `c` and determines if it is a digit. A character constant is identified as a specific value by placing the character value in single quotes. Therefore, the expression `c>='0'` determines if the character in the location `c` is greater than or equal to the character 0. If it is, the result of this expression is TRUE. Otherwise, the result is FALSE. The expression `c<='9'` determines if the input character is less than or equal to the character 9. If both of these logical expressions are TRUE, then the AND of the two will be TRUE, and the statement `nn++` will be executed to count the numbers found in the input stream. Program control will then skip to the end of the `if` statement and continue to execute the while loop. If, on the other hand, either of these expressions is FALSE, then the AND of the two results will be FALSE and the statement `no++` will be executed. This statement keeps count of the number of characters that are not digits found in the input stream.

At the conclusion of the program, the `getchar()` will return an EOF character and the program will fall out of the `while` loop. It will then execute the following statement:

```
printf("Digits=%d and other characters=%d\n",nn,no);
```

The string contained within the double quotes in this argument causes a combination of text plus calculated values of variables to be printed out. Suppose that the program found 51 numbers and 488 other characters. The printout from the program would then be:

```
Digits=51 and other characters=488
```

Each `%d` is associated with its corresponding argument and converted to a numerical value before it is sent to the screen.

The If-Else If Statement

Sometimes it is necessary to select among several alternatives. One of the methods that C offers is the `if-else if` sequence. Examine the following program that counts the number of occurrences of each vowel in an input. The program also counts all other characters found in the input.

```
/* Count the number of occurrences of each vowel
found in an input and also count all other charac-
ters. */

#include <stdio.h>

int main(void)
```

```
{
   int na=0,ne=0,ni=0,no=0,nu=0;
   int nother=0,c;
   while ((c=getchar())!=EOF)
      if(c=='A' || c=='a')
         na=na+1;
      else if(c=='E' || c=='e')
         ne=ne+1;
      else if(c=='I' || c=='i')
         ni=ni+1;
      else if(c=='O' || c=='o')
         no=no+1;
      else if(c=='U' || c=='u')
         nu=nu+1;
      else
         nother=nother+1;
   printf( "As=%d, Es=%d, Is=%d, Os=%d, Us=%d and"
           " Others=%d\n",na,ne,ni,no,nu,nother);
   return 0;
}
```

This program shows several new features of C. The first is found in the program lines

```
int na=0,ne=0,ni=0,no=0,nu=0;
int nother=0,c;
```

When the variables na and so forth are defined, they are assigned initial values of 0. Such an initialization is always possible when variables are defined. The next statement of the program is

```
while ((c=getchar())!=EOF)
   if(c=='A' || c=='a')
      na=na+1;
   else if(c=='E' || c=='e')
      ne=ne+1;
   else if(c=='I' || c=='i')
      ni=ni+1;
   else if(c=='O' || c=='o')
      no=no+1;
   else if(c=='U' || c=='u')
```

```
        nu=nu+1;
  else
        nother=nother+1;
```

This single statement has quite a few lines of code associated with it, and there are some new concepts here. First, the arguments of the ifs are combinations of two logical expressions. The expression

```
c=='A' || c=='a'
```

says that if c is equal to uppercase a OR if c is equal to lowercase a the argument is TRUE. The vertical bars || are the logical operator OR.

The first if statement is evaluated. If its argument is TRUE, the statement following the if is executed and program control moves to the end of the if statements. Otherwise, the first else if statement argument is evaluated. If this argument is TRUE, the following statement is executed and program control moves to the end of the if statements. This process is repeated until one of the arguments is found to be TRUE, or all of the else if statements are evaluated. At that time, the final statement following the else entry is evaluated. The final else is not required.

In the above statement, please note that the while statement itself and all that follows it form a single statement to the compiler. Likewise, the combination of all of the if-if else constructs also form a single statement. Furthermore, each of the statements following either an if or an if else form single statements. The formatting of this statement helps you understand what is going on, but remember, the format of such a statement is completely up to the programmer. The language is completely free format. The while statement above and its statement following is indeed confusing to observe, and probably the one thing that the programmer can do to reduce the confusion is to block the statements following both the while and the if and else key words. In this case the while statement would look like

```
while ((c=getchar())!=EOF)
{
   if(c=='A' || c=='a'*
   {
```

```
        na=na+1;
    }
    else if(c=='E' || c=='e')
    {
        ne=ne+1;
    }
    else if(c=='I' || c=='i')
    {
        ni=ni+1;
    }
    else if(c=='O' || c=='o')
    {
        no=no+1;
    }
    else if(c=='U' || c=='u')
    {
        nu=nu+1;
    }
    else
    {
        nother=nother+1;
    }
}
```

This block of source code is longer than the original form, but it is exactly the same to the compiler. This code is indeed longer than the original form, but it is also much less open to misinterpretation and misunderstanding. Writing code that cannot be misinterpreted is as much the responsibility of the programmer as writing code that works. Therefore, when writing production code, you should seriously aim to generate code that is not open to any misinterpretation, even though it will make your source code somewhat longer. Usually when you write code that is easy to maintain, it will not affect the size of your object code.

The `printf` function call

```
printf("As=%d, Es=%d, Is=%d, Os=%d, Us=%d and"
 " Others=%d\n",na,ne,ni,no,nu,nother);
```

has the normal `printf` arguments. Note however that the string is not confined to one line. C compilers will not allow a string to be split among several lines in a program. However, an ANSI C compliant compiler will cause two adjacent strings to be concatenated into a single string, so the above code will compile without error.

Break, continue, and goto

These commands will cause a C program to alter its program flow. If a `break` statement is encountered, the program will exit the loop in which it is executing. `Break` can be used to exit `for`, `while`, `do-while`, and `switch` statements. An example of the use of the break statement is shown in the next section.

The `continue` statement causes the next iteration of a `for`, while, or a `do while` loop to be started. In the case of the `for` statement, the last argument is executed, and control is passed to the beginning of the `for` loop. For both the `while` and the `do-while`, the argument of the `while` statement is tested immediately, and the program proceeds according to the result of the test.

The `break` statement is seen frequently, and the `continue` statement is rarely used. Another statement that is even more rarely used is the `goto`. In C, the programmer can create a label at any location by typing the label name followed by a colon. If it is necessary, the `goto <label>` can be used to transfer control of the program from one location to another. In general, C provides enough structured language forms that the use of the `goto <label>` sequence will rarely be needed. One place where the `goto` can be used effectively is when the program is nested deeply and an error is detected. In such a case, the `goto` statement is an effective means of unwinding the program from a deep loop to an outer loop to process the error. In general, you should avoid `goto` statements whenever possible.

That said, there is an excellent alternative to a `goto`. The reach of a `goto` is limited to the function in which it is defined. In fact, the reach should probably be confined to the block in which it is defined. Since new variables can be defined at the beginning of any block, undefined behavior can be introduced when a `goto` branches into a block where new variables have been defined. Also, you can introduce undefined behavior when you branch out of a block where variables have been defined. In both cases, the branch is around operations of defining or deleting local variables. The alternative is to make use of the `setjmp()` and `longjmp()` functions.

These handy library functions are declared in the `setjmp.h` header file. Also declared in this header is a type called `env`. In your program, you must define an external instance of `env`. This parameter is used as an argument to `setjmp()`. `setjmp()` saves the status of the computer in the instance of `env` when it is called. When called originally `setjmp()` returns a zero or a FALSE. The function `longjmp()` takes two arguments. The first is the `env` variable corresponding to the return location in the program. The second is an integer that is returned. Execution of the `longjmp()` returns control to within the `setjmp()` function. When control is returned to the function that called the `setjmp()` the parameter that was passed from the `longjmp()` is returned. Therefore, when control returns from `setjmp()` a simple test determines whether the function `setjmp()` or `longjmp()` was called.

These functions restore the status of the computer to that which existed when the `setjmp()` was called. This restoration automatically unrolls all function calls between the execution of the `setjmp()` and the `longjmp()`. Also, there are no block or function limits on the use of these functions. As long as `env` is a globally accessible variable, `longjmp()` can pass control from any location in a program to any other.

The Switch Statement

A second approach to selection between several alternates is the switch statement. This approach is sometimes called the switch/case statement. Following is a program that accomplishes exactly the same as the above program. In this case, the switch statement is used.

```
/* Count the number of occurrences of each vowel
found in an input and also count all other charac-
ters. */

#include <stdio.h>

int main(void)
{
```

```
int na=0,ne=0,ni=0,no=0,nu=0;
int nother=0,c;

while ((c=getchar())!=EOF)
switch( c)
{
   case 'A':
   case 'a': na=na+1;
         break;
   case 'E':
   case 'e': ne=ne+1;
         break;
   case 'I':
   case 'i': ni=ni+1;
         break;
   case 'O':
   case 'o': no=no+1;
         break;
   case 'U':
   case 'u': nu=nu+1;
         break;
   default:
         nother=nother+1;
}
printf("As=%d, Es=%d, Is=%d, Os=%d, Us=%d and"
     " Others=%d\n" ,na,ne,ni,no,nu,nother);
return 0;
   }
```

This program performs exactly the same function as the earlier one. The data are read in a character at a time as before. Here, however, the switch statement is used. The statement switch(c) causes the argument of the switch to be compared with the constants following each of the case statements that follows. When a match occurs, the next set of statements to follow a colon will be executed. Once the program starts to execute statements, all of the following statements will be executed unless the programmer does something to cause the program to be redirected. The break instruction does exactly this operation for us. When a C program encounters a break, it jumps to the end of the current block. Therefore, the breaks following the executable statements above will cause the program to jump out of the executing sequence and return to get the next character from the input stream.

When all options have been exhausted without a match, the statements following the default line will be executed. It is not necessary to have a default line.

EXERCISES

1. Write a program that counts the number of lines, words, and characters in an input stream.

2. Extend the program from the exercise above to calculate the percentage usage of each character in the alphabet.

3. A prime number is a number that cannot be evenly divided by any number. For example, the numbers 1, 2, and 3 are all prime numbers. Write a program that will calculate and print out the first 200 prime numbers.

   Write this program without the use of either a modulo or a divide operation.

Functions

The function is the heart of a C program. In fact, any C program is merely a function named main. The purpose of a function is to provide a mechanism to allow a single entry of a code sequence that is to be repeated many times. A function is the most reusable element in the C language. Properly written and debugged functions can be collected into a program when needed. Therefore, the use of functions will allow the programmer to write smaller programs and it is not necessary to rewrite common functions that are used often.

A function can have many arguments or none whatsoever. Function arguments are contained in parentheses following the function name. The values of the arguments are the parameters needed to execute the function. A function can return a value, or perhaps it will not have a return value. An example of a function that returns a value is getchar() which returns a character from the input stream.

A function may not be nested inside another function. Therefore, any function must be created outside of the boundaries of any other function or program structure. Functions have only one entry point, and they return only one return item. The return can be of any type that C supports. Functions can have several arguments. The arguments can be of any valid C type.

In ANSI C, the use of a function requires the use of a function prototype. A *function prototype* is a statement of the following form:

```
type function_name(type, type, type,...);
```

The first `type` preceding the function name is the type to be returned from the function. It is also called the type of the function. The several types found in the argument are the types of the corresponding arguments that are sent to the function. Variable names may or may not be used for the arguments of a function prototype. The type list is the important item. The types in the argument list are separated by commas.

Thus far, it might seem that we have been blindly using functions like `printf()`, `getchar()`, and `putchar()` without the benefit of function prototypes. Not so! The header file `stdio.h` contains the function prototypes of all input/output related functions, so it is not necessary for you to put a function prototype in your code for these functions. Other library functions have their prototypes in their own header files.

Compilers will differ. If a programmer attempts to send the wrong type of data to a function through its argument, the compiler might consider it an error or it might well convert the argument to the correct type prior to calling the function. In either case, the compiler will not let a program use the wrong type of data as an argument to a function. The standard allows that the parameter type be corrected to the correct type and proceed. Some embedded systems compilers will require that the type of each parameter be correct before the program can compile.

One item that is important. Copies of parameters are passed to any function. Copies are placed on the system stack or in registers or both before the function call is executed. Therefore, the program can use these parameters in any way without altering the calling program. In fact, after you have done what is needed with a parameter, you may use the parameter as a storage location for your function.

Data returned from a function will always be converted to the correct type before it is passed back to the program. If you wish to have a different type returned, the cast operator can be used to change the return data type to any type desired.

ANSI C defined the type `void`. This type is used in several different ways. If there is no function return, the prototype must identify the function as type `void`. Also when there are no function arguments, the argument list must contain the type `void`. This use of the keyword `void` will prevent problems in function calls.

Note that the function prototype above is terminated with a semicolon. The semicolon is needed in the function prototype, but it is not to be used after the name of the function in the code where the function is defined. The function prototype is a declaration statement that merely provides information to the compiler while the function prologue, the first line of the function, is a function definition which opens the code for the function.

The philosophy in C is to use functions with little provocation. Using many functions produces code that is easy to read and follow. Often it is easier to debug many small functions rather than a larger program. One must temper these ideas somewhat when writing code for small microcontrollers. Calling a function requires some overhead that is repeated each time the function is accessed. If the total overhead is more than the length of the function, it is better to use in-line code. In-line code implies that the function code is repeated in-line every time that it is needed. If the function code is much greater than the calling overhead, the function should be used. In between these limits, it is difficult to determine a hard-and-fast rule. In microcontroller applications, it is probably best to use function calls to a single function if there is a net savings of memory as a result. This savings is calculated by first determining the code needed prior to calling the function, the code needed to clean up the process after the function call, the number of times the function is called, and the length of the function. The in-line code will be smaller than the corresponding function code. Therefore, if the total code for the number of function calls listed first exceeds the total in-line code required to accomplish the same operations, then use the in-line code. Otherwise, use function calls.

The above argument is valid for microcontroller applications code. It does not necessarily follow for code written for large computers. When writing for a large computer, there are usually few memory constraints. In those cases, it is probably best to use more function calls and not be worried about the memory space taken up by function calls unless there is a serious speed constraint. When speed is a problem, the programmer must go through an analysis similar to that above with the dependent parameter being time rather than memory space. In small computers where several registers can be saved and restored when a function is entered and exited, single instructions can require many clock cycles. When deciding whether to use a function or in-line code, the programmer must assess the total time lost to entering and exiting a function each time it is entered, and weight that time lost as a fraction of the total time the program resides in the function. If this time is large, and the program requires too much execution time, consider the use of in-line functions.

It is always good to write small functions and create simple calling programs to exercise the small functions. These programs are used to debug the functions, and they are discarded after the functions are debugged. If later, the program constraints dictate that in-line code should be used, the essential code of the function can be written into the program wherever it is needed. Another approach that will be discussed in the next chapter is to use a macro definition to specify a small function. With a macro definition, the function code is written in-line to the program whenever the function is invoked.

Let us revisit an example used earlier. Write a program to calculate and display the square root of each integer less than 11:

```
/* Calculate and display the square roots of
numbers
1 <= x <=10 */

#include <stdio.h>

#define abs(t)  (((t)>=0) ? (t) : -(t))
#define square(t)  (t)*(t)

double sqr ( double );
int main(void)
```

```
{
   int i;
   double c;

   for(i=1;i<11;i++)
   {
      c=sqr(i);
      printf("\t%d\t%f\t%f\n",i,c,square(c));
   }
   return 0;
}
/* the square root function */

double sqr( double x )
{
   double x1=1,x2=1,c;

   do
   {
      x1=x2;
      x2=(x1 + x/x1)/2;
      c=x1-x2;
   }
   while(abs(c) >= .00000001);
   return x2;
}
```

The result of this calculation is shown below:

```
1     1.000000          1.000000
2     1.414214          2.000000
3     1.732051          3.000000
4     2.000000          4.000000
5     2.236068          5.000000
6     2.449490          6.000000
7     2.645751          7.000000
8     2.828427          8.000000
9     3.000000          9.000000
10    3.162278          10.000000
```

The second line of code

```
#define abs(t) (((t)>=0) ? (t) : -(t))
```

is called a macro definition. In this case, the macro definition has the appearance of a simple function. This function will calculate the absolute value of the argument t. The absolute value of the argument is a positive value. If the argument is positive, it is returned unchanged. If it is negative, it is multiplied by −1 before it is returned. A macro definition is a type of character expansion. Whenever the function abs(x) is found in the code, the character string (((x)>=0) ? (x) : -(x)) is put in its place. The argument x can be any valid C expression. This function returns the absolute value of its argument. The macro definition

```
#define square(t) (t)*(t)
```

returns the square of t. Since these arguments can be any valid C expression, it is necessary to be cautious when writing the macro definitions. Suppose that the parentheses were left out of the above expression, and the macro were written

```
#define square(t) t*t
```

Also suppose that the code using this function were as follows:

```
x=square(y+3);
```

The character expansion of this expression would be

```
x=y+3*y+3;
```

The result of this calculation is 4*y+3 and not (y+3)*(y+3) as expected. When writing macro definitions, surround all arguments and functions created by the macro with parentheses so that all arguments are evaluated prior to use in the macro definition function.

Another problem can sneak into your code through improperly written macros. Suppose that you want a macro that doubles the value of its argument. Such a macro could be written

```
#define times_two(x)   (x)+(x)
```

This macro when expanded in the following expression

```
x = 7*times_two(y);
```

will yield

```
x = 7*(y) +(y);
```

which is the wrong answer. The problem can be easily corrected by wrapping parentheses around the whole macro as

```
#define times_two(x)  ((x)+(x))
```

Remember, always place any argument of a macro within parentheses and always place the entire macro definition in parentheses.

Note that there is no semicolon at the end of the macro definition. There should not be. If a semicolon were placed at the end of a macro definition, extra semicolons would be entered into expressions containing macros with unpredictable results.

The function prototype

```
double sqr( double );
```

notifies the compiler that the function returns a double and takes a double argument.

Inside of the main function, the `for` loop

```
for(i=1;i<11;i++)
{
   c=sqr(i);
   printf("\t%d\t%f\t%f\n",i,c,square(c));
}
```

is used to calculate the several results. The variable c is of the type double, and i is an `int`. The expression `c=sqr(i)` will be accepted by the compiler. This function returns a double which can be stored in c. The argument is an `int`, but the compiler recognizes that `sqr` requires a double argument and converts i to a double before it is sent to the function `sqr()`.

The code in the function `sqr()` is a restatement of a square root operation that we saw earlier. In this case, the function processes floating-point numbers rather than the integers used before. Three double variables are needed. The variables x1 and x2 are the current and last values found in the Newton iteration. As the result converges to the correct value for the square root, several things happen. Variables x1 and x2 become equal. The square of x2, or x1 for that matter, becomes equal to x. The product of x1 and x2 becomes equal to x. Any of these tests can be used to determine if the estimate has been through enough iterations to be accurate.

The macro definition abs(x) is used to test for the end of the loop. You will note that the argument is evaluated three times for the expansion of the macro. If we place a lot of calculation within the argument of a macro definition, the expansion of the macro may cause the code to calculate the argument to be repeated several times. For this reason, the expression

```
c = x1 - x2;
```

is placed inside of the while loop, and the test to determine loop termination uses abs(c).

At the end of a function, a return statement will cause the value of the expression following the word return to be evaluated and returned to the calling function. If this expression is not of the type specified by the function prototype, it will be converted to the correct type prior to being returned to the calling function. The expression following the return statement can be enclosed in parentheses or not.

Another example will show the use of static external variables.

```
/* Read in a string from the keyboard and print it
out in reverse order. */

#include <stdio.h>

#define null 0
/* some function prototypes */
void push(int);
int pull(void);

int main(void)
{
    int c;
    push(null);
    while((c=getchar())!='\n')
        push(c);

    printf("\n");
    while((c=pull())!=null)
```

```
      putchar(c);
   printf("\n");
   return 0;
}
```

The following code is to be compiled in a separate file from the code above:

```
/* the stack function */
#define MAX 100

static int buffer[MAX];
static int sp;

void push(int x)
{
   if(sp < MAX)
      buffer[sp++]=x;
   else
   {
      printf("stack overflow\n");
      exit(1);
   }
}
/* the unstacking function */

int pull(void)
{
   if(sp > 0)
      return buffer[-sp];
   else
   {
      printf("stack underflow\n");
      exit(1);
   }
}
```

A stack is a last in, first out (LIFO) structure. Therefore, a stack can be used to reverse the order of data sent to it. The above program uses a stack operation. The functions push and pull identified in the function prototypes perform the stacking operations for the main program. In the main program, a `null` is pushed onto the stack to identify the end of the data as it is pulled off of the stack a character at a time. Data are read in a character at a time, and as each character is read in, it is pushed onto a stack. When a new line character is detected, the input phase is stopped, and the data written to the stack is pulled off and printed. When the `null` is detected, the data have all been pulled off the stack, and the program is ended.

In the function above, the function `exit()` is used. This function is similar to the `return` operation. Whenever a call to the function `exit` is executed, the argument is evaluated, any files open for write are flushed and closed, and the control of the computer is returned to the operating system. The evaluation of the argument is returned to the operating system. Whenever a `return` is encountered, the expression following the `return` call is evaluated and returned to the calling function. If control is in `main()` when the `return` is encountered, the evaluation of the expression is returned to the operating system. Also, from `main()` all files open for write are flushed and closed. The function `exit()` and `return` work the same in `main()`, but `exit()` exits a program and returns control to the operating system from anywhere in the program.

In a separate compilation, the stack functions are compiled. In that function, the macro definition `MAX` is defined as 100. Macro definitions can be used to define any character string that is needed in a program. They are not limited to defining pseudo functions. Two external variables are defined in this file: an array of `MAX` integers named `buffer` and an `int` called `sp`. These variables are declared to be static. As such, these variables can be accessed by any function in the file, but they are not available to any function outside of the file. The variable `sp` is used as an index into the array buffer. When a push is executed, a test to determine if `sp` is less than `MAX` is completed. If `sp` is less than `MAX`, the data are stored at the `sp` location in buffer and `sp` is then incremented. Otherwise, an error message indicates that a stack overflow has occurred and the program is exited.

The `pull()` function is the reverse of the `push()` operation. First a check is made to see if there are some data on the stack to be pulled off. If there are data, the stack pointer is decremented, and the content of the buffer at that location is returned to the calling program. In the event that sp is 0 when the pull operation is executed, a stack underflow message is sent to the screen prior to exiting the program.

The advantage to our making the buffer and the stack pointer in the stack functions static can be easily seen. Suppose that these variables could be accessed from anywhere in the program. In that case, it would not be necessary for the programmer to call the functions push or pull to stack and unstack data. If several different programmers were using the same stack for different tasks in one large program, it would be possible for different programmers to access the stack as expected, or from their own tasks. Suppose a programmer made the mistake of pre-decrementing the stack pointer on stacking and post-incrementing the stack pointer on unstacking. The whole program would suddenly be in chaos. Therefore, masking these variables from the rest of the program can reduce serious potential debugging problems.

Recursion

A recursive routine is one that calls itself. The C language is supposed to produce recursive code. Compilers for large machines usually support recursion, but recursion is often one of the first casualties on small microcontrollers. Automatic variables are created when a function is entered, and they are stored on the stack. Therefore, each time a function is called, a new stack frame is created, and variables in place from an earlier execution of the function are unaltered. Such a function is called re-entrant, and re-entrant functions are also recursive. An example of a simple recursive function is the factorial:

```
n! = n*(n-1)*(n-2)*....*2*1
```

An interesting observation that can be made of factorial is that

```
n! = n*(n-1)!
```

or n factorial equals n times n-1 factorial. Also, the factorial of 0 is defined as 1. With these definitions, it is possible to write the following recursive function to calculate the factorial of a number:

```
long factorial( int n)
```

```
{
   if(n==0)
      return 1;
   else
      return n*factorial(n-1);
}
```

This surprisingly simple function calculates the factorial. Whenever you write a recursive routine, it is important to have means of getting out of the routine. In the above case, when the argument reaches zero, the function returns a result rather than calling itself again. At that time the routine will work itself back a level at a time until it reaches the initial factorial call, and the calculation will be done.

Recursion can create some elegant code in that the code is very simple—often too simple. There is a cost in the use of recursive code, and that is stack space. Each time a function call is made, the argument is placed on the stack and a subroutine call is executed. As a minimum, the return address is two bytes, and the value of the argument is also two bytes. Thus, at least four bytes of stack space are needed for each function call. That is no problem when the factorial of a small number is calculated. (The factorial of 13 is larger than can be held in a `long`, so only small numbers can be considered for a factorial.) However, if a recursive function is written that calls itself many times, it is possible to get into stack overflow problems.

Another interesting recursive routine is the function to calculate a Fibonacci number. A Fibonacci number sequence is described by the following function:

```
long fib(int n)
{
   if(n==1)
      return 1;
   else if(n==0)
      return 1;
   else
      return fib(n-1) + fib(n-2);
}
```

This sneaky function calls itself twice. Some interesting characteristics of this function are left to the exercises that follow.

EXERCISES

1. Write a function to calculate the Fibonacci number for 10, 20, 30, and 40.

2. Devise a means for determining the number of times the `fib` function is called in the above program. What is this number for `fib(20)`?

3. A separate problem from the number of times the function is called is the number of times the function is called without exiting through the bottom of the function. This term is called the depth of the function. Determine the maximum depth of the `fib()` function in calculating `fib(20)`.

4. Repeat problem 1, but rewrite the Fibonacci number function so that it does not employ recursion. How does the time to execute this version of the `fib(30)` compared to that above?

Summary

The basics of writing programs in C have been discussed in this chapter. Several important concepts have been skipped over in this presentation and will be covered in Chapter 2.

If you have not done so, it is recommended that you enter and compile each example shown. These programs will all compile and run under the MIX PowerC Compiler, the Cosmic compiler for the M68HC11, the M68HC16, and the M68300 series of chips. They also compile on the DIAB MCORE compiler. With the exception of the MIX PowerC compiler, all of the compilers listed are cross compilers that run on a PC platform, but compile code for another computer.

The ANSI version of the language is the current standard, and none of the classical C constructs have been introduced in this text. It is not to the programmer's advantage to use the classical version of the language, even though programs that conform to classical C will compile on an ANSI compliant compiler. Any version of a C++ compiler structured to compile C code will also compile ANSI C code. The DIAB compiler listed above is a C/C++ compiler.

Chapter 2

Advanced C Topics

Pointers

The use of pointers sets the C language apart from most other high-level languages. Pointers, under the name of *indirect addressing*, are commonly used in assembly language. Most high-level languages completely ignore this powerful programming technique. In C, pointers are simply variables that are the addresses of other variables. These variables can be assigned, be used as operands, and treated much the same as regular variables. Their use, however, can simplify greatly many of the truly difficult problems that arise in day-to-day programming. Let's discuss pointers with the view that they offer us a powerful new tool to make our programming jobs easier.

How To Use Pointers

A pointer to a variable is the address of a variable. Thus far in our discussion of variables, the standard variable types were found to occupy 8-, 16-, or 32 bits depending on the type. Pointers are not types in the sense of variables. Pointers might occupy some number of bits, and their size is implementation dependent. In fact, there are cases of pointers to like types having different sizes in the same program depending on context.

If the variable px is a pointer to the type int and x is an int, px can be assigned the address of x by

```
px = &x;
```

The ampersand (&) notifies the compiler to use the address of x rather than the value of x in the above assignment. The reverse operation to that above is

```
x = *px;
```

Here the asterisk (*) is a unary operator that applies to a pointer and directs the compiler to use the integer pointed to by the pointer px in the assignment. The unary * is referred to as the *dereference* operator. With these two operators, it is possible to move from variable to pointer and back again with ease.

A pointer is identified as a pointer to a specific type by statements like

```
int *px,*pa;
long *pz;
float *pm;
```

In the above declarations, each of the variables preceded by the unary asterisk identifies a pointer to the declared type. The pointers px and pa are the addresses of integers. pz is the address of a long, and pm is the address of a floating point number. Always remember: if pk is a pointer to an integer, *pk is an integer. Therefore, the declarations in the above examples are correct and do define ints, longs and floats. However, when the compiler encounters the statement

```
int *pi;
```

it provides memory space for a pointer to the type int, but it does not provide any memory for the int itself. Suppose that a program has the following declaration:

```
int m,n;
int *pm;
```

The statement

```
m = 10;
pm = &m;
```

will assign the value 10 to m and put its address into the pointer pm. If we then make the assignment

```
n = *pm;
```

the variable n will have a value 10.

Another interesting fact about pointers is shown in the following example:

```
int able[10];
int *p;
```

After it has been properly declared, when the name `able` is invoked without the square brackets, the name is a pointer to the first location in the `array`. So the following two statements are equivalent:

```
p = &able[0];
```

and

```
p = able;
```

Of course, the `nth` element of the array may be addressed by

```
p = &able[n];
```

The unary pointer operators have higher precedence than the arithmetic operators. Therefore,

```
*pi = *pi + 10;
y = *pi + 1;
*pi += 1;
```

all result in the integers being altered rather than the pointers. In the first case, the integer pointed to by `pi` will be increased by `10`. In the second case, `y` will be replaced by one more than the integer pointed to by `pi`. Finally, the integer pointed to by `pi` will be increased by 1. The statement

```
++*pi;
```

causes the integer pointed to by `pi` to be increased by 1. Both `++` and the unary `*` associate from right to left, so in the following case

```
*pi++;
```

the pointer `pi` is incremented after the dereference operator is applied. Therefore, the pointer is incremented in this case, and the integer `*pi` remains unaltered. If you wish to post-increment the integer `*pi`, use

```
(*pi)++;
```

At times, it is necessary to pre-increment the pointer before the dereference. In these cases, use

```
*++pi;
```

Pointers work with arrays. Suppose that you have

```
int *pa;
int a[20];
```

As we have already seen,

```
pa = a;
```

assigns the address of the first entry in the array to the pointer pa much the same as

```
pa = &a[0];
```

does. To increment the pointer to the next location in the array, you may use

```
pa++;
```

or

```
pa = pa + 1;
```

In any case, the pointer that is 1 larger than pa points to the next element in the array. This operation is true regardless of the type that pa points to. If pa points at an integer, pa++ points to the next integer. If pa points at a long, pa++ points to the next long. If pa points to a long double, pa++ points to the next long double. In fact, if pa points to an array of arrays, pa++ points to the next array in the array of arrays. C automatically takes care of knowing the number that must be added to a pointer to point to the next element in an array.

In the above case, since a is an array of 20 integers, and pa points to the first entry in the array, it follows that pa+1 points to the second element in the array, and pa+2 points to the third element. Similarly, *(pa+1) is the second element in the array and *(pa+2) is the third element.

There is a set of arithmetic that can be applied to pointers. This arithmetic is different from the normal arithmetic that can be applied to numbers. In all cases, the arithmetic applies to pointers of like types pointing to within the same array only. When these conditions are met, the following arithmetic operations can be completed:

- *Pointers can be compared.* Two pointers to like types in the same array can be compared in a C logical expression.

- *Pointers can be subtracted.* Two pointers to like types in the same array can be subtracted with a C arithmetic expression. The result of the subtraction will be the number of elements between the two pointers, not the difference in the values of the pointers.
- *Pointers can be incremented and decremented.* A pointer into an array can be either incremented or decremented. The result will be a pointer that points to an adjacent element in the array.
- *Pointers can be assigned.* A pointer can be assigned to another pointer of the same type.

Pointers cannot be added, multiplied, divided, masked, shifted, or assigned a pointer value of an unlike type. Pointers can be assigned to another pointer of the type `void*`.

The name of an array is a pointer to the first element in this array. This type of variable occupies a special place in C. The name can always be used as a pointer, and assigned to another pointer, but it cannot be assigned to. Any attempt to assign a new value to the array name would upset the beginning of the array to the program.

Therefore, an array name as a pointer can be used as a value on the right side of an assignment equal sign, but it cannot be used on the left side of the equal sign. C variables are broken into the types `rvalue` and `lvalue`. All C variables, with the exception of the names of functions and arrays, are both `rvalues` and `lvalues`. They can be used on either side of an equal sign in an assignment statement. Function names and array names are `rvalues` and can be used only on the right side of the equal sign in an assignment statement. Variables declared as constants and constants created by the `#define` statement are also `rvalues`.

The type `void*` has a special meaning when applied to pointers. A `void` pointer (sometimes referred to as a *generic* pointer) is a pointer that does not point at any specific type. Some functions will return `void` pointers, and to use these pointers, they must be cast onto the type that they represent. Therefore, a pointer of the type `void*` can be assigned the value of a typed pointer. However, unless the `void*` pointer is cast onto the specified type, the `increment`, `decrement`, `subtraction`, and so forth will not work. A void pointer can be used in expressions, but it is impossible to alter its value.

Pointers and Function Arguments

Values passed to functions as arguments are copies of the real values. The data to be passed to a function are pushed on the stack or saved in registers prior to the function call. Therefore, if the function should modify any of the arguments, this modification would not propagate to the calling function. Therefore, a function like

```
void swap(int x, int y)
{
    int temp;

    temp = x;
    x = y;
    y = temp;
}
```

does nothing but swap two passed values, and those values are never returned to the calling program. This performance is good as well as bad. The bad situation is shown above when the function simply does not work as expected. The good side is that it is not easy to inadvertently change variable values in the calling program. The use of pointers permits this problem to be avoided. The technique is called passing parameters by *reference*. Consider the following function:

```
void swap(int* px, int* py)
{
int temp;

temp = *px;
*px = *py;
*py = temp;
}
```

Here the integers pointed to by px and py are swapped. These integers are the values in the calling program. The pointer values in the calling program are unaltered.

C makes extensive use of passing parameters by reference. Recall the first program in Chapter 1. That program has the line

```
printf("Microcontrollers run the world!\n");
```

We now know that the string "Microcontrollers run the world!\n" is nothing more than a character array terminated by a zero. The compiler creates that array in the memory of the program. That array is not passed to printf. A pointer to that array is passed to printf. Then printf prints the characters to the standard output and increments through the array until it finds the 0 terminator. Then it quits. In other instances, arguments like s[] were used. In these cases, C automatically knows to pass a pointer to the array. It is interesting. If you have an array int s[nn] and pass that array to a function as s, you can use an argument int* p in the function and then in the function body deal with the array p[]. The C language is extremely flexible in the use of pointers. An example of this type of operation is as follows:

```
/* Count the characters in a string */

#include <stdio.h>

int strlen(char *);

int main(void)
{
   int n,s[80];

   fgets(s,80,stdin);
   n=strlen(s);
   printf("The string length = %d\n",n);
   return 0;
}

int strlen( char *p)
{
   int i=0;

   while(p[i]!=0)
      i++;
   return i;
}
```

The above version of `strlen()` will return the number of characters in the string not including the terminating 0. The function `gets()` is a standard function that retrieves a character string from the keyboard. The prototype for this function is contained in `stdio.h`. It is interesting to note that the function `strlen()` can be altered in several ways to make the function simpler. The index `i` in the function can be post-incremented in the argument so that it is not necessary to have the statement following the `while()`. This function will be

```
int strlen(char *p)
{
   int i=0;

   while(p[i++]!=0);
   return i;
}
```

Because the `while()` argument is evaluated as a logical statement, any value other than 0 will be considered TRUE. Therefore,

```
int strlen(char *p)
{
   int i=0;

   while(p[i++]);
   return i;
}
```

provides exactly the same performance. Yet another viewpoint is found by

```
int strlen(char* p)
{
   int i;

   for(i=0;*p++;i++);
   return i;
}
```

Each of the above functions using `strlen()` show a different operation involving passing arguments by reference.

Let us examine another use of passing arguments by pointers.

The following function compares two strings. It returns a negative number if the string p is less than the string q, 0 if the two strings are equal, and a positive number if the string p is greater than the string q.

```
int strcomp(char* p, char* q)
{
   while( *p++ == *q++)
   if(*(p-1) == '\0') /* The strings are equal */
   return 0; /* zero */
   return *(p-1) - *(q-1); /* The strings are
                    unequal, return the correct value*/
}
```

Here the while() statement will examine each character in the string until the argument is not equal. Note, that the pointers p and q are incremented after they are used. The test for equality will not occur until after the pointers are incremented, so it is necessary to decrement the pointer p to determine if the last equal value in the two strings is a zero character. If the last value is a zero, the two strings are equal, and a 0 is returned. If the last tested character is not a zero, the two strings are not equal and the difference between the last character in p and the last character in q is returned. This choice will give the correct sign needed to meet the function specification.

Another approach can be used that eliminates the increments and decrements within the program and confines them to the argument of a for construct. Consider

```
int strcomp(char* p, char* q)
{
   for( ; *p==*q ; p++, q++)
      if( *p == '\0')
         return 0;
   return *p - *q;
}
```

The pointers p and q are both incremented after the test in the if statement. Therefore, the pointer values are correct for the if statement as well as for calculation of the return value when the strings are not equal.

C is completely unforgiving if you exceed the boundaries of the array in your calculations. C does not have intrinsic boundary checks, and it is possible to increment pointers right off the end of an array. For that matter, the array index can be decremented to addresses below the beginning of the array. When a program makes such a mistake, it will destroy other data, perhaps destroy the program being executed, or in the worse case, the program can crash the system. The simple single-tasking computers that run MS-DOS or similar operating systems have no memory protection feature that protects one task from another. In such cases, a program that runs wild and overwrites memory not assigned to it can destroy other tasks that are loaded in memory whether they are running or not.

EXERCISES

1. Write several versions of a function that copies one string into another.

2. Write a function that *concatenates* two strings. Concatenation of two strings means that one string will be written at the end of the other. Write this function with and without the use of pointers. What is the advantage of the pointer version of the function?

Let us consider a problem that will be analyzed in more detail later. Suppose that a function is needed that will sort the contents of an array into ascending (or descending) order. There are several sort functions, and each has its own set of advantages and disadvantages. The simplest and probably most intuitive is called the *bubble* sort. In a bubble sort, a swap flag is reset. The first entry in the array is compared with the second, and if they are in the wrong order, they are swapped. Then the second entry is compared with the third, and they are swapped or not. If a swap ever occurs, the swap flag is set. The elements of the array are successively compared and swapped until all of the elements in the array have been compared. If a swap has occurred during the scan of the array, the swap flag is reset, and the whole process is repeated. This process repeats until the scan of the array causes no swaps to occur. At that time, the array is in order.

If the array contains n elements, this approach requires on the order of n squared compares and swaps. For large arrays, the time to sort an array with a bubble sort is inordinate. Another approach was discovered by D. L. Shell. The *Shell* sort allows the array to be sorted with *n* times the logarithm base two of *n*. This difference can be substantial. For example, if the array contains 1000 elements, the bubble sort will require on the order of 1,000,000 compares and swaps while the Shell sort will need only 10,000. Therefore, the bubble sort should not be used to sort large arrays.

Another sort technique that is used was discovered by C. A. R. Hoare. This technique is called the *quick* sort. It uses a recursive approach to accomplish the sort, and it also requires *n* log base 2 of *n* compares and swaps. The shell sort and the quick sort usually require about the same time to accomplish the sort. We will demonstrate a shell sort now.

The code for a shell sort follows. Note that extensive use of pointers is used in this version of the shell sort.

```
/* shell_sort(): sort the contents of the array *v
into ascending order */

void shell_sort(int* v, int n)
{
   int gap, i, j, temp;

   for( gap = n/2; gap > 0; gap /= 2)
      for( i=gap; i < n; i++)
         for(j=i-gap;j>=0&&*(v+j)>*(v+j+gap);
            j -= gap)
         {
            temp = *(v+j);
            *(v+j) = *(v+j+gap);
            *(v+j+gap) = *(v+j);
         }
}
```

The shell sort receives as arguments a pointer to an array of integers and an integer *n* which is the length of the array.

This sort works by successively dividing the array into subarrays. The first operation divides the whole array into two parts; the second divides each of the two subarrays into two new subarrays, for a total of four arrays; and so forth.

The corresponding elements in each subarray are compared, and if they are out of order, they are swapped. At the close of this operation, a new set of subarrays are created, and the process is repeated over these subarrays. Eventually, the gap between elements in the subarrays will be reduced to one, and the array contents will be sorted.

The outer `for` loop above controls the splitting of the arrays into the subarrays. The second loop steps along the array pairs. The innermost loop successively compares the elements that are separated by gaps in the subarrays. If elements are found that are out of order, they are reversed or swapped in the array.

EXERCISES

1. Restate the shell sort above to use arrays rather than pointers to arrays.

2. Write a program that reads in characters from the input stream and record the number of occurrences of each character. Calculate the percentage occurrence of each character, and print out the result in ascending order of percentage of occurrence.

Functions can return pointers. For example, prototype to a function that returns a pointer is:

```
int *able(char* );
```

Here `able` returns a pointer to an integer.

In C, a NULL pointer is never used. A NULL pointer implies that something is to be stored at the address zero. This address is never available for data storage, so no function can return a valid NULL pointer. The NULL pointer can be used as a flag or an error return. The programmer should never allow a NULL pointer to be dereferenced, which implies that data are read or stored at 0.

If C will support a pointer to a variable, it requires but little imagination to reason that C will support pointers to pointers. In fact, there is no practical limit in the language to the depth of dereference C will support. C will also allow arrays of pointers, and pointers to functions. (We discussed pointers to arrays in the preceding section.) An array of pointers can be very useful when needed. A most obvious use for an array of pointers is to read the contents of a command line to a program. So far in our discussion of programs, there have been no provisions for reading the content of a command line that is written to the screen when the program is executed. A command line can be read by the program. The definition of the program name `main` when extended to read in a command line is as follows

```
void main ( int argc, char *argv[])
```

The integer variable `argc` is the number of entries on the command line. The array of pointers to the type `char argv[]` contains pointers to strings. When entering arguments onto the command line, they must be separated by spaces. The first string pointed to by `argv[0]` is the name of the program from the command line. The successive pointer values point to additional character strings. These strings are each 0 terminated, and they point to the successive entries on the command line. The value of `argc` is the total number of command line entries including the program name. It must be remembered that each entry in `argv[]` is a pointer to a string. Therefore, if a number is entered on the command line, it must be converted from a string to an integer, or floating point number, prior to its use in the program. Let us see how this concept can be used. Earlier, we wrote a function to calculate a Fibonacci number. Let's use this function in a program in which the argument for the Fibonacci calculation is read in from the command line:

```
#include <stdio.h>
#include <stdlib.h>

long fib( int ); /* Fibonacci number function
                prototype */

int main( int argc, char* argv[] )
{
   int i;
   i = atoi(argv[1]);
   printf("The Fibonacci number of %d  = %ld\n", i,
      fib(i));
   return 0;
}
```

We will not repeat the code for `fib(i)`. A new header file, `stdlib.h`, is included with this program. The function prototype for `atoi()` is contained within this header file. The standard command line arguments are used in the call to `main()`. The line

```
i = atoi(argv[1]);
```

causes the function atoi—ASCII to integer conversion—to be executed with a pointer as an argument. This pointer points to the first argument following the program name on the command line. In this case, it will be pointing to an ASCII string that contains the number to be used as an argument for the fib() call. This string must be converted to an integer before fib() can operate on it, which is exactly what the atoi() function accomplishes. The final line in this program prints out the result of the calculation.

Another example of use of the command line arguments is to print out the command line. The following program will accomplish this task.

```
#include <stdio.h>
int main( int argc, char* argv[] )
{
   int i;
   for(i=0; argc--; i++)
   printf("%s ",argv[i]);
   printf("\n");
   return 0;
}
```

The arguments to main() are the same as before. This program enters a for loop that initializes i to zero. It decrements argc each time it tests its value, and executes until the loop in which argc is decremented to 0. The printf call

```
printf("%s ",argv[i]);
```

prints out the string to which argv[i] points. Notice the space in the string "%s ". This space will force a space between each argument as it is printed. The program is written so that there are no new line characters printed. Arguments will all be on one line, and they will each be separated by a space. The printf() statement after execution of the for() loop will print out a single new line so that the cursor will return to the next line after the program is executed.

Command line entry is but a simple example of use of arrays of pointers. Another area in which arrays to pointers are needed is in ordering strings of data. For example, it is possible to collect a large number of words in memory, say from an input stream. Suppose that it is needed to alphabetize these words. We saw earlier, that the shell sort will order the contents of an array, so it might be possible to simply modify the shell sort to do this job.

First, a comparison is needed that will determine if the lexical value of a one word is smaller, equal, or larger than that of another word. Such a compare routine was outlined above. Second, a swap routine that will swap the words that are in the wrong order. Here is a case where an array of pointers can be quite useful. Assume that the program that reads in the data will put each word into a separate memory location and keep an array of pointers to the beginning of each word rather than just the array of the words themselves. Then in the shell sort, when a swap is required, rather than swapping the words, swap the pointers in the array. Swap routines that swap pointers in the array are very easy to implement. On the other hand, swap routines to swap two strings in memory are difficult and slow. Therefore, we can create a sort routine that is much more efficient if we use an array of pointers rather than an array of strings.

```
/* shell_sort(): sort the contents of the array
char* v[] into ascending order */

#include <string.h>
void shell_sort(char* v[], int n)
{
   int gap, i, j;
   char* temp;
   for( gap = n/2; gap > 0; gap /= 2)
      for( i=gap; i < n; i++)
         for( j=i-gap; j>=0 &&
                 strcmp(v[j],v[j+gap]); j -= gap)
         {
            temp = v[j];
            v[j] = v[j+gap];
            v[j+gap] = v[j];
         }
}
```

Here the strcmp() routine is used to determine if a swap is needed. strcmp() is identified in the header file string.h. When needed, the contents of the array of pointers to the beginning of the words is swapped rather than swapping the words themselves.

An important point of style: Recall that a few pages back a function `strcomp()` was written as an example. It did the exact same operations as `strcmp()` above. Why should we choose one over the other? Well, library functions have been written, rewritten, debugged, and worked over for years. They work correctly, and you can count on their robust construction. As a general rule, use a library function if you can find one to do the job that you are attempting. Most of the time, programmers who write duplicates of library functions do it to satisfy their own egos. The reward is poor when a bug is discovered, especially in production code, that could have been avoided by using a standard library function.

Multidimensional Arrays

C supports multidimensional arrays. Programmers often find that much of the need for multidimensional arrays will go away with the availability of pointers. Multidimensional arrays in C are thought of as arrays of arrays. This idea can be extended to more than two dimensions. A two-dimensional array is identified as

```
array[x][y]; /* [row][column] */
```

The first argument to the right can be thought of as the row dimension, and the second the column dimension. Elements specified by the rightmost argument are stored in adjacent memory locations.

An array can be initialized at declaration time. For example:

```
int array [3][4] = { {10,11,12,13},
                      {14,15,16,17},
                      {18,19,20,21} };
```

It is equally valid to initialize the array as follows:

```
int array [3][4]={10,11,12,13,14,15,16,17,18,19,20,21};
```

Either form of initialization will place the proper numbers in the proper location in memory, and the two-dimensional indices will work properly in either case.

Frequently, it is needed to know the size of a variable in C. This variable can be a basic type, an array, a multiple dimension array, or even a structure that will be introduced later. C provides an operator that has much the appearance of a function called `sizeof`. To determine the size of any variable, use the `sizeof` operator as follows:

```
a = sizeof array;
```

which will return the number of bytes contained in `array[][]` above.

There are several important different ways that you can use the `sizeof` operator. First, the value of the return from `sizeof` is in characters. The type of the return is called a type `size_t`. This special type is usually the largest `unsigned` type that the compiler supports. For the MIX compiler, it is an `unsigned long`. If you should want the size of a type in your program, you should enclose the parameter in parentheses. If you want the size of any other item, do not use the parentheses. One other item. If the `sizeof` operator is used in a module where an array is defined, it will give you the size of the array as above. If you should pass an array to a function, the array name degenerates to a pointer to the array, and in that case, the return from the `sizeof` operator would give you the size of a pointer to the array.

A common example program using two-dimensional arrays is to determine the Julian date. The Julian date is simply the day of the year. The following function is one that allows counting the number of days that have passed in a year.

```
int month_days[2][13] = {
        {0,31,28,31,30,31,30,31,31,30,31,30,31},
        {0,31,29,31,30,31,30,31,31,30,31,30,31}};
int Julian_data(int month, int date, int year)
{
   int i,leap;
   leap = year%4==0 && year%100!=0 || year%400==0;
   for(i=1;i<month;i++)
      day += month_days[leap][i];
   return day;
}
```

The declaration

```
int month_days[2][13] = {
        {0,31,28,31,30,31,30,31,31,30,31,30,31},
        {0,31,29,31,30,31,30,31,31,30,31,30,31}};
```

types `month_days` as an array of 26 integers. This array is a two-dimensional array of two rows of 13 columns each. The values assigned are shown. The extra 0 entry at the beginning of each array is to allow the conventional month designations 1 through 12 to be used as indices and not have to worry about the fact that arrays in C start with a 0 index.

The introduction of the program is normal. The function returns an `int` and expects to receive three `int` arguments; one for the month, one for the day of the month, and one for the year. Note that the year must be the full year, like 2013, rather than merely 13. The first executable statement is the logic statement:

```
leap = year%4==0 && year%100!=0 || year%400==0;
```

Leap years are usually every four years. However, a small discrepancy still exists in the length of the year with the "once each four years" correction. To further correct the error, the calendar makers have decided that years divisible by 100 will not be a leap year unless the year is divisible by 400. The above statement is a logical statement that determines first if the year is divisible by 4. If it is divisible by 4, it is then checked to determine if it is divisible by 100. The result of this much of the analysis will be TRUE for any year divisible by 4, and not divisible by 100. If this portion of the calculation is TRUE, `leap` will be assigned a value TRUE, or 1, and the evaluation will terminate. If the result of the first portion of the calculation is FALSE, it will be necessary to evaluate the last term to determine if the whole statement is TRUE or FALSE. The variable leap will be assigned the result of

```
year%400==0
```

in this case.

`leap` is assigned a value of 1 or 0 according to the result of the logic evaluation. This value can be used as an index into the two dimensional array to determine if the number of month days in a leap year or a nonleap year will be used in the calculation of the Julian date.

Pointers and Multidimensional Arrays

Perhaps one of the most widely misunderstood and therefore mysterious aspects of pointers and C has to do with multidimensional arrays. These problems are really not difficult. A multidimensional array must always be understood as being arrays of arrays of arrays and so forth. For example, the declaration

```
int ar[3][5];
```

defines three arrays of five elements each. We have already seen that data stored in this array is column major. The second argument points to a column in the two-dimensional array, and `ar[n][0]` and `ar[n][1]` are stored in adjacent memory locations. Following the logic of array names and pointers, the array name `ar` is a pointer to the first element in the array. The value obtained when using `ar` as an `rvalue` is `&ar[0][0]`. The order of evaluation of the square brackets is from left to right so that `*(ar+1)` is a pointer to the element `ar[1][0]` in the array. Think of the two-dimensional array as being `*(ar+n)[i]` where n has a range from 0 to 2 and i has a range from 0 to 4. An increment in n here will increment the absolute value of the pointer by `5*sizeof(int)`.

These ideas can be carried to the next level. The element `*(ar+n)` is a pointer to the first element of a five-element array. Therefore, the evaluation of `*(*(ar+n)+i)` is the value found in the location `ar[n][i]`. The important item is that the right-most argument in multiple dimensional arrays point to adjacent memory locations, and the increments of the left arguments step the corresponding pointer value from array to array to array.

These ideas can be extended to arrays of more than two dimensions. Had the array been

```
double br[3][4][5];
```

then `*(*(*(br+1)+2)+3)` would be the element `br[1][2][3]` from the above array, and `*(*(br+1)+2)+3` is a pointer to this element.

EXERCISES

1. Write a function that will receive the year and the Julian date of that year, and calculate the month and date.

2. Write a function that will calculate the product of a 4 by 4 matrix and a scalar. The scalar product requires multiplication of each element in the matrix by the scalar value.

3. Write a function that will calculate the vector product of two 1 by 4 matrices. The vector product of two arrays is the sum of the products of the corresponding elements of the two arrays.

4. Write a function that will calculate the matrix product of two 4 by 4 matrices. The matrix product is a matrix whose i, j element is the vector product of the ith row of the first matrix and the jth column of the second matrix. As such, the product of two 4 by 4 matrices is a 4 by 4 matrix.

C has pointers to functions. A common use for pointers for functions is in creating vector tables for microcontrollers. Most microcontroller applications involve the use of interrupts. When an interrupt occurs, the machine status is saved, and program control is transferred to an interrupt service routine. At the close of the interrupt service routine, the machine status is returned to the earlier condition, and control is returned to the interrupted program. An address table in memory called the vector table contains the addresses of each interrupt service routine. It is the programmer's responsibility to fill the vector table with the proper addresses for the various interrupt service routines. Pointers to functions allow the programmer access to the addresses of the interrupt service routines. We will see several different methods to create the vector tables in the chapters on the individual microcontrollers that follow.

A pointer to a function is identified by

```
int (*function_ptr)();
```

The above declaration says that `function_ptr` is a pointer to a function that returns a type `int`. The arguments are not declared here. The parentheses surrounding `*function_ptr` are required. If they were not included, the declaration would declare `function_ptr` as a function that returns a pointer to the type `int`. If `function_ptr` is a valid pointer to a function, the function can be accessed by

```
(*function_ptr)(args);
```

The above declaration form can be used for arrays as well as functions. The declaration

```
char (*array_ptr)[];
```

states that `array_ptr` is a pointer to an array of char. The declaration

```
char* array[];
```

states that `array` is an array of pointers to the type `char`. Although you will rarely find it used, these declarations can be combined to create very complicated declarations.

One construct from the general area of complicated declarations is so important to microcontroller code that it must be covered. C supports variable types called `lvalues`. As mentioned earlier, an `lvalue` is a type of variable that can be the destination for an assignment. Most variables in C are `lvalues`. Notable exceptions are function names and array names. If a program deals with a number that can be a memory address, it can be made accessible to the language by casting the address to an appropriate type. For example, suppose that a special table is located at the address 0x1000 in memory. Further, suppose that the type to be stored at that address is an integer. Here, the code sequence

```
(int *) 0x1000
```

forces the number 0x1000 to be a pointer to a type `int`. Often this idea must be carried further, and the programmer wants to put a value into a specific address. The above representation is a pointer to the type `int`. Therefore, a value can be assigned to that `int` by

```
*(int *) 0x1000 = integer_value;
```

which will place `integer_value` into the location 0x1000 in the computer memory.

Frequently, control registers, data registers, and input/output port registers are placed at specific locations in memory. These register locations can be converted to tractable C names by use of the `#define macro` capability of C. Suppose that an I/O port is located at the address.

```
#define PORTA (*(char *) 0x1000)
```

allows the use of the name `PORTA` in the computer program. I define `PORTA` as a pseudo-variable. It is created by a macro expansion and such things are usually constants or function-like expansions. The above is neither. `PORTA` can be assigned to, or its value read. You can even do operations like

```
PORTA |=0X80;
if(PORTA & 0x6==0)
   do something;
```

You can perform any operations with PORTA that you would want with any normal variable in the language. While PORTA has been generated by a macro expansion it also can be used as a variable. Thus, the title pseudo-variable. Unless there is a specific reason, though, I will call these macros "variables."

The above pseudo-variable is of the type char, and its address is 0x1000. This capability is very useful in programming microcontrollers.

When programming microcontrollers, there are two reasons why it is necessary to be able to manipulate direct addresses in memory. Most high-level languages will not allow the programmer direct access to specific memory locations. As seen above, C does allow the programmer to bend these rules enough to be able to store data into a specific memory address. Another feature that is highly desirable is to be able to place the address of a function at a specific address. This capability is necessary when implementing an interrupt service routine. When an interrupt occurs, the computer will stop its current operation, save at least the values contained in the status register and the program counter, and begin execution at an address contained in a vector location. Each interrupt will have a vector address, and each interrupt will require its own interrupt service routine. Program initialization when interrupts are involved will require that the program place the addresses of any interrupt service routines into the specific vector addresses for each interrupt.

A continuation of the above approach can be used in this case. There is a direct address in memory that is to receive the interrupt service routine address. Let's think for a moment about what this address is. The address is going to contain the address of a function. The address itself is a pointer to a memory location. The contents of this location are a pointer that points to a function that is the interrupt service routine. All interrupt service routines are of the type void. Therefore, the vector address is a pointer to a pointer to a type void. To be able to place the value of the pointer to a type void into this location, we must assert one additional level of indirection to access the content of the specific memory location. Therefore, the following line of code

```
*(void **) 0xfffc = isr;
```

> places the beginning address of the function isr into the memory location 0xfffc.

> A convenient method of executing this operation is to create a macro. The following macro definition works:

```
#define vector(isr, address) (*(void **)(address)=(isr))
```

> Now the function call

```
vector(timer, 0xffd0);
```

> will place the address of the function named timer into the location 0xffd0 in the computer memory map. It is important that timer be defined as a function that returns a type void.

Structures

> Another feature of C not found in many high-level languages is the structure. A structure is similar to an array, but far more general. A structure is a collection of one or more variables identified by a single name. The variables can be of different types. Structures are types in the sense that an int is a type. Therefore, if you have properly declared a structure, you may declare another structure of the same type. Examine the following structure:

```
struct person
{
    char *name;
    char *address;
    char *city;
    char *state;
    char *zip
    int height;
    int weight;
    float salary;
};
```

> This structure contains some of the features that describe a person. The person's name and address are given as pointers to character strings. The person's height and weight are integers, and the salary is

a floating-point number. The structure declaration must be followed by a semicolon. This combination of variables is created every time a structure of the type `person` is declared. The name `person` following the `struct` declaration is called the structure tag or merely the tag. Tags are optional. If a tag is used, it may be used to declare other structures of the same type by

```
struct person ap,bp,cp;
```

Here `ap`, `bp`, and `cp` are structures of the type `person`. A single instance of a structure can be declared by

```
struct {....} a;
```

In this case `a` is a structure with the elements defined between the braces. The elements that make up a structure are called its members. Members of a structure can be accessed by appending a period followed by the member name to the structure name. For example, the name of the person represented by `ap` is accessed by

```
ap.name
```

and that person's salary is

```
ap.salary
```

A pointer to a structure can be used. The pointer `pperson` is created by

```
struct person *pperson;
```

If a pointer to a structure is used, members of the structure can be accessed by the use of a special operator `->`. This operator is created by use of the minus sign (`-`) followed by the right angle bracket character (`>`). The height of a person identified by the pointer `pperson` is accessed by

```
pperson->height
```

Arrays of structures are used, and when dealing with pointers to arrays of structures, to increment the pointer will move the pointer to the next structure. If a program has

```
struct person people[20], *pp;
```

and `pp` is made to point at `people[0]` by

```
pp=people
```

then

```
people[1].name
```

is the same as

```
++pp->name
```

A structure can have another structure as a member element. Consider

```
struct point
{
   int x;
   int y;
};
```

where `point` contains two integer elements, `x` and `y`. A circle can now be defined by

```
struct circle
{
   struct point center;
   int radius;
};
```

In this case access to the members of center is by

```
circle.center.x
```

or

```
circle.center.y
```

Of course the radius is accessed by

```
circle.radius
```

Functions can take structures as arguments and they can return structures. For example, the function `make_point()` that follows returns a structure.

```
struct point make_point(int x, int y)
{
   struct point hold;

   hold.x=x;
```

```
    hold.y=y;
    return hold;
}
```

Observe that the `struct point` is treated as a type with no difficulty in this function. The return type is `struct point`, and within the body of the function `hold` is also a type `struct point`. The `x` argument passed to the function is placed in the `x` member of `hold` as is the `y` argument placed in the `y` member. Then the `struct hold` is returned to the calling function. All of these operations are legal.

Since structures create types, structures can have members that are structures. For example, suppose that the `struct rect` for a rectangle is defined as

```
struct rect
{
   struct point p1;
   struct point p2;
};
```

Let's outline a program that will inscribe a circle within a rectangle. The circle is tangent to the sides that make up the narrowest dimension of the rectangle.

```
/* Inscribe a circle in a rectangle */
/* first declare some useful structures */

struct point
{
   int x;
   int y;
};

struct circle
{
   struct point center;
   int radius;
};

struct rect
{
```

```
     struct point p1;
     struct point p2;
};

/* here is a function to make a circle */
struct circle make_circle ( struct point ct, int
rad )
{
   struct circle temp;

   temp.center = ct;
   temp.radius = rad;
   return temp;
}

/* some useful function like macros */
#define min(a,b) (((a)<(b)) ? (a) : (b))
#define abs(a) ((a)<0 ? -(a) : (a))
/* function prototypes */

void draw_rectangle(struct rect);
void draw_circle(struct circle);

int main ( void )
{
   struct circle cir;
   struct point center;
   struct rect window = { {80,80},{600,400} };
   int radius,xc,yc;
   center.x = (window.p1.x+window.p2.x)/2;
   center.y = (window.p1.y+window.p2.y)/2;

   xc = abs(window.p1.x -window.p2.x)/2;
   yc = abs(window.p1.y -window.p2.y)/2;
   radius = min(xc,yc);

   cir=make_circle(center,radius);
```

```
draw_rectangle(window);
draw_circle(cir);
return 0;
}
```

At the beginning of the program, several important `struct` types are defined.These include a point, a rectangle, a circle, and a function to make circle given its center and its radius. Two macro definitions are needed. The first is the calculation for the minimum value of `a` and `b` and the second returns the absolute value of the argument. Two function prototypes are included. These functions will draw a rectangle and a circle to the screen, respectively.

Inside the `main` program `cir` is declared to be of the type `struct circle`, `center` is `struct point`, and `window` is of the type `struct rect`. When `window` is defined, it is initialized to the values shown. This type of initialization is acceptable to structures as well as arrays. The rectangle is defined by two points. The point {80,80} is the lower lefthand corner of the rectangle, and the point {600,400} is the upper right corner. These locations are implementation dependent, and in some cases might represent the upper left corner {80,80} and the lower right corner {600,400}.

The center of the window is calculated by determining the average value of the `x` members of each point along with the average value of the `y` members. These values are the exact center of the rectangle. The center of the inscribed circle will lie at this point. The radius of the inscribed circle will be one-half the length of the shortest dimension of the rectangle. The two potential value are calculated as `xc` and `yc`. Here the absolute value is used, because in general, it is impossible to know that the rectangle will be specified by the lower left hand corner in `p1` and the upper right hand corner in `p2`. If these points were interchanged, negative values would be calculated. The selection of the positive result through the `abs()` macro avoids this problem. The final choice for radius is the minimum value of `xc` or `yc`.

The above calculations provide enough information to specify the circle, so `cir` is calculated as the return value from `make_circle()`. Finally, two compiler specific functions, `draw_rectangle()` and `draw_circle()`, are used to draw the calculated figures to the screen.

While it would not be difficult to execute these calculations without the use of structures, it is obvious that the structure formulation of the program makes a much simpler and easier to follow program. The variables and program elements are objects here rather than mere numbers.

C has a command called `typedef`. This command can rename a specified type for the convenience of the programmer. It does not create a new type, it merely renames an existing type. For example,

```
typedef int Miles;
typedef char Byte;
typedef int Word;
typedef long Dword;
```

are all valid `typedef` statements. After the above invocations, a declaration

```
Miles m;
Byte a[20];
Word word;
Dword big;
```

would make m an `int`, a an array of 20 characters, `word` the type `int`, and `big` a `long`. All that has happened is that these types are a redefinition of the existing types. New types defined by `typedefs` are usually written with an upper case first letter. This is a tradition, not a requirement of the C language.

Structures used earlier could be modified by use of the `typedef`. Consider

```
typedef struct
{
   int x;
   int y;
} Point;
```

This typedef redefines the earlier struct point as `Point`. The judicious use of `typedefs` can make a program even easier to read than the simple use of `structs`. The program that follows is the same as that above where all of the structures are typedef new names.

```
/* Inscribe a circle in a rectangle */
```

```c
typedef struct
{
   int x;
   int y;
}Point;

typedef struct
{
   Point center;
   int radius;
}Circle;

typedef struct
{
   Point p1;
   Point p2;
}Rect;

Circle make_circle( Point ct, int rad )
{
   Circle temp;
   temp.center=ct;
   temp.radius = rad;
   return temp;
}

/* some useful macros */
#define min(a,b) (((a)<(b)) ? (a) : (b))
#define abs(a) ((a)<0 ? -(a) : (a))

/* function prototypes */
void draw_rectangle(Rect);
void draw_circle(Circle);

int main ( void )
{
   Circle cir;
   Point center;
   Rect window = { {80,80},{600,400} };
```

```
int radius,xc,yc;

center.x = (window.p1.x+window.p2.x)/2;
center.y = (window.p1.y+window.p2.y)/2;
xc = abs(window.p1.x - window.p2.x)/2;
yc = abs(window.p1.y - window.p2.y)/2;
radius=min(xc,yc);

cir=make_circle(center,radius);
draw_rectangle(window);
draw_circle(cir);
return 0;
}
```

The function make_circle() returns a type Circle and requires arguments Point and int. Circle is used to declare the variable temp in make_circle(). Within the main program Circle, Point, and Rect are used as types in the definition of the several structure type variables used in the program. With the typedef declarations, there is no basic change to the program, but it is easier to read and follow.

The two functions draw_rectangle() and draw_circle() are not standard C functions. These functions are programmed and the listing of the final program is shown in Appendix B.

Self Referential Structures

A structure cannot contain itself as a member. Structure definitions are not recursive. However, a structure can contain a pointer to a structure of the same type. This capability has proven quite useful in dealing with complicated sort or listing problems. One sort problem that can be easily treated is the *binary tree* sort. A tree sort receives data, such as a word. Within the tree there is a *root node* that contains a word. The node also contains a count of the number of times the word has been seen and two pointers to additional nodes. These nodes are called *child* or *descendent* nodes. Traditionally, the node to the left contains a word that is less than the root word, and the node to the right contains a word that is greater than the root node. As data are read into the tree, the new word is compared with the root word. If it is equal to the root word, the count in the node is incremented, and a new word is received. If the word is not the same and is less than the root word, the node to the left is accessed and the comparison is repeated. This process is repeated until a match is found or a node with no descendants is found. If a match is found, the count of that node is incremented. If no match is found a new node is created and placed at the bottom location, and the word is inserted into the new node. This process is repeated with the smaller words always going to the left and the larger words always going to the right.

Eventually a tree that contains all of the different words entered into the program will have been created. The tree has several properties. Since each node has two child nodes, the tree builds in a binary manner. The root level has exactly one entry, the second level has two entries, the third level has four entries, and so forth. If the tree is balanced, each level will have a power of two entries. However, if word data are taken in randomly, it is possible that some tree branches will terminate early and others will extend to a depth or level that exceeds some.

Once the data are placed into the tree, it is possible to sort it or to arrange the words in alphabetical order. If we traverse the tree from the root node to the extreme left, we will find the word that is smallest. Immediately above that word will be the next larger word. To the right of the second word will be words larger than itself but smaller than the word in the next node above. Therefore, the right path must be traversed to the left to find the next larger words. All of this—right and left, larger and smaller—sounds complicated. It is not.

Here is a case where a little thought and recursive code will help things along easily. The code that follows is a complete function to alphabetize and count the number of times that each word is used in a document. Several new concepts will be shown in this program, so it will be broken into short blocks and described in these small pieces of code rather than trying to bite into the whole program at one time. The structure tnode is listed below.

```
typedef struct tnode
{ /* the tree node */
    char *word; /* points to text */
    int count; /* occurrences */
    struct tnode *left;/* pointer to left child */
    struct tnode *right; /* pointer to right child*/
}Tnode;
```

The first two elements to this structure are a pointer to the word contained in the node, and the number of times that the word has been seen. The last two elements are pointers to structures of the type `struct tnode`. Note that in the `typedef` of the `struct tnode`, the structure tag was used. It is necessary to use the tag in this case because the self-referential pointers inside of the structure needs the tag to find the correct type. For the remainder of the program, the `typedef Tnode` is used.

Some new files are included in the include files. These files will be discussed in a following section. The first is `ctype.h`. This program uses several character tests that are identified in `ctype.h`. There are string operations found in `string.h`, and there are standard functions defined in `stdlib.h`.

```
#include <stdio.h>
#include <ctype.h>
#include <string.h>
#include <stdlib.h>
```

The list of function prototypes for the functions written in this program follow:

```
Tnode *addtree(TNODE *, char *);
Tnode *talloc(void);
void treeprint( TNODE *);
int getword(char *, int);
char *strsave(char *);
```

The first function is the function that is used to add a tree to the program. `*talloc()` is a function that allocates memory for a `Tnode`. The function `treeprint()` prints out the tree, and `getword()` reads in a word from the input stream. There are two function prototypes defined within the program `getword()`. These functions are used by `getword()` only. The final function above saves the string pointed to by the argument in a safe place and returns a pointer to the word location in memory.

The `main` program is relatively simple. The constant `MAXWORD` is the maximum number of characters that can be allowed in a word. Within `main()` a pointer to a structure `Tnode` named `root` is declared along with a character array named `word` and an integer `i`.

```
/* word frequency count */

#define MAXWORD 100

int main(void)
{
   Tnode *root;
   char word[MAXWORD];
   int i;

   root=NULL;
   while(getword(word,MAXWORD) != EOF)
      if(isalpha(word[0]))
         root=addtree(root,word);
   treeprint(root);
   return 0;
}
```

A NULL value is assigned to root, and the program enters a loop that reads words from the input stream. It is assumed that if the first character of the word is a letter, that a word has been read in. The function isaplha() returns a TRUE if its argument is a letter, and a FALSE if it is not. If the input is a letter, the routine addtree is executed with the arguments root and word. The first time that this function is executed, root is a NULL. This loop is repeatedly executed until getword() receives an EOF character from the input stream. At that time the input loop terminates, and the function treeprint() is executed. When treeprint() is completed, main() returns a 0 to the calling program or the operating system. This signal can be used to notify the operating system that the program has executed correctly.

Here is a good point to introduce the concept of a NULL pointer. Often people use the term NULL to mean zero or a character containing a zero. This idea is incorrect. In this text, the word NULL means a NULL pointer to a type void. Such a pointer can be defined in a header file and the definition will look like

```
#define NULL ((* void) 0)
```

Any place you see the name NULL in this text, it has the above meaning.

The function addtree() is the most complicated function in

this program. This function receives a pointer to a Tnode and a pointer to a character string as arguments. It returns a pointer to a Tnode. If the Tnode pointer argument is a NULL on entry to the function, the first if loop is executed. Within that loop, talloc() returns a pointer to a new Tnode. The function strsave() copies the string into a safe place and returns a pointer to this location. This pointer is put into the word pointer location in the new Tnode. A value of 1 is put into the count location, and the pointers to the left and right child Tnodes are set to NULL. At this time, the word has been put into the Tnode , and the pointer to this Tnode is returned to the calling program.

```
/* addtree: add a node with w, at or below p */

Tnode *addtree(Tnode *p, char *w)
{
   int cond;

   if(p == NULL) /* new word has arrived */
   {
      p=talloc(); /* make a new node */
      p->word=strsave(w);
      p->count=1;
      p->left=p->right=NULL;
   }
   else if((cond = strcmp(w,p->word)) == 0 )
      p->count++; /* repeated word */
   else if(cond <0) /* less than into left subtree*/
   p->left = addtree(p->left,w);
   else /* greater than into right subtree */
      p->right = addtree(p->right,w);
   return p;
}
```

Suppose now that at a later time addtree() is called and this time p is no longer a NULL. In this case, a string compare test will be executed to determine if the word that has been passed is equal to that in the Tnode pointed to by p. If it is equal, it is a repeated word for that node, so the word count is incremented and control is returned to the calling program. If it is not equal (say, it is lexically less than the word of the node), it is necessary to either traverse to the node on the left or add a new node on the left if there is none there. The code

```
p->left = addtree(p->left,w);
```

does exactly what is needed in this case. This recursive call to addtree() will descend to the left Tnode if one exists. If one does not exist, the pointer to the left Tnode will be NULL, so addtree() will create a new node for this location. addtree() returns a pointer to the new node, so it will be placed in the left pointer of this node.

Had the lexical value of the word been greater than that of the word stored in the node, the addtree() call would work on the pointer to the right child node. Therefore, the function addtree() will start at the root node and traverse down the tree to the right or left child nodes depending on the size of the word relative to the sizes of the words in the tree. If the word is found in the tree, its count is incremented. If control proceeds down the tree, and the word is not found, eventually, a Tnode with a NULL pointer to the direction that the path must move. A that time a new Tnode is created and the word is assigned to that Tnode.

When all of the data to be input into the tree are read in, control is passed to the function treeprint(). The argument of treeprint() is a pointer to the root node and treeprint() returns nothing to the calling program. Efficient printing out of the data requires another recursive routine. The function treeprint() shown below shows this routine. treeprint() is called with the root pointer as an argument. The root pointer will not be a NULL, so the code following the if statement will be executed. The first

```
/* treeprint: in-order print of tree p */

void treeprint(TNODE *p)
{
   if(p !=NULL)
   {
      treeprint(p->left);
      printf("%4d%15s\n",p->count,p->word);
      treeprint(p->right);
   }
}
```

statement of this code is a recursive call to treeprint() with the pointer to the left child pointer as an argument. This recursive call will cause control to propagate to the lowest and leftmost level of the tree. At this time treeprint() will return normally, and the word pointed to in the Tnode will be printed out by the printf() call. The program will then start a recursive treeprint() to the right of this node. Control will immediately go to the left side of the right branch and will descend the lowest level and print out the word "found." This routine will repeat up and down until the whole tree content has been printed.

The function strsave() copies the word passed to it as an argument into a save place and returns a pointer to this memory location to the calling program. C provides for dynamic allocation of memory.

```c
char *strsave(char *s) /* make a duplicate of s */
{
    char *p;

    p = (char *)malloc(strlen(s)+1);
    if(p != NULL)
        strcpy(p,s);
    return p;
}
```

Up to this point, all memory access was to memory allocated by declaration statements. With dynamic allocation, the program can go to the operating system and request memory at any time. This memory is from a memory area called the program heap. The first call to allocate memory is malloc() shown above. The function prototype for this function is found in stdlib.h , and the function requires an argument that is the length of the memory space needed. The program returns a pointer to the base type of the system, chars, to the required block of memory. If there is not enough memory available, the function returns a NULL pointer. C will never return a valid NULL pointer. Therefore, if you have a C program call that returns a pointer, the program can always test for a NULL pointer to determine if the call succeeded. In the above code, it is assumed that the calling program will check for the NULL pointer and determine what to do.

The function prototype of the function `malloc()` is contained in `stdlib.h`. This function returns a pointer to a type `void`. If you will recall the rules of arithmetic for pointers, a pointer to a type `void` can be assigned to any other pointer type, and vice versa. Therefore, it is not necessary to `cast` the return from the `malloc` call to the type `char*` as was done above. In fact, there are people that say that you should not perform this `cast`. If you make a mistake and do not include the header `stdlib.h`, then the compiler would assume that the return from the `malloc` function call is an `int`. The `cast` operation would introduce a bug into your code if the header were not included. The error of the missing header file would be identified by the compiler if there is no `cast` because the program would be loathe to assign an integer to a pointer. Therefore, the code without the `cast` is more robust.

I personally object to this logic. The current trend is to insist that all operands in expressions that are not of the correct type be `cast` to the correct type prior to execution of any operators. This trend largely ignores the automatic type promotion that takes place when there are mixed types within a single expression. In fact, many programming standards insist that all mixed types be cast explicitly. In such an environment, it seems rather silly that one case be singled out and done differently.

The function `talloc()` also makes use of the `malloc()` function. In this case, the argument of the `malloc()` call is the `sizeof` operator. `sizeof` is a C operator that returns the size of the argument. `sizeof` is an operator in C and requires no prototype.

```
/* talloc : make a tnode */
TNODE *talloc(void)
{
    return (TNODE *) malloc(sizeof(TNODE));
}
```

The memory allocation function `malloc()` returns a pointer to the basic memory size of the system. It is always necessary to cast the return from `malloc()` onto the type of variable that the return must point to. In this case, the `cast` is to the type pointer to TNODE. In `strsave()`, the `cast` was to the type pointer to `char`. Therefore, the statement

```
return (TNODE *) malloc(sizeof(TNODE);
```

will return to the calling function a pointer of the type TNODE to a free memory space the size of a TNODE. This function is called by the function addtree(). You will note that addtree() does not test the return from talloc() to determine if the return is a NULL pointer. Good programming practice would dictate that the return from talloc() should be tested to make certain that malloc() did not return a NULL pointer.

The next function that must be incorporated into the program is getword(). getword() returns the first character of the word or an EOF in the case that an EOF is detected. It requires two arguments. The first is a pointer to a character array into which the input data are to be stored. The second argument is the length of the array and hence the maximum length of any word that can be read into the program by getword(). Two functions are accessed by getword(). The first is getch() which returns a character from the input stream. The second is ungetch() which restores a character back onto the input stream. We will see later that for getword() to work correctly, it must pull one more character than the length of the word in some cases. When that happens, the extra character must be put back onto the input stream so that it will be available for the next getch() call.

```c
/* getword: get next word or character from input */
int c, getch(void);
void ungetch(int);
int getword(char *word, int lim)
{
   char *w=word;
   while(isspace(c=tolower(getch())));
   if(c!=EOF)
      *w++=c;
   if(!isalpha(c))
   {
      *w='\0';
      return c;
   }
   for( ; -lim >0 ; w++)
```

```
            if(!isalnum(*w=tolower(getch())))
            {
                ungetch(*w);
                break;
            }
        *w='\0';
        return word[0];
    }
```

The first executable statement

```
while(isspace(c=tolower(getch())));
```

includes two standard C functions. The function isspace() has its prototype in the header file ctype.h. This function returns a TRUE if its argument is a space and FALSE otherwise. The second function, tolower(), is also prototyped in ctype.h. It tests the argument and returns a lower case letter if the argument is upper case. Therefore, this operation will loop until it receives a nonspace input, and the lower case version of the letter input will be stored in c.

If c is not an EOF, it is put into the next open location of the word array and the pointer into this array is incremented. If the return is an EOF, the second if statement will execute. The if statement

```
if(!isalpha(c))
{
    *w='\0';
    return c;
}
```

tests to determine if the character from the input stream is a letter. isalpha() returns a TRUE if its argument is a letter and a FALSE otherwise. If the character taken from the input stream is an EOF, isalpha() will return a FALSE and the statement following the if will be executed. In this case, a zero character is written to the word, and the EOF is returned to the calling function.

If the return is a letter, the following sequence will be executed:

```
for( ; --lim >0 ; w++)
    if(!isalnum(*w=tolower(getch())))
    {
        ungetch(*w);
```

```
        break;
    }
```

The central loop here is the `if` statement. Its argument executes `getch()` to read in characters from the input stream. These inputs are converted to lower case and stored into the array location pointed to by `w`. The result is then checked to determine if it is a letter or a number. If it is not, it is put back onto the input stream and the loop is exited by the break instruction. If it is a letter or a number, the pointer to the output array `w` is incremented and the maximum length of the array is decremented. If this last result is greater than zero, the next character is read in. Otherwise, the `if` statement is skipped.

The last two statements in the function are

```
*w='\0';
return word[0];
```

The last entry of the character array is made a `'\0'` to satisfy the C requirement that a string must terminate with a 0, and the first character of the word is returned to the calling program. This value is returned just to guarantee that an `EOF` is not returned.

There are two final functions required for this program. These functions work together to form the `getch()`/`ungetch()` pair. A global buffer with an argument `BUFSIZE` is created, and a global integer `bufp` is declared. `bufp` is initialized to 0. Since these variables are global, they are initialized at the beginning of the program, and they will not be changed when control is returned to a calling function.

In `getch()`, the value of `bufp` is tested. If it is greater than 0, the character returned is taken from the buffer and the `bufp` is decremented. Otherwise, a new character is read from the input stream by `getchar()`.

```
#define BUFSIZE 100

char buf[BUFSIZE]; /* buffer for ungetch */
int bufp=0; /* next free position in buffer */

int getch(void)/* get the next character from the
buffer */
```

```
{
   return(bufp>0) ? buf[—bufp] : getchar();
}

void ungetch(int c) /* put character back onto
input */
{
   if(bufp>=BUFSIZE)
     printf("ungetch: too many characters \n");
   else
   buf[bufp++]=c;
}
```

When ungetch() is executed, a test is made to determine if this store would exceed the buffer length. If bufp will be greater that BUFSIZE after it is incremented, an error return is executed. Otherwise, the return character is put in the buffer and it is there to be read the next time getch() is executed.

This program has introduced several C functions. In fact, C has a ungetchar() that does the same function as ungetch() above. It has given examples of structures containing pointers to like structures, recursive functions to process data, and dynamic memory management.

EXERCISES

1. With an editor, enter the tree program and compile it. To run this program, use an input redirection such as c:>tree < (filename)

2. Modify the above program to determine the number of levels, and record the level number with each word. Print out the levels when the output is printed. Hint: the levels can be determined in the addtree() function, and the printout of the levels require a modification of treeprint().

3. Add a histogram to the above program that shows the number of entries in each level. Count the total number of different words in the document tested.

More Structures

There are two more important considerations that should be placed under structures. The first of these is the *union*, and the second is *bit manipulations* and *bit fields*.

Unions

A union is defined much the same as a struct. There is a significant difference, though. A union can have several different arguments, each of which is a different type. The compiler, when it sees a union declared, provides enough memory to hold the largest argument of the union. When different arguments are used, the different types occupy the same memory location. Consider the following sequence:

```
struct bothints
{
    int hi,lo;
};
union both
{
    long l;
    struct bothints b;
}compound;
```

This sequence will cause the compiler to generate a structure that contains two ints. The union both will provide space for whichever is larger, a long or a struct bothints. Of course, a long is the size of struct bothints, so enough memory will be provided to store a long. In use, a sequence like

```
compound.b.hi = a;
compound.b.lo = b;
```

will place the int a into the upper location of compound, and b will go into the lower location of compound. After these operations, compound.l will contain a in its upper word, and b in its lower word. If compound.l were used as a variable, it would be this combination.

Unions are most often thought of as a method of saving memory. If several variables are completely independent and never used at the same time, a `union` that contains these several variables will allow the programmer to store each in the same memory location.

Bitfields

The concepts of *bit fields* fall loosely under structures. The first built-in operations that allow bit manipulations from C involve an `enum`. Consider the following `enum`:

```
enum {PB1=1,PB2=2,OUT1=4,OUT2=8};
int PORTA;
```

Notice that the different elements of the `enum` are each powers of 2. We can then use an expression like

```
PORTA |= PB1 | PB2;
```

to turn on bits corresponding to `PB1` and `PB2` in the integer `PORTA`. Or, these corresponding bits might be turned off in `PORTA` by

```
PORTA &= ~(PB1 | PB2);
```

Tests can be executed like

```
if(PORTA & (PB1 | PB2) == 0)
```

Here the argument of the `if` call will be `TRUE` if both bits corresponding to `PB1` and `PB2` are turned off in `PORTA`.

These manipulations are not really special bit manipulations. What is seen here is merely creation of bit-like operations using normal C. C does support bit fields. Bit fields are created in the form of a `struct`. The following `struct` defines several bit fields:

```
struct
{
   unsigned int PB1 : 1;
   unsigned int PB2 : 1;
   unsigned int OUT1: 1;
   unsigned int OUT2: 1;
   unsigned int ALL : 4;
} FLAGS;
```

This `struct` consists of several bit fields. The colon that follows the field name designates the bit field whose size is the number following the colon. The first four fields are each 1 bit wide, and the final field `ALL` is 4-bits wide. These bits can be turned on by:

```
FLAGS.OUT1 = FLAGS.OUT2 = 1;
```

or off

```
FLAGS.OUT1 = FLAGS.OUT2 = 0;
```

and they can be tested

```
if(FLAGS.PB1 == 0 && FLAGS.PB2 == 1)
    . . .
```

Some of the compilers have special bit constructs. These constructs are usually `structs` that have either 8- or 16-bits within the field. These `structs` are useful as Boolean variables.

When setting up a microcontroller program, the programmer will frequently want to have bit fields at specific locations in memory. These bit fields can be used as I/O ports, control registers and even arrays of bits to be used internally as flags. An approach to this problem is found in the bit array.

```
typedef struct
{
    bit_0 :1;
    bit_1 :1;
    bit_2 :1;
    bit_3 :1;
    bit_4 :1;
    bit_5 :1;
    bit_6 :1;
    bit_7 :1;
} BITS:
```

A macro definition is used to create a variable:

```
#define PORTA (*( BITS *) 0x1000)
```

With these definitions, instruction statements like

```
PORTA.bit_7 = 0;

if(PORTA.bit_3 == 1 && PORTA.bit_2 == 0)
```

etc. can be used in dealing with the bits within this memory location.

Input and Output

C has no provision for either input or output within the language. Any such operations must be programmed as functions that are called from the programs. There are several standard I/O functions that are always included with a complete C compiler. However, in many instances, compilers for very small microcontrollers will have no built-in input/output capability.

File accesses are through a set of functions and a special structure. The `struct FILE` is a structure that contains all parameters needed to access a file. The file pointer is given a value by

```
FILE *fp;
FILE *fopen(char* name, char* mode);
```

Here `name` is a pointer to a character array that contains the name of the file. This `name` can be simply the file name if the file resides in the default directory, or it can be the complete path name-file `name` combination for files elsewhere in the file system. The string pointed to by mode is a one or two character string. The various modes are

```
"r" read only
"w" write only
"a" append
```

and sometimes

```
"b" binary
"rw" read/write
```

To open a file, the program must contain a statement

```
fp=fopen(name,mode);
```

where `fp` is declared as a pointer to the type `FILE`. Once the file is opened, all file accesses will use `fp` as an argument in some way. There are two single character file access functions:

```
int getc(FILE* fp);
int putc(int c, FILE* fp);
```

The first function returns a character from an open file `fp` and the second puts a character `c` onto an open file `fp`. There are three special file pointers created by the operating system when a program is loaded. These three file pointers are `stdin`, `stdout`, and `stderr`. `stdin` is the standard input, and defaults to the keyboard. `stdout` in the standard output and defaults to the terminal screen, and `stderr` is the standard error output which defaults to the terminal screen. These defaults can be redirected by the operating system through the use of redirection operators and pipe operators when the program is executed.

The functions `getchar()` and `putchar()` are macro definitions:

```
#define getchar() getc(stdin)
#define putchar(c) putc(c,stdout)
```

Therefore, `getchar()` will read a single character from the keyboard, and `putchar()` will put a single character to the terminal screen.

There are string I/O with the file functions. A string can be put to a file by

```
int fputs(char* string, FILE* fp);
```

This function returns an EOF if an error occurs. Otherwise, it returns a zero when there is no error.

A string can be read from a file by

```
int fgets(char* string, int maxline, FILE* fp);
```

`fgets` returns the next input line from the file `fp`. These data are written into `string`, and can contain at most `maxline-1` characters. The string is 0 terminated. `fgets` will return a zero when an error occurs. If there is no error, `fgets` returns the pointer string.

Another important file access function is

```
int fclose(FILE* fp);
```

This function releases the file pointer `fp` and it causes any data written to the file, but not written to the final destination, to be written. For example, most disk file systems use a buffer to store up a significant amount of data before it is written to the disk. If data are buffered when the `fclose()` routine is executed, any buffered data is written to the disk and the program connection to the file system is dissolved.

Different compilers have several file handling routines. To obtain the maximum advantage of these routines, you should consult the manual that comes with your specific compiler.

We have seen several instances of input/output functions. The most common is the `printf()` function. This function is but one of several functions that can be grouped together under the category of formatted input/output. The formatted output functions include

```
int printf(char* format, arg1, arg2,...);
int sprintf(char* string, char* format,
arg1,arg2,...);
int fprintf(FILE* fp, char* format, arg1,
arg2,...);
```

The function `printf()` has been used throughout the text so far, and there should be little question as to its use. This function prints data to the terminal screen. The pointer format points to a character string that contains the information to print out. Within the format, there can be conversion commands identified by a percent sign %. These commands will be discussed in detail later. The number or arguments, `arg1`, `arg2`, etc. depend on the number of commands within the format string.

The second function above, `sprintf()`, performs exactly the same conversion as `printf()`. Rather than printing the result to the screen, it is written into the character array string in memory designated by `char* string`.

Data can be printed to a file with the third function `fprintf()`. Here `fp` is the file pointer of an open file, and the remainder of the arguments are exactly the same as with the `printf()` function.

The following is a list of all the conversion commands and their meanings. If the character following the % in a format string is not found in the table, function behaviors will be undefined.

Character	Printed as
d,i	`int`: decimal number
o	`int`: unsigned octal number
x,X	`int`: unsigned hexadecimal number
u	`int`: unsigned decimal number
c	`int`: single character
s	`char*`: print characters from string until '\0' is detected.
f	`double`: floating point [-]m.dddd where the number of d's is given by precision. Default precision is 6.
e,E	`double`: floating point [-]m.ddddExx where the number of d's is given by the precision, and the power of 10 exponent can be plus or minus. Default precision is 6.

`g,G`	Uses e or f format, whichever requires least space.
`%`	No argument is converted. Print as a `%`

Additional formatting capability exists. The complete form of the % command is as follows:

`%[flags][width][.precision][modifier]type`

where the various optional fields have the following meanings.

flags

-	Left-justify the value if there is a width field specified.
+	Print leading + character for positive outputs

width

Width	*Precision*
Minimum width of print zone. Pad the field with 0 if the leading width specifier is 0.	The number of digits to the right of the decimal point. For strings, the maximum number of characters.

Modifier

Modifier	*Meaning*
`h`	Used with any integral type, data type is short.
`l`	Used with any integral type, data type is long.
`L`	Used with any floating point type, the data type is long double.

This formatting approach works with strings as well as numerical outputs. If a string output contains a field width or a precision, it works much the same as a numerical output. If the field width specified is not long enough to hold the output string, the specified field width is ignored. The complete string is printed out. With a string, precision specifies the number of output characters. If the precision is smaller than the field width, the width of the output field will be the precision specification. A minus sign applied to field width will cause the data to be printed to the left edge of the field. Otherwise, the data will print to the extreme right edge of the field.

Formatted input is also provided in C. The three formatted input functions are

```
int scanf(char* format,arg1,arg2,...);
int sscanf(char* string,char*
format,arg1,arg2,...);
int fscanf(FILE* fp,char* format,arg1,arg2,...);
```

The first of these functions reads formatted data from the standard input, the second reads formatted data from a string in memory, and the third reads formatted data from an open file fp. There is a significant difference between the formatted inputs and outputs. Recall that the arguments in the formatted outputs are data values. In the formatted inputs, all arguments must be pointers to locations in memory that can hold the data types specified in the data string. If the program should contain

```
int i;
.
.
.
scanf("%d",i); /* BAD */
```

the program will probably compile, the compilers will not note an error, and the program will crash because the scanf() argument is not a pointer. It is generally recommended that you avoid use of the scanf() function.

Memory Management

There are two functions in C that allow dynamic memory management. The function

```
void* malloc(size_t n);
```

returns a pointer to a block of memory that is n bytes long. This memory is uninitialized. malloc() (and calloc() below) returns a pointer to a block of memory of the type void. Therefore, before it can be used, the return pointer must be cast onto the data type that the program requires. In this case, an assignment of the return value to a pointer of the correct type will work. If this step is not taken, none of the usual pointer arithmetic will work. If the memory is not available to be allocated, the allocation program will return a NULL.

A second dynamic memory allocation function is

```
void* calloc(int number, size_t n);
```

This function returns a pointer to a type `void`. It takes two arguments. The first argument is the number of items and the second argument is the size of the specific item. Therefore, if you wished to allocate space for ten TNODEs from the above problem, you should use

```
(TNODE *) calloc(10, sizeof(TNODE));
```

`calloc()` differs from `malloc()` in that it returns a pointer to initialized space. The `calloc()` function initializes all memory assigned by the function to zero. When memory is allocated by either `malloc()` or `calloc()`, this memory must be returned to the system. If allocated memory is not returned to the system, eventually all of the available memory will be allocated. Further attempts to allocate memory will fail. If the return pointer from the allocation function is not tested and properly handled when a NULL is returned, the program will exhibit undefined behavior. This problem can be avoided by deallocating memory when it is no longer needed by the program. The function

```
void free(void* );
```

will release allocated memory. The argument in this case must be a pointer that was returned from an allocate function, either `malloc()` or `calloc()`.

An additional memory allocation function is

```
void *realloc(void *,size_t);
```

The first parameter here is a pointer to a previously allocated memory block, and the second is the new size that you want for the allocated memory. This function will change the size of the allocated memory block. If you are reducing the size of the block, the pointer returned will probably be the same as that passed to `realloc`. If you are increasing the size of the block, and there is not enough contiguous memory available, the function will search for a proper sized block.

If one is found, the data stored in the original block are copied to the new block and the proper space is allocated. The pointer returned in this case will not be the same as that passed.

Miscellaneous Functions

C provides several sets of miscellaneous functions that are very useful. As mentioned earlier, the input/output functions that are prototyped in `stdio.h` are not parts of the language, but instead are separate functions. The miscellaneous functions are prototyped is several different header files. Many of these functions are macros, and if they are, the macro definitions are contained in the header file.

Following is a list of several header files specified by the ANSI Standard. With each header file is a list of functions or function prototypes contained within these files. You should look to the documentation for your compiler to get the detailed descriptions of each of these files.

As mentioned earlier, these files are there, tested, work and are probably as robust and reliable as any code written. They do what they are supposed to, and they are free. It makes little sense to re-write these programs just to satisfy an ego trip. I have seen shop coding standards that did not allow the use of standard library files because the programmers could not examine the source code. Such a standard is absolutely silly. If you do not trust the compiler to provide safe, satisfactory library code, you should not trust the compiler to generate any code. Therefore, you are using the wrong compiler. Get a compiler that meets your standard.

Most embedded programs that employ microcontrollers do not have standard connections to input/output devices. For example, there is rarely a terminal or a keyboard attached to an embedded system. More likely you would expect to find a numeric, or special, keypad and an LCD special purpose output display. None of the standard i/o programs can connect with such devices. In such cases, it will be your responsibility as the programmer to write drivers to interface with these devices. Also, you will rarely see disk files in your embedded product. Therefore, you will rarely need to include the header file `stdio.h` in a microcontroller program.

STRING OPERATIONS—Defined in `string.h`

`strcat(s,t)`	concatenate t to s
`strncat(s,t,n)`	concatenate n characters of t to s
`strcmp(s,t)`	return negative, 0, or positive for s< t, s==t, or s>t respectively
`strncmp(s,t,n)`	compare n characters of t and s
`strcpy(s,t)`	copy t to s
`strncpy(s,t,n)`	copy n characters of t to s
`strchr(s,c)`	point to the first c in s or '\0'
`strrchr(s,c)`	point to last c in s or '\0'

CHARACTER TESTS—in `ctype.h`

`isalpha(c)`	non-zero if c is alphabetic, 0 if not
`isupper(c)`	non-zero if c is uppercase, 0 if not
`islower(c)`	non-zero if c is lowercase, 0 if not
`isdigit(c)`	non-zero if c is a digit, 0 if not
`isxdigit(c)`	non-zero if c is a hexadecimal digit, 0 if not
`isalnum(c)`	non-zero if c is a digit or alphabetic, zero if not
`isspace(c)`	non-zero if c is a blank, tab, newline, return, formfeed, vertical tab, 0 if not
`ispunct(c)`	non-zero if c is a printing character except space, letter, or digit
`iscntrl(c)`	non-zero if c is a control character, 0 if not
`isgraph(c)`	non-zero if printing character except space
`isprint(c)`	non-zero if printing character including space
`toupper(c)`	return c converted to upper case
`tolower(c)`	return c converted to lower case

MATH FUNCTIONS—in `math.h`

`sin(x)`	sine of x, x in radians
`cos(x)`	cosine of x, x in radians
`tan(x)`	tangent of x, x in radians
`asin(x)`	arcsine of x, $-1<=x<=1$
`acos(x)`	arccosine of x, $-1<=x<=1$
`atan2(x,y)`	four quadrant arctan of y/x, in radians
`atan(x)`	two quadrant arctan of x, in radians
`sinh(x)`	hyperbolic sine of x

`cosh(x)`	hyperbolic cosine of `x`
`tanh(x)`	hyperbolic tangent of `x`
`exp(x)`	exponential function of `x`
`log(x)`	logarithm base `e` of `x`
`log10(x)`	common logarithm of `x`
`pow(x,y)`	raise `x` to the `y` power
`sqrt(x)`	square root of `x`
`fabs(x)`	absolute value of `x`
`ceil(x)`	smallest integer not less than `x`
`floor(x)`	largest integer not greater than `x`
`ldexp(x,n)`	x*2^n
`frexp(x,int *exp)`	split double into mantissa and exponent
`modf(x,double *ip)`	Splits `x` into integral and fractional parts. Puts integral in `ip` and returns fractional part.
`fmod(x,y)`	floating-point remainder of `x/y`

These functions all require double arguments and return double values.

Utility functions—in `stdlib.h`

```
double atof(const char *s);
integer atoi(const char *s);
long atol(const char *s);
double strtod(char *str,char **scanstop);
long strtol(char *str,char **scanstop,int base);
unsigned long strtoul(char *str,char **scanstop,int base);
ldiv_t ldiv(long numer,long denom);
div_t div(int numer,int denom);
long labs(long n);
int abs(int n);
void *bsearch(void *key, void *base,size_t number,size_t size,
    int(*compare)(void*,void*));
void qsort(void *base,size_t number, size_t size,
    int(*compare)(void*,void*));
int rand(void);
int srand(unsigned int seed);
void *calloc(size_t nobj, size_t size);
void *malloc(size_t size);
void *realloc(void *p, size_t size);
```

```
void free(void *p);
void abort(void);
int atexit(void(*fcn)(void));
void exit(int status);
int system(const char *name);
```

Non-local Jumps—in `setjmp.h`

```
int setjmp(env);
int longjmp(env,int);
```

Diagnostics—in `assert.h`

The macro `assert()` is defined in `assert.h`

Signals— in `signal.h`

```
void (*signal(int sig, void (*sighandler)(int)))(int);
int raise(int sig);
```

Date and Time Functions —in `time.h`

```
char *asctime(struct tm *timeptr);
clock_t  clock(void);
char *ctime(time_t *timer);
double difftime(time_t time2, time_t time1);
struct tm *gmtime(time_t *timer);
struct tm *localtime(time_t *timer);
time_t mktime(struct tm *timeptr);
size_t strftime(char *buffer, size_t bufsize, char *format,
    struct tm *timeptr);
time_t time(time_t *timer);
```

Standard input and output—in `stdio.h`

```
void          clearerr(FILE *fp);
int           fclose(FILE *fp);
int           feof(FILE *fp);
int           ferror(FILE *fp);
int           fflush(FILE *fp);
int           fgetc(FILE *fp);
int           fgetpos(FILE *fp, fpos_t *pos);
char          *fgets(char *buffer, int n, FILE *fp);
FILE          *fopen(char *filename, char *access);
```

int	fprintf(FILE *fp, char *format, ...);
int	fputc(int c, FILE *fp);
int	fputs(char *string, FILE *fp);
size_t	fread(void *buffer, size_t size, size_t number, FILE *fp);
FILE	*freopen(char *filename, char *mode, FILE *fp);
int	fscanf(FILE *fp, char *fs, ...);
int	fseek(FILE *fp, long offset, int origin);
int	fsetpos(FILE *fp, fpos_t *pos);
long	ftell(FILE *fp);
size_t	fwrite(void *buffer, size_t size, size_t number, FILE *fp);
int	getc(FILE *fp);
char	*gets(char *buffer);
void	perror(char *string);
int	printf(char *format, ...);
int	putc(int c, FILE *fp);
int	puts(char *string);
int	remove(char *filename);
int	rename(char *oldname, char *newname);
void	rewind(FILE *fp);
int	scanf(char *format, ...);
void	setbuf(FILE *fp, char *bufptr);
int	setvbuf(FILE *fp, char *bufptr, int buftype, size_t bufsize);
int	sprintf(char *s, char *format, ...);
int	sscanf(char *s, char *format, ...);
FILE	*tmpfile(void);
char	*tmpnam(char *buffer);
int	ungetc(int c, FILE *fp);
int	vfprintf(FILE *fp, char *format, va_list arglist);
int	vprintf(char *format, va_list arglist);
int	vsprintf(char *s, char *format, va_list arglist);

As you scan through the above listings, you will see many types that are not from the usual list of available types. Whenever you find such a type, the type will be either a `typedef` of a structure or a `typedef` of a standard type. These `typedefs` will be found in the header file where they are used.

Summary

In this chapter, you have seen how pointers are used in C, learned about structures, and viewed a summary of the contents of the standard C library. You have also been shown the essentials of creating header files that can attach peripheral devices to your programs. These areas will be explored in detail in later chapters.

Chapter 3

What Are Microcontrollers?

A microcontroller differs from a microprocessor in several important ways. The early name for a microcontroller was *microcomputer*. The big difference between a microprocessor and a microcomputer/microcontroller is the completeness of the machine each represents. A microprocessor was simply the "heart" of a computer. To put a microprocessor into use, the designer required memory, peripheral chips, and serial and parallel ports to make a completely functional computer. By contrast, the microcomputer was designed to be a complete computer on a single chip. Necessary memory and peripheral components were integrated onto the chip so that a complete computer-based system could be built with a minimum of external components. A basic microcontroller is shown in block diagram form in Figure 3-1.

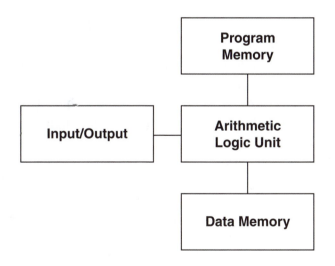

Figure 3-1: A Typical Microcontroller Block Diagram

The central control unit of the microcontroller is the *arithmetic logic unit* (ALU). Figure 3-1 shows that the ALU is connected to three different blocks. The first is the *input/output* block (I/O), the second is the *program memory*, and the third is the *data memory*. Most Motorola microcontrollers combine the last three blocks into one block. The architecture shown in the figure is known as a *Harvard* architecture, as opposed to the more common *Von Neumann* architecture. The Harvard architecture is a computer configuration in which the memory area that contains the program instructions for the computer is separated from the memory area in which data are stored. By contrast, the Von Neumann architecture has just one memory space where both program and data are stored.

The main functional difference between Harvard and Von Neumann architectures is in their ultimate operating speeds. Both architectures require that the ALU access memory once each instruction to get the next instruction to execute. Often the instruction being executed will also require an access to memory. Reading data into a register, storing data in a memory address, and accessing a location in memory that is in fact an input/output register are examples of operations that require memory accesses in addition to the normal memory fetches. As seen in Figure 3-1, the Harvard architecture has two or more internal data busses over which these different accesses can take place. There are usually two such internal busses: one for instruction access, and one for other data access. The processor can easily tell which data bus to use. If the access is to fetch an instruction, it is relative to the program counter. These accesses will go to the program memory area. All other memory accesses will go to the data memory area. It is entirely possible to have two or more memory accesses simultaneously with a Harvard architecture.

The Von Neumann architecture is somewhat simpler than the Harvard architecture. A Von Neumann processor has only one memory bus. All memory accesses must go through this single path on the system. With such a system, the processor can never process more than one memory access at a time and all memory accesses—instruction, data, or input/output—must pass through a single data bus. This is the origin of the term "Von Neumann bottleneck." The multiple accesses to memory for each instruction ultimately limit the maximum speed of a Von Neumann architecture processor. However, the speed of such processors can be many millions of instructions per second, so there are numerous excellent, fast microcontrollers constructed with the Von Neumann architecture. The Von Neumann architecture has been the mainstay of microcontrollers and will be the only microcontroller configuration available for the foreseeable future.

A microcontroller has its program stored internally, and the ALU reads an instruction from memory. This instruction is decoded by the ALU and executed. At the completion of the execution of the instruction, the next instruction is fetched from memory and it is executed. This procedure is repeated until the end of the program is found, or the program gets into a loop where it is instructed to branch back to a beginning point. In this case, the machine will stay in the loop forever or until something happens to release it from the never-ending loop.

There are three ways for a machine locked in a loop to be removed from the loop so it can execute code outside of the loop. These operations are called *exceptions*. The first is to reset the part with a reset signal. A reset signal usually requires connecting the reset pin of the part to a logic low signal. A logic low is usually ground. When this condition is detected, several internal registers are set to predetermined values, and the microcontroller fetches the address of the reset routine from a specific memory location. This address is placed in the program counter, and the program starts to execute. There is a table in memory that contains the addresses of several routines accessed when exceptions occur. These are the addresses of the interrupt service routines, reset routines, etc. This table is called the *vector table*, and the addresses are called *vectors*.

A second means of forcing the part out of the loop is for the part to detect an external interrupt. An external interrupt occurs when the *interrupt request* (IRQ) pin on the part is set low. This pin is tested at the beginning of the execution of each instruction. Therefore, if an instruction is being executed when an IRQ is asserted, the instruction will complete before the IRQ signal is processed. Processing for the IRQ consists of first determining if IRQs are enabled. If they are, the status of the machine is saved. All interrupts are disabled by setting the interrupt mask bit in the status register of the microcontroller. Then the address stored in the IRQ vector location is fetched. This address, the address of the *interrupt service routine* (ISR), is placed in the program counter. The ISR then executes.

The process of saving the status of the machine is to push the contents of all machine registers onto the machine stack. Therefore, the ISR can safely use any of the central machine resources without disrupting the operation of the main line of code when control is returned. When exiting an ISR, it is necessary to use a special instruction called a *return from interrupt* or a *return from exception*. This instruction restores the status of the machine from the stack and picks up execution of the code from the instruction following the one where the interrupt occurred.

The third means for exiting the main loop of the program is from internal interrupts. The microcontroller peripherals can often cause interrupts to occur. An internal interrupt causes exactly the same sequence of operations to occur as an external interrupt. Different interrupt vectors are used for each of the several internal peripheral parts so the cause of the interrupt is generally known and control is directed to the specific ISR for each of the several possible internal interrupts.

Data are transferred, information is passed, or events are handled either synchronously or asynchronously. The difference between these two methods of data transfer has mainly to do with how the clocking of the data is handled. The most common form of synchronous data transfer is with a three-wire serial link. One of the wires is a clock, and the other two are input data and output data, respectively. For a synchronous transfer, the value of the input is usually sampled at one edge of the clock signal (such as the fall of the clock) and the value of the bit to be sent out is guaranteed to be correct at the fall of the clock signal. Any synchronous system must set its output at such a time that it will be stable while the clock is high, and hold it in that condition until the clock signal falls. To receive a bit, the condition of the input line must be latched into the system as the clock signal falls from high to low.

Within the computer, there is another distinction for synchronous. Often an input is allowed to set a bit when it occurs. If this happens, the program will not expeditiously observe the fact that the bit is set. In fact, the program will test the state of the bit according to the program timing requirements. This type of operation is also called *synchronous* because the test is synchronized with the program.

Asynchronous operation, on the other hand, usually depends on a prearranged series of events to cause the data transfer. Serial data communications is a common example of asynchronous data transfer. Here, an input line can have two states: mark and space. A line is held at the mark state whenever no data are being transferred. When data are to be transferred, the data line is transitioned to the space state and held there for a specified time. This period is called the *start bit*. From that time onward, the data bits are placed on the line a bit at a time so that at the specified time intervals the receiving device can examine the data line and determine the bit sequence.

Asynchronous operation means that there is no computer clock-related specification as to the time that events will occur. Another example of asynchronous transfer occurs within the computer. Generally, events and data transfers that are initiated by interrupts are considered to be asynchronous. Most of the peripheral devices that are found on Motorola microcontrollers will allow either synchronous or asynchronous notification of the program that the peripheral business is completed.

The following sections contain brief discussions of microcontroller memory and several of the standard peripherals found on these devices. These discussions are intended to be qualitative and provide a broad overview of these parts of the microcontroller. Detailed descriptions of how to access and use these several peripherals will be found in later chapters.

Microcontroller Memory

In a microcontroller, the program instructions are usually stored in a memory type called *read-only memory* (ROM). ROM is usually programmed by a special mask during the manufacture of the microcontroller and is called *masked ROM*. ROM is the least expensive means of storing a program in a microcontroller, especially for high-volume manufacturing.

There are at least two means for the end user of the microcontroller to place the program memory into the chip. The first is called *erasable programmable read-only memory* (EPROM). EPROM is a memory technology that can be erased by exposing it to high-energy ultraviolet light. The EPROM requires the application of a high voltage to be programmed. The memory can be programmed with either a development system or a special programming board designed specifically to program the microcontroller.

Packages that contain EPROM have a quartz glass window through which the ultraviolet light can pass with minimum attenuation. These packages are quite expensive, and therefore, microcontrollers with EPROM are usually too expensive to use. EPROM was used for development purposes in the past, but it is just too expensive in light of more recent developments to be used for that purpose today.

For limited production purposes, a less expensive version of the EPROM chip is available. This is the *one time programmable* (OTP) chip. An OTP chip has the exact same silicon component as an EPROM, but it is packaged in a standard plastic package. This package is much less expensive than the windowed package discussed previously. However, once the chip has been programmed, the program contents cannot be changed.

There is yet another means of storing programs or, in some instances, data in a microcontroller. This technique is called electrically erasable programmable read-only memory (EEPROM). EEPROM is programmable from instructions within the microcontroller. EEPROM also requires a high programming voltage. If there are large blocks of EEPROM on the chip, the programming voltage is usually applied during the programming cycle through a pin connected to an external voltage source. In cases where the amount of EEPROM to be programmed is relatively small, a charge pump on the microcontroller chip will allow the EEPROM to be programmed with no externally applied voltage. The amount of EEPROM that can be programmed with an on-board charge pump is usually so small that it is not useful for storing program instructions. But on-board EEPROM can be quite useful in the storage of data generated by the program that must be saved through a power-down cycle. Sometimes in the execution of the program, some data are generated that must be saved for later use. These data are called volatile data or variables, and are usually stored in random access memory (RAM). Careful design of a program will usually result in the need for much less RAM than ROM. In most microcontrollers, the amount of RAM is usually 60 to at most a few hundred bytes. The amount of ROM, EPROM or EEPROM usually runs from 1000 bytes upwards to a few tens of thousands of bytes.

EEPROM is quite expensive, and has been replaced by a newer technology called FLASH memory. FLASH programs in a manner similar to EEPROM and it is inexpensive enough to allow rather large amounts of programmable memory on a microcontroller chip. You will find chips with 30,000 bytes and more of FLASH and the intent is to use these chips for production runs. The FLASH is programmed as part of the production cycle. We will see many details on programming FLASH memory in later chapters.

The architecture of Motorola microcontrollers is strictly Von Neumann. That is, within the microcontroller chip, there is only one data bus over which all program, data, and input/output must pass. In a Harvard architecture system, each of these different data types will have a dedicated bus over which the information will pass. Therefore, the Harvard architecture microcontroller is able to access data, program, and I/O simultaneously. The simultaneous availability of these different data paths can result in a significant increase in overall processor speed. It also increases the area of the microcontroller die and, hence, the cost of the microcontroller. In general, most of the applications to which the microcontrollers are directed do not require extreme speed. Thus, the Von Neumann architecture is completely satisfactory.

Input/Output

The Motorola microcontrollers use an architecture called memory-mapped I/O. Each I/O device input and output registers, its control registers, and status registers are mapped into memory locations. I/O transactions require no special computer instructions. It is merely necessary to know the memory locations of the pertinent registers and the uses of the register bits to be able to handle any I/O function. Listed below are brief descriptions of several microcontroller I/O peripherals found on Motorola microcontrollers. Not all of these peripheral systems are found on each microcontroller. It is possible to pick and choose between needs for the several peripheral systems and select a microcontroller that has exactly those peripherals required.

Timer Subsystems

There are four popular timer systems that you will find on different microcontrollers. The first is a general-purpose timer. Motorola refers to the general-purpose timers as either 8- or 15-bit timers. These timers are different. The 8-bit system contains a prescaler that counts down from system clock. The output from the prescaler is fed into a counter that counts down from it the value stored in it. When the counter underflows, a flag is set, and an interrupt can be executed.

The 15-bit timer is a strictly Motorola name, and it is even simpler than the 8-bit timer. This timer has a 15-bit minimally programmable prescaler. An interrupt can be taken from two locations in this ripple counter.

A second class of timer is the 16-bit timer. This timer is often called a general purpose timer. These timers contain a 16-bit counter that is clocked by the system clock. There are two associated subsystems: the first is called an *input capture* system, and the second is the *output compare*.

The input capture system simply captures the value of the system timer counter when an input occurs. These inputs can set a flag or request an interrupt so the input can be processed either synchronously or asynchronously. The important fact is that the exact time of the input relative to the 16-bit clock is saved when the input occurs. Applications for input capture systems are interpulse period measurements or frequency measurements.

The output compare system allows the programmer to specify a time relative to the 16-bit counter when an output is to occur. This time is calculated by adding the time offset value to the current value of the 16-bit counter. This result is stored in the output compare register. When the 16-bit counter counts to the value in the output compare register, the output occurs, a bit is set, and an interrupt can be processed if desired.

Input capture and output compare functions are sometimes called high-speed inputs and outputs. The number of input captures and output compare systems vary from as few as one each to as many as 16 programmable timers, each of which can be either input capture or output compare.

There is another style of timer subsystem that is used on high-end microcontrollers. This system is called the *timer processor unit* (TPU). In most conventional computers, the contents of a memory location are called an operand, and the processor has built-in operators that operate on the operands. A TPU is also a computer, but rather than using memory location contents as operands, time is the main operand used by the TPU. Most TPUs contain many complex systems to implement their operation. The TPU of the M68300 family and the M68HC16 family contains sixteen registers, each of which can be operated as either an input capture or an output compare. Each output compare can have its events coupled to other registers to control intricate timing events with fine time resolution. We will not see the direct programming of a TPU in this text, but we will see some of the types of events that are controlled by the TPU programmed with the usual 16-bit timer.

On the newer computers, such as the MCORE architecture, a time-of-day (TOD) clock has been introduced. This clock is based on a 32768-Hz watch crystal. These crystals are readily available, small, very accurate, and quite inexpensive. Their only problem is that they are slow, and are not very good for fine time measurements unless the crystal is used as a time base to a frequency synthesizer.

Another timer function found on most microcontrollers is the computer operating properly (COP) or watchdog timer. Most microcontrollers are placed in embedded controls. That is, the microcontroller is a part of a larger system, and usually an operator never deals directly with the microcontroller. Even though great care has been taken in the design of the microcontroller, it is possible to cause these devices to get lost from the program that they are executing. The power might dip, or a large transient magnetic field might cause the part to go into abnormal operation. In such a case, the easiest way to restore normal operation is to send the part through a reset sequence. Such a sequence will restore all of the initial internal status of the microcontroller, execute the initialization code procedure of the program, and restart the execution of the application loop. A COP timer provides just this function. A COP timer is a timer with a relatively long period. Once the COP timer is started, it is necessary for the main program to reset the COP periodically prior to the expiration of the COP period. The COP timer is never allowed to time out. If the computer gets lost, the program no longer resets the COP, so the timer will eventually overflow, and this operation causes the microcontroller to reset. Therefore, if the part ever gets lost from its normal program sequence, the COP will force a reset and restore the normal operation of the system.

Digital Input/Output

Most microcontrollers have several digital I/O ports. Usually a port consists of eight or fewer bits, and the bits in these ports can be outputs, inputs, or often bit programmable as either input or output bits. If a port has programmable I/O, it will have an associated data direction register—DDRA, DDRB, and so forth. The ports are usually named PORTA, PORTB, and so forth. DDRA is associated with PORTA. Each bit in DDRA has a corresponding bit in PORTA. If a bit in DDRA is set, the corresponding bit in PORTA is an output. The same is true for PORTB, PORTC, PORTD, PORTE, and so forth if these ports exist on the part.

A port pin can be made into an output. When this occurs, this pin becomes a latched output. In other words, when this bit is set it will remain set until it is reset by the program, and vice versa. Just because a port pin is designated to be an output does not mean that its state cannot be read by the computer. When a port is read in, the state of all of the outputs as well as the state of the inputs will be shown in the result.

Some I/O pins are multiplexed and serve multiple functions. For example, microcontrollers with analog-to-digital converters, ADC, usually allow the ADC pins to serve as digital input pins as well. In that case you need merely read the input port, and those pins that are above the high threshold will indicate one, and those below the low threshold will indicate zero. Reading the port does not affect the ADC operation at all.

Analog-to-Digital Converters

The ADC subsystem on most microcontrollers consists of a single successive approximation analog-to-digital converter preceded by an analog multiplexer that can switch the converter to any of several input pins. The program controls this switching. The electromagnetic environment of the surface of a microcontroller die is about as bad as can be found anywhere. Therefore, attempts to do fine resolution measurements of analog voltages in these parts is fraught with problems. Most ADCs use a resistive ladder to act as a digital-to-analog converter. The inputs to this ladder are sequenced in a prescribed manner to build a voltage that matches the voltage being measured. The input to the D-to-A is then the digital equivalent to the voltage being measured.

Precision resistors are very difficult to manufacture on silicon, and even precision matching between resistors is extremely difficult. While making precision capacitors is very difficult on a silicon die, it is possible to make several capacitors with highly accurate ratios between the capacitor values. Therefore, the approach is to use a set of matched capacitors and a charge balance technique to accomplish the successive approximation of the analog voltage. This method works well, and 8- and 10-bit systems are available on microcontrollers with plus or minus one-half bit accuracy.

Serial Input/Output

Where would the computer be without serial input/output? The serial system was probably the first direct human interface with any computer system. It has expanded, and today, relatively low-speed asynchronous serial interfaces are used for terminal and modem and network interfaces. High-speed synchronous serial links are used for all of the above plus inter-computer connections, hardware peripheral communications, and other types of devices where high-speed, secure communication is required.

Many microcontrollers have both asynchronous and synchronous communications peripherals built in. Usually, an asynchronous interface is called a serial communications interface (SCI) while the synchronous interface is called a serial peripheral interface (SPI).

Typically SCI systems can communicate at any of the popular asynchronous serial bit rates. These systems have built-in baud rate generators, double buffered input and output registers, and all of the error detection found on a universal asynchronous receiver-transmitter (UART) chip. These I/O devices can be either polled or interrupt-driven by the computer portion of the microcontroller.

The SPI is designed to communicate at high speeds with other microcontrollers or perhaps with hardware devices with a synchronous serial interface. These devices typically run at megabit per second rates. Since synchronous systems require a system clock, each microcontroller SPI can act as either a master or a slave. The main difference between the master and the slave is which chip generates the system clock. The master generates the system clock, and the data are clocked into and out of the slave by the system clock. Communications with the microcontroller and the SPI can be either polled (synchronous) or via interrupt controller (asynchronous).

Different Controllers

Not all of these peripheral systems are found on each microcontroller. It is possible to pick and choose between needs for the several peripheral systems and select a microcontroller that has exactly those peripherals required. The smallest microcontroller has only a 15-bit timer, and the most complete MC68HC05 part has everything but a SPI system. All varieties in between these extremes exist. In the larger chips, some of the basic requirements for the microcontroller change. We will see these larger chips in later chapters.

Programming Microcontrollers

The preceding brief description of what microcontrollers are gives a rather bleak picture of a potential programming environment from a computer standpoint. Most programmers are used to having an operating system that handles such mundane things as I/O, memory management, time management, program loading, error processing, interdevice or intertask communications, and so forth. Be prepared for a giant step backwards when you address the microcontroller. There is usually no operating system, no libraries of useful functions, no I/O handling, nothing but a bare-bones computer with a bunch of hard-to-tame peripheral components onboard the single-chip device.

C compilers for the microcontrollers have been available long enough that they are thoroughly tested and do a good job of creating proper code. Anyone who has programmed a microcontroller in assembly language knows that the programs must be very direct and have no fancy overhead. Memory is strictly limited, and the compiler must generate assembly code that is as resourceful as can be created by any thoroughly qualified assembly language programmer for the machine.

The development environment, while quite sophisticated in terms of how it works, does little for the programmer in terms of direct help in debugging a program. There are two different types of development systems that are in common use. Both of these systems require a host computer to run the device. The simplest of these systems goes by names like evaluation module, evaluation system, or evaluation board. These devices are usually board-level products that require a power supply in addition to a host computer.

The software to run the development boards is merely a good terminal emulator. Assemblers and linkers for the different chips are provided as part of the development board. The programmer writes the code for the part in the host computer. This code is assembled, compiled, and linked in the host computer. The code is then down-loaded to the development board through either a serial or a parallel link depending upon the individual system.

The development board has the microcontroller that is to be emulated on board. This microcontroller sometimes operates in a nonuser mode that allows internal bus access. A second computer on the development board controls the operation of the microcontroller. Code delivered from the host is put into memory accessed by the microcontroller, and the microcontroller can operate as if the code were contained within its internal memory. All of the I/O lines associated with the microcontroller are brought to a header on the development board, and a cable can be attached to this header to a plug-in device that plugs into a target board. This target system then operates as if it had a programmed microcontroller plugged into its socket.

The microcomputer on the development board has a complete monitor system in its firmware. This monitor provides communications with the host, down-loading and up-loading capability and, most important, complete debugging firmware for the microcontroller.

There is a single line assembler and disassembler in the firmware. This package allows the programmer to examine and change memory in assembly mnemonics. The microcontroller program can be single stepped, run, address breakpointed, and the memory can be displayed in normal hexadecimal format. The microcontroller runs at full speed when emulating operation in a target board.

An experienced programmer will be able to debug code in a microcontroller with the help of such a development board. There is additional software available that provides a nice display of all pertinent information in a single screen on the host computer. In this area, you will also find that the microcontroller can be controlled from a display of C source code on the host computer. This technique is called source level debugging.

On later chips, another feature is incorporated to help the development environment. This feature is called Background Debug Mode, or ONCE. Both of these similar operations allow debug to take place in an external computer without any access to the microcontroller resources such as interrupts or memory. When a chip is put into BDM, certain pins become a special serial input/output port. There are several commands that can be delivered to this port from an external computer. These commands allow the computer to set memory, examine memory, examine registers, set and clear registers, execute code, set and clear break points, and so forth. All of the operations normally needed to debug a program can be executed through this special serial port. There is no need for an on-board monitor on the microcontroller, and it is not necessary to make use of the chip interrupts by the debugger during the debug operation. All of the programming needed for the debugger can reside in a host computer. Most modern chips have this type of interface, which greatly simplifies debugging of microcontrollers.

All of the above capabilities are available with the development boards. Another level of capability is available. These devices are box level, and usually have a built-in power supply. Most development systems require a host computer, and usually they come with special software to interface with the host computer. These systems have all of the capabilities outlined above plus some significant improvements. The breakpoint capability of these systems is much improved over the simple address breakpoint above. Here a complicated breakpoint can be employed that will break the program operation on read or write, at any data or address location, on access of data or program, or access of a range of data or address locations. Also, the breakpoint can occur after a specified number of occurrences of the breakpoint conditions.

Another major difference in the development systems is the trace buffer. A trace buffer is a memory that is as many as 48 or 64 bits wide. Each clock cycle of the microcontroller, the condition of all address bits, the data bus, the internal microcontroller control bus, and as many as 16 external test point lines are captured in the trace buffer.

Usually, the trace buffer is 4 to 16 kilowords deep, so it can hold a significant number of microcontroller clock cycles. Even if the microcontroller is running slowly, one million clock cycles per second, such a trace buffer represents an insignificant execution time. To help make the data contained in the trace buffer, trace buffer capture can be controlled by a system that is the same as the breakpoint operation. Therefore, the portion of the program that is traced is under the detailed control of the programmer.

The data in the trace buffer can be displayed in several different manners. The simplest, of course, is to print to the computer screen the I/O pattern of all the lines captured. This type of display is extremely difficult to interpret, but it is useful in some cases. To help the programmer determine where the microcontroller is operating, it is possible to read the data bits and display a disassembled version of the code being read into the microcontroller. This display is also quite useful in debugging the code.

Yet another display is called a logic analyzer. A logic analyzer is an oscilloscope display that shows the logical status of the various lines captured in the trace buffer. A logic analyzer is a separate device, but it can have a built-in disassembler that displays the disassembled code along with the condition of the designated lines.

The devices with logic analyzers and trace buffers are quite a bit more expensive than the development boards discussed earlier. Some of the development systems provide source level debugging capability for high-level languages like C.

Another approach to development systems has been made available in some of the newer microcontrollers. The microcontrollers from the MC68HC16 family and those from the MC68300 family all have a background mode of operation. When operating in the background mode, these chips stop their normal computing and start serial communications with an external computer. The background mode can be entered as the result of an internal command or an external signal. There are enough debug commands that can be communicated over this port to allow complete debug of any program that the microcontroller might be running. Minimum external circuitry is needed to support the debug mode, so these high-powered chips can operate as their own development environment. Here, the development support is mostly software contained within the host computer, and the deliverable system can contain all of the essential components of a development system.

Coding Tips for Microcontrollers

One of the major tasks facing a programmer when writing code in a high-level language for any microcontroller is to make the resulting program as readable as possible. Other people who might later need to read or modify your code must be able to understand what is going on in your program. It is extremely important that mnemonics be used as much as possible when dealing with various registers, their bit contents, and special memory addresses throughout your programs; otherwise, the resulting code will be a "quasi-C" program filled with many numbers and funny-looking cast operations that will be largely incomprehensible to others trying to maintain your program.

Modern large 8-bit systems are very nearly as complete as yesterday's mini-computers. Programs for microcontrollers are also getting large, and they are increasingly being written by a team of programmers instead of a single programmer. The management of the programming task is much more important when a team is involved because the person-to-person interface is the most "dangerous" in all programming. It is easy for people to make mistakes; misspelled words, interchanged order of parameters in a function argument, passing incorrect data to a function, etc., are all problems that can (and often will) arise when several people try to write a single program. Each person on the team can probably keep a maximum of six or seven function interfaces in mind, and even then mistakes will be made. When the program is really large, there will be dozens of function interfaces and sources of possible error will abound. For example, most of the time programmers will try to use what is in mind rather than look up a questionable function call.

How can these problems be minimized? Fortunately, there are several steps you can take to keep errors down to a manageably low level. A big one is to make certain that an ANSI-certified compiler can be used for any program and that the computer will enforce strict type checking in the program. This will cause most of the errors due to carelessness or simple accidents to be caught at compile time and not during the debugging phase. If such errors are not caught by the compiler, they are among the most difficult to find when they show up at run time. Programs can be improved by following these conventions:

1. Use a consistent constant definition naming convention. Use all capital letters for the names of constants defined in `#define` statements.

2. Make all function names thoroughly descriptive. Use several words for each name if possible, and capitalize the first letter of each word so that the words can be distinguished easily. For example, a function that returns the time should have a name like `TimeOfDay()` or `Time_Of_Day()`. The important thing is that the name describe the function in some detail. Be consistent with naming conventions. If you start with names having no underscores, keep this convention throughout the entire program.

3. Make use of `typedef` and the fact that structures create new types. Where possible, create a type that is descriptive to your program. For example, in the C language the interface to all input/output and

files is through the type `FILE`. `FILE` in this case is simply a type that has been made available to the program through a `typedef` statement of a structure that contains all of the essential elements needed for file input/output. It is recommended that types created by `typedef` statements involving a structure should have names with the first letter capitalized. The names of these types should be brief and should be one word only.

4. The C macro gives us the ability to create functions that will be compiled in-line. This is very useful, but remember there is no type checking for macro defined functions and you do not know what will be returned from a macro function. This does not mean to avoid macro functions, but it does mean that with no type checking we have a part of the program that does not get the normal scrutiny we expect with most function calls. Be careful when using macros—you, not the compiler, must make certain that the argument types and the return type are all correct.

5. Use as few global variables as possible. Global variables seem to be an easy way to avoid using function arguments, but they also can make debugging extremely difficult. Parameters should be passed to functions as arguments. This approach will give the programmer some assurance that compile time checking will catch any typing errors, and will eliminate the problems of "side effects" found with global variables.

6. Be consistent. When designing function calls, make the order of the parameters in the argument list logical and be consistent throughout the whole program.

7. Avoid complicated argument lists. If it becomes necessary to send many variables to a function, create a structure that contains all of the arguments and send a pointer to the argument structure as a parameter rather than the arguments themselves. Since the members of the structure are all mnemonics with—hopefully!—meaningful names, the filling of the structure should result in a more accurate creation of the parameter list than would be found if the parameters are all listed in the function argument.

8. Be courteous—document your code. There is no need for a comment for every line of code, but it is unconscionable that a program should go page after page with no comments. Every function

should have a comment header that explains what the function does in the program. The header should contain a list of the function arguments and a record of modifications to the function. If there is some nonobvious code in the program, this code should be documented to explain what is happening.

9. Kernighan and Ritchie, the developers of C, gave us a language that is rich with shorthand notation. While often useful, this is a tool that can be badly misused. Don't write code in a convoluted manner using tricky shorthand notations so that nobody can understand what you're trying to do. You might be the one who has to maintain the resulting code a decade later!

10. Always create mnemonics for addresses and bit variables used in the program.

11. Use assembly language only when C cannot accomplish necessary operations.

12. Microcontrollers usually work with unsigned variables. Often the microcontroller can create more efficient code when doing compares with unsigned variables. Therefore, use unsigned variables everywhere unless it is necessary to use signed variables. A few `typedef` operations will help keep the use of unsigned variables in mind:

```
typedef WORD unsigned int;
typedef BYTE unsigned char;
```

In writing code for a microcontroller, you must be able to deal with ports at some specific address location, the bits within these ports, control registers, certain data registers, and other important control functions. The program can be easily written if the names of all registers, ports, bits, and so forth are taken from the component specification and not made subject to the whim of the programmer. Therefore, if all of the ports and bits have the name given in the microcontroller reference manual, then any variable used will be defined in one place and the programmer need only find the appropriate register along with the definition of its bits in the reference manual to understand what is being done in the program. The reference manual becomes part of your documentation.

We saw back in Chapter 2 how to create mnemonics for address locations. It is necessary for programmers to write a series of define statements which identify all of the register locations, port addresses, and bit names for the microcontroller. It is best if all of these preliminary data are placed in a header file specific to a device and then it can be included at the beginning of programs for the device. This approach is convenient because it gives the programmer the ability to use mnemonics throughout the program. It also provides some limited portability. If a second device in the same family has somewhat different register and port locations, it is possible to write a new header file for this second device and code written for the original device will probably work with the second device when the proper header file is included. We will discuss header files further as we examine the programming of specific microcontrollers.

In this book, you will see the C programming language used to develop code for embedded systems products. C is a powerful language that can be abused and is often the subject of such abuse. Don't abuse this fine language. It is your responsibility as a programmer to write code that is easy to read, easy to understand, easy to debug, easy to maintain, and easy to use. Any fancy antics in coding are uncalled for and not needed. Any group that is writing project code should implement a shop coding standard that will direct the way code is written. There are several organizations that have prepared such standards. I cannot agree with all of the elements of these coding standards that I have seen, but any coding standard is better than none. Research the matter and find a coding standard that comes close to suiting your needs and then modify it to meet your needs exactly. And then use it.

I previously mentioned "group development." I have seen many small projects in which the software development was an afterthought. There are still many projects that can fit into a few hundred bytes or even a few thousand bytes. Such projects might not seem to require the careful oversight implied by a shop standard. But they do! In the development of software, you will find that the initial cost of the software is dwarfed by the cost of maintenance. Therefore, any effort put forth to help the maintenance of the software into the future is effort well spent regardless of the size of the project. You will often find that newer chips have massive memory spaces and with these memory spaces available, the code will expand to fill them. This expansion of the code is not necessarily bad if the expansion accompanies improved operation and features. However, one implication is that the one hardware engineer in the lab can no longer be expected to develop the code for such projects. More and more teams are coming together to develop embedded systems software today. Such teams almost demand that a coding standard be in place if the code thus generated is to ever be useful.

Testing

In future chapters we will see many programming examples developed. The code in such examples has been tested and runs on the microcontroller it was designed for. The hardware interface used to test these programs has been a series of evaluation boards built by Motorola. The several different boards used each represent a different approach to an inexpensive development environment. For the MC68HC05 family, I used the MC68HC05EVM and MC68HC05EVS series. The EVM is an evaluation module for many of the older MC68HC05 family of microcontrollers. This family grew so large and so many different devices were introduced into the product line that it was impossible for a single board to provide the development services needed for the whole family. A new series designated as the EVS family was developed to meet these expanding needs. The EVS is comprised of a base board with many of the needed interface components. A daughterboard containing all of the "component unique" capability of the system is plugged into the base board.

These systems provide excellent development environments. Each system contains a microcontroller of the type being developed. In the case of the MC68HC05 family, the on-board components are operating in a special mode called the non-user mode. In the non-user mode, the device reorganizes its pins to provide an expanded bus operation so that memory and external control can be used on what is usually a single-chip microcontroller. For these parts, a special chip is used called a port replacement unit, or PRU, which is put onto the expanded bus, and its outputs are exactly the same as what would be seen on the ports if the chip were operating in the normal mode. With this capability, RAM can be substituted for internal ROM on the microcontroller and the software implications of this change are unlimited. For example, programs can be loaded into the memory at will, memory access breakpoint can be implemented, the program can be started at any point, the contents of the memory can be changed, and so forth.

A firmware monitor is placed on the evaluation board. A serial port is added, and the board, under the control of the monitor, communicates with a host computer. As a matter of fact, the host need only be a terminal in the most basic case. The monitor provides several important debug and development functions. Code can be downloaded from the host computer into the memory of the evaluation board. The monitor contains a single line assembler/disassembler module. With this software available, the user can examine the code in the microcontroller memory in either hexadecimal or mnemonic form. Also, memory contents can be changed in either hexadecimal or mnemonic form. Breakpoints can be inserted into the code, and the operation of the program can be observed from the terminal of the host computer.

In addition to the evaluation capability of these boards, they also each have extension headers that allow the board to be plugged into a target system. When operated in this manner, the target system operates as if it has a programmed microcontroller in its socket. Of course, the code in the microcontroller is that in the memory on the evaluation module. The emulation of the microcontroller in this case is excellent. The microcontroller on board the evaluation module is executing the code for the target system. The microcontroller resources used in the target system are those found on board the microcontroller in the evaluation module. In most cases, the lines connected to the target system are connected directly to either the evaluation module microcontroller or PRU. The PRU is designed to emulate the pin operation of the microcontroller as well as possible. Therefore, the loads presented to the target system by the emulator and the signal responses of these pins very nearly duplicate those of the microprocessor being emulated.

Figures 3-2 and 3-3 are photographs of the MC68HC05EVM and the MC68HC05EVS, respectively. Note that Figure 3-2 shows a single-board system while Figure 3-3 shows a two-board device. All of the development devices discussed here require an external power supply. Any device suffixed "EVM" needs a +5 volt supply as well as +12 and -12 volt sources. The higher voltage supplies are needed only to drive the RS232 communications signals necessary to communicate with the host computer. On the EVS and the MC68HC16EVB boards, only a +5 volt supply is needed. Most of the components being emulated by these boards can have some on-board nonvolatile memory, such as EEPROM or EPROM. The evaluation boards each provide a means of programming these memories. Usually, a high voltage is needed. This voltage will vary from part-to-part and it must be applied as a separate voltage on the board.

Figure 3-2: MC68HC05EVM Single Board Development System.

Figure 3-3: MC68HC05EVS Two-Board Development System.

The previous discussion is aimed at the MC68HC05 development boards. The MC68HC11 also has an EVM. This EVM operates exactly the same as that discussed above. There are several EVMs for the different parts in the MC68HC11 family. Figure 3-4 shows a photograph of the MC68HC11EVM.

Figure 3-4:
MC68HC11EVM Development System.

In all of the devices we've discussed, communication with the host computer is through an RS232 serial link from the board. The development work can be done with a terminal emulator on the host computer. Such a terminal emulator might be found in the software programs PROCOMM or KERMIT. Certainly, any terminal emulator can do the job, and those mentioned here are just two of many.

Development through a simple terminal emulator is usually not the easiest approach. Motorola ships a software package with each evaluation device. This software package is written by P & E Microcomputer Systems, Inc. It provides a "windowed" environment that shows the conditions of many internal features of the device being evaluated. All of the internal microcontroller registers are displayed, a listing of the breakpoints that have been set are shown, a block of the code being debugged is displayed, and a control screen is also available. The display will differ with various devices, but in general the screen interface is easy to use and understand.

In some cases, a complete symbolic debug environment can be established. P & E Microcomputer Systems, Inc., has a file system that allows transfer of source code to the program. This source code is displayed in the code window. It helps during the debug procedure to see the source code while stepping through it. This capability is available with the compiler used for the MC68HC05, but not for the MC68HC11 and MC68HC16 families.

All of the software written by P & E runs under the DOS operating system. Today, there are fewer computers that run under DOS than in the past, so you might want to find a later development system that runs under a more modern operating system. I have always used DOS by itself or running under either Windows 3.1 or one of the later Windows operating systems. During the availability of OS/2, it was an ideal operating system to develop microcontrollers under these DOS-based development systems. More recent systems require Windows 95 or later.

Development is enhanced if the host computer has multitasking capability. This capability can come from any of the popular operating systems, such as Microsoft Windows or X Windows. In such a case, it is possible to keep an editor with the listing file of the program being debugged in a window along with the P & E display in another window. This approach provides both insight into the code being developed along with the condition of the hardware as the execution of the code proceeds. It is recommended that this approach be used when debugging programs on evaluation boards.

Unfortunately, none of these evaluation boards allows access to the operation of the component being debugged when the device is executing the test code. To achieve this level of operation would increase the complexity of the development board significantly. Also note there is no provision for a trace buffer on these boards. The newer boards, EVS and MC68HC16EVB, provide headers that can be connected to a logic analyzer that can act as a trace buffer and provide many of the functions usually found in the more comprehensive development devices.

One last development device used is the MC68HC16EVB, shown in Figure 3-5. This board differs from the others in one significant way. The MC68HC16 family of devices (as well as those found in the MC68300 family) provide a capability not usually found in most microcontrollers. These devices can be placed into a background debug mode that facilitates development of the code. In all of the above development boards, the monitor software must execute on the microcontroller being developed. This operation means there must be either memory bank switching or the monitor must occupy some of the normal memory space of the microcontroller. In either case, the presence of the monitor software can be seen in the operation of the microcontroller. Also, development with the above systems will require that other resources be used for the development operation. For example, if control of the development is through a serial port to a terminal, the SPI of the microcontroller will probably be used. In the background debug mode (BDM) of the larger devices, this memory and resource sharing between the monitor and the code being debugged is unnecessary. When a device is placed into BDM, its normal operation ceases and all communications with the device occurs through a special serial port. Through this serial port, the condition of the device can be examined, memory can be accessed and changed, breakpoints can be established, etc., just like the invasive technique. However, when the device is returned to operation, there is no way the presence of the monitor can be seen by the running program.

Figure 3-5: MC68HC16EVB Development System.

With BDM, all debug software can be placed on a host computer. Therefore, it is possible to write much better debugging systems that connect through the BDM system. Most recent microcontroller chips employ some type of background debug mode operation. It is not always called BDM. It might be called ONCE or JTAG, but these operations are basically similar and allow background debugging of the microcontroller and a host to contain all of the necessary debug software.

Another advantage to using the BDM is that all communication between the host and the device is through an 8-wire bus. This bus can be accessed in any device, so these devices become their own development systems and no development system is really needed to work on the final target system.

The MC68HC16EVB uses the BDM operation. Communications with the host computer are through the parallel, or printer, port on the computer. P & E has written an interface for "PC clone" computers that uses the parallel port. Its operation is essentially as for the MC68HC05EVM and MC68HC11EVM devices. P & E has also made an additional device which can connect to the 8-wire BDM bus and connect directly to the host computer parallel port. It requires a +5 volt power source and takes its voltage through the lines to the target system.

Three other chips will be examined in the following chapters. The M68HC08 family and the M68HC12 family are extensions of the M68HC05 and the M68HC11 families, respectively. Yet another recent chip family is the MCORE. These RISC chips are very fast and run at extremely low power. The MMC2001 chip from this family will be examined. The development system for the MCORE chips is called an EBDI—Extended Background Debug Interface. This package is interfaced to an evaluation board through an 8-wire serial port.

All of the programs in the chapters that follow were tested on the appropriate development boards. These programs are relatively small because each is designed to show some feature of either the microcontroller or the language as applied to the microcontroller; larger programs were not appropriate to the tutorial aims of this book. With judicious use of the boards along with the development environment on the host computer, these programs were easy to develop and debug.

Chapter 4

Small 8-Bit Systems

Not surprisingly, writing code for any microcontroller—whether in assembly language or a high-level language like C—requires a detailed knowledge of the microcontroller being programmed. Usually, a high-level programming language requires little knowledge of the underlying computer on the part of the programmer. This approach allows the programmer to concentrate on the nature of the problem being solved rather than how to squeeze the problem into a specific computer. A C abstract computer has been designed and you write code for this computer when you write C code. The abstract computer has no registers, control registers, index or address registers, or any other of the normal resources found on a typical computer. The language is sufficient to allow proper creation of code needed to run the core computer. However, the essence of any microcontroller is the special on-board peripherals that it provides. These peripherals are not directly available from the C language either.

Programming techniques must allow use of these peripherals or the high-level language is valueless. Three distinct levels of microcontrollers will be covered in different sections of this text. The simplest microcontroller is embodied in the M68HC05 family. These 8-bit devices are usually completely self-contained and do not support an expanded bus. Another level of complexity is found in the M68HC08, the M68HC11 and the M68HC12 families. These computers are also 8-bit machines, but they have more registers than the M68HC05 family and support an expanded data bus. With the expanded data bus, these families can have external memory and peripherals in addition to those within the chip itself. (The peripherals on these chips are not very different from those found on the M68HC05.) The step up in computer power is the M68HC16 microcomputer. This computer is a

16-bit machine and its peripheral components are nearly all different from those found on the 8-bit machines. The M68HC16 is a superset of the M68HC11; it will execute M68HC11 code, but the hardware computer extensions and new peripheral components are significant.

To successfully program a microcontroller using a high-level language, the programmer must be able to access various control and status registers in the computer. The program must force the language to place both program and data memory addresses in the proper locations in the memory map. Vectors associated with interrupt service routines, and the service routines themselves, must be handled directly by the program. These tasks are difficult to accomplish with most high-level languages, but C allows access to these things without extensions. However, most C compilers for microcomputers have extensions that allow such special features to be easily treated.

The compiler used in this chapter is called C6805[1]. It was written to support the M68HC05 family of devices. Be forewarned: some M68HC05 microcontroller instructions have no counterpart in the standard C language. Special directives identify unique microcontroller characteristics to the compiler. Listed in Table 4-1 below are nine assembly instructions available to the 68HC05. These instructions have no equivalent C call. They can be accessed as either a single instruction (all uppercase) or as a function call as shown. The function call requires a pair of closed parentheses to follow the name of the instruction.

Function	Operation
CLC or CLC ( )	clear carry bit
SEC or SEC ( )	set carry bit
CLI or CLI ( )	clear interrupt flag (interrupts on)
SEI or SEI ( )	set interrupt flag (interrupts off)
NOP or NOP ( )	no operation
RSP or RSP ( )	reset stack pointer
STOP or STOP ( )	STOP instruction
SWI or SWI ( )	software interrupt
WAIT or WAIT ( )	WAIT instruction

Table 4-1: Assembly Codes Directly Callable By C6805

[1] Byte Craft Limited, 421 King Street North, Waterloo, Ontario, Canada N2J 4E4

A *pragma* is a C preprocessor command not defined by the language. As such, the compiler writer can use the `#pragma` command to satisfy a need not specifically identified by the language. C6805 uses pragmas to identify microcontroller-specific characteristics. Table 4-2 contains a list of pragmas used by C6805. The format of a pragma directive here is

```
#pragma portxx portname @ address
```

where `portxx` can be `portr`, `portw`, or `portrw` which shows whether the port is read, write, or both. `portname` is the name used in the program for the port. The at symbol (`@`) identifies a memory address. `#pragma mor` identifies the contents of the masked option register used on field programmable chips. There are some instructions that are not found across the whole M68HC05 family. In particular, some devices may not have the `MUL`, the `DIV`, the `STOP`, and the `WAIT` instruction. The `#pragma` has a preprocessor call that identifies the instructions from this set in the particular microcontroller.

pragma	*Function*
`#pragma portxy`	I/O port definition
`#pragma memory`	RAM/ROM definition
`#pragma mor`	mask option register
`#pragma has`	instruction set options
`#pragma options`	compiler directives
`#pragma vector`	interrupt vector definitions

Table 4-2: C6805 pragma Directives

This compiler has certain options that can be inserted from the command line or, if needed by the programmer, the `#pragma options` preprocessor command can also be used to set the appropriate compiler options. Finally, the `#pragma vector` identifies a given function name as an interrupt service routine. When the compiler compiles the name specified, it will place the address of the function into the defined vector location. Another modification in the compiled code will take place when `#pragma vector` is used. All returns from a function identified by a `vector` pragma will use the return from interrupt instruction rather than the usual return from subroutine.

Another useful directive pair is the #asm and the #endasm. The code enclosed in a block that starts with #asm and ends with #endasm must be in standard assembly language. Variables defined in the C program can be used safely.

C can accomplish almost everything that the assembly language program can. You will find that the C6805 compiler will create tight, efficient code that is probably as good as can be written by a competent assembly programmer. There are, however, some items that are absolutely foreign and inaccessible to a compiler. A compiler creates code for an abstract machine that does not exist in reality. The usual registers found in the real machine are nonexistent in the abstract machine. For example, it is not possible to access the status register of the microcontroller with compiled code. Usually, status register contents are not directly important to the conduct of the program. But later we'll see an example where the ability to manipulate the carry bit of the status register can save many bytes of code. Therefore, it is important to be able to use some assembly code as well as C.

This chapter will concentrate on small 8-bit microcontrollers. Subsystems such as timers, analog-to-digital converters, computer operating properly (COP) timers, etc., found on the 8-bit systems will be outlined and their programming discussed. While the main details of the central processor in the microcontroller are important to the assembly language programmer, they are of little interest to the C programmer. This observation is true at least at the C level. If it becomes necessary to enter an assembly language program for optimization of code size or other considerations, then the programmer is required to have detailed knowledge of the programming model and the internal architecture of the computer.

Let's start by discussing important microcontroller peripheral components that you can expect to find. We'll begin with what is probably the most important single consideration in the selection of a microcontroller to do a job—the device memory. This discussion will be followed by sections on other important peripherals such as timers, analog-to-digital and digital-to-analog converters, serial communications devices, and simple digital input/output lines.

Microcontroller Memory

Most microcontrollers have memory on-board. The memory is in the form of *random access memory* (RAM), *read-only memory* (ROM), *erasable programmable read-only memory* (EPROM), *electrically erasable read-only memory* (EEPROM), and a newer type of EEPROM memory called *FLASH*. These memory types are discussed in the following paragraphs. The discussion of FLASH memory will be deferred until the chapter on the M68HC08 family.

Random Access Memory (RAM)

In a microcontroller, onboard RAM is static random access memory. It is always volatile—when the power to the microcontroller is removed, the contents of this memory disappear. Sometimes, special provisions are made to deliver power to RAM when the processor is in the "off" state. This provision is called battery backed- up RAM, and it is one of the alternative ways that a small amount of important data can be saved when power is removed from the main system.

The requirement for RAM in typical microcontroller applications is modest to small. Available RAM is usually limited to a few hundred bytes, and often there will be as little as a few tens of bytes of RAM. In the design of the microcontroller, price is a major consideration. The total silicon area of the computer die often drives the final price of the component. In most computers, a base page is the first 256 bytes of memory. This page is unique because it requires only 8-bits of address to reach any location. Silicon area needed to construct the address decoding for the upper address bits is not required to address base page memory. Therefore, onboard RAM is usually located in the computer base page. There are some other functions that are usually assigned to the base page. Generally, you will find that the amount of RAM is limited to less than 256 bytes.

Read-Only Memory (ROM)

Programs and other data that can never be changed are stored in ROM. ROM is programmed during the manufacture of the chip, and its contents cannot be changed once the microcontroller is delivered to the customer. The ROM program is installed as a mask layer and is called *masked* ROM.

Most microcontroller applications require more program memory space than RAM space. The smallest microcontroller usually has about 512 bytes of ROM, while the largest can contain as much as 32,000 bytes (32 kilobytes, or 32k) or more. Sometimes, the programmer will find it desirable to have a small amount of ROM that can be accessed from the computer base page. To meet these requirements, the microcontroller designers will place a few bytes of ROM in the base page memory map.

Erasable Programmable Read-Only Memory (EPROM)

EPROM is a form of programmable memory that permits the programmer to change the program contents and, if necessary, return and change it later after testing. As the name implies, it is possible to reprogram EPROM. First, this memory must be erased. The erasing procedure involves allowing ultraviolet light to fall upon the memory area of the die. This high-energy light removes stored charge that is placed on each memory gate during programming.

EPROM programming requires that a higher than normal voltage be applied to the chip, and the code be systematically placed in each memory location. The procedure is slow because the code must be left in place for several milliseconds for each memory location stored. Often, a separate programmer board is used to transfer code from an EPROM to the microcomputer EPROM. These programming boards can program as many as one to eight parts at a time.

EPROM requires a larger silicon die area than the corresponding amount of ROM. Therefore, it is somewhat more expensive. Also, the window package that allows the EPROM to be erased is expensive. This additional expense makes it impractical to use normal EPROM for production volumes. The window package EPROM devices are excellent for development purposes, though. The modestly higher cost of these devices is not a serious impediment to their use in development programs.

The economics of production sets the smallest production volume for a masked ROM microcontroller at about one to five thousand units. An alternative to the use of masked ROM at smaller levels of production is called the *one-time programmable* (OTP) chip. These devices use the standard EPROM technology for their program memories. They are programmed in the same manner as EPROM chips.

Their packages, however, have no windows to allow erasure of the program once it is put in place. These devices cost somewhat more than masked ROM, but they are sufficiently less expensive than the EPROM parts to allow economic production of rather small quantities. They do have the disadvantage that, once programmed, they can never be used for a different program.

Electrically Erasable Programmable Read-Only Memory (EEPROM)

EEPROM is a technology that uses a memory cell similar to the standard EPROM cell. These cells are somewhat larger than the standard EPROM, and are therefore more expensive. It is possible to erase an EEPROM electrically without the high-energy ultraviolet light. EEPROM requires a high voltage in programming and erasing the memory. Some microcontrollers have EEPROM that can be programmed without an externally applied high voltage. This programming is accomplished by the use of an onboard charge pump to generate the programming voltage. Such charge pumps are not capable of delivering much current, so the amount of EEPROM that can be programmed from an onboard system is usually limited to a maximum of 512 bytes. This EEPROM is used for the storage of information gathered after the microcontroller has been placed into a system. This memory is not often used for the storage of program.

The smaller block of EEPROM can be programmed with the use of the onboard charge pump, and can be programmed "on the fly" during the normal execution of program. Devices with EEPROM are moderately expensive because EEPROM requires the largest silicon area of any memory technology.

Other Memory Considerations

Not all microcontrollers have enough onboard memory to suffice in some jobs. In these cases, an *expanded bus* part can be used. Expanded bus parts allow the programmer to access memory that is external to the microcontroller. None of the small microcontrollers currently provide for expanded bus operation. The larger microcontrollers—large 8-bit, 16-bit, or 32-bit—provide expanded bus. In some instances, they provide no onboard memory at all. As we will see later, pins on a microcontroller are at a premium. An expanded bus operation means that some of the component pins must be used to access memory and will not be available for other microcontroller features. (Pin usage, bus expansion, and pin multiplexing will be discussed in later sections.) The important consideration at this point is that the limited program memory area usually associated with a microcontroller should not cause serious concern. If the program grows to exceed the available size of onboard memory for a microcontroller family, it is always possible to get a larger microcontroller that can handle any additional memory requirements. The programming goal, though, is usually to confine the program in the smallest possible program memory space so that the least expensive microcontroller will do the job.

Using Microcontroller Memory

In our discussion on variables in Chapter 1, it was shown that C treats all automatic variables as local to the block in which they are declared. The scope of these variables is the block where they are declared. Since these variables exist only in the block where they are declared, the memory locations dedicated to the storage of these variables can be freed when the variables go out of scope. These rules create an ideal situation for storage on the program stack. Memory space is easily created on the stack at the beginning of a block, and it is equally easily destroyed at the close of the block. This operation is exactly what is needed, but it cannot be used in a typical small microcontroller. Most microcontrollers have very limited RAM, and the stack arrangement in them is completely different from that you will find on a large computer. On the M68HC05 family of parts, for example, the chip has a hardware stack and no stack pointer into memory that the compiler writer can access. Therefore, it is impractical to even attempt to use the system stack to store local variables. The hardware stack on these chips is used only for storage of the processor status when an interrupt occurs or to store the return address from a jump to a subroutine. The stack pointer is set to its initial value on microcontroller reset, and the occurrence of an interrupt or a jump to subroutine instruction are the only ways that the stack pointer can be changed.

In the larger machines, the stack pointer is set to a value that points to a memory location. This pointer will be automatically incremented and decremented by the equivalent of stack push or pull operations. The program can arbitrarily change the stack pointer value so that room for automatic variables can be easily provided or eliminated. In the small microcontrollers, automatic variables are stored in RAM and their scope is not limited to the block in which they are defined. Their access is limited to their block, however. Consider the following code segment:

```
main()
{
    int i;

    .

    .

    .

}


void able(void)
{
    int i;

    .

    .

    .

}
```

The two occurrences of the variable i in this case will cause no trouble because each i will be given a unique location in RAM and the scoping arrangement will insure that any reference to i in main() will not be confused with the i in able() and vice versa.

An important implication of this change in storage: recursion is no longer available! Only one memory location is available for each variable in the program. When a stack is used to store automatic variables, a function can call itself and a new block is created each time the function is entered. Thus, each time a function calls itself, a new stack frame that contains space for all automatic storage in the function is created. The function can call itself repeatedly as long as there is space on the stack to create new stack frames for the successive calls. Without stack space for variable storage, recursion is impossible.

A second limitation that occurs is in the available arguments for function calls. The compiler C6805 for the M68HC05 family defines an int as an 8-bit number and a long as a 16-bit number.

This definition is not compliant with the ANSI Standard, which requires that an `int` be at least 16 bits wide and a `long` be at least 32 bits. Since the stack cannot be used to pass arguments, they must be passed in either registers or as global variables. If they are passed in registers, only two bytes can be passed. The arguments can be either two `int`s or one `long`. Function return values have the same limitations. Of course, the program can use global variables to pass information to or from a function. A global variable defined external to any function can be accessed by any function in the program.

Most C compilers for the M68HC05 family provide automatic placement of variables in the available RAM of the part. Specific memory addresses are identified to the compiler by the `#pragma` memory directives. The following code segment shows an example of how the memory is defined within an M68HC05 program:

```
#pragma memory ROMPAGE0 [48] @ 32;
#pragma memory ROMPROG [5888] @ 2048;
#pragma memory RAMPAGE0 [176] @ 80;
#pragma memory RAMPROG [256] @ 256;
```

This sequence of code will be used to identify the memory map of the M68HC05B6. This part has 48 bytes of ROM in page zero starting at address 32. There are 5888 bytes of program ROM starting at address 2048. The 176 bytes of page zero RAM starts at address 80. There are 256 bytes of EEPROM in this part that begin at the address 256. Here we treat EEPROM as program RAM because it is programmable and is outside of the base page.

Inclusion of the above code lines will identify the necessary memory locations for the compiler, and further concerns about memory locations should be unnecessary. The compiler will automatically place the code in the ROMPROG area and the RAM requirements will fall at the starting address identified by RAMPAGE0. Programmers who wish to make use of the ROMPAGE0 memory can do so by a command like

```
const int table[]={-,-,-,...,-} @ 32;
```

This instruction will place the specified array of data in the ROMPROG0 area and will start it at the address 32.

Programming EEPROM

EEPROM is read like any other memory in the microcontroller. Two different types of EEPROM can be found on a microcontroller: *program memory* and *data storage memory*. Program memory usually cannot be programmed without the aid of an externally applied programming voltage. Data storage memory can be programmed from within the program and requires no externally applied programming voltage. In the case of the M68HC805B6, there are 5888 bytes of program EEPROM and 255 bytes of data storage EEPROM. Other than the reduced size of the data storage memory, this memory is no different from the program memory. It is possible to write code to the data storage EEPROM and execute this code.

One additional byte of data storage EEPROM exists and is called the OPTION register. This register content is saved in EEPROM which is read into a latched register during the initialization of the microcontroller. The address of this register is 0x100. The bits in this register control the security option of the part and control a block protect region in the data storage EEPROM that will prevent accidental writing of data into the protected memory area. A description of these bits follows:

Options Reg 0x0100	Bit 7	Bit 6	Bit 5	Bit 4	Bit 3	Bit 2	Bit 1	Bit 0
							EE1P	SEC

SEC   Bit 0   Security Bit. When the SEC bit is programmed to zero, the contents of the EPROM are secured by preventing access to the test mode. The only way to erase the SEC bit to a one state is to enter the self-check mode. In this event, the data on EEPROM will all be erased. When the SEC bit is changed, its new value will have no effect until after the next chip reset.

EE1P   Bit 1   EEPROM Block Protect Bit. The EEPROM is in two parts: 0x101 to 0x11f is part 1 and 0x120 to 0x1ff is part 2. The EE1P bit allows part 2 to be protected. If this bit is in the erased state, (1), part 2 of the EEPROM will be protected. This memory area can be read as usual, but any attempt to write to this area will fail. The protection remains in effect after this bit is erased until after the next chip reset.

Control of the EEPROM programming is through the EEPROM CTL register found at address 0x07. The bits in this register are as follows:

EEPCTL/CLK 0x07	Bit 7	Bit 6	Bit 5	Bit 4	Bit 3	Bit 2	Bit 1	Bit 0
	0	0	0	0	ECLK	E1ERA	E1LAT	E1PGM

E1PGM   Bit 0   EEPROM Program Bit. This bit turns the internal Vpp charge pump on and off. When this bit is 0, the charge pump is turned off, and when it is at 1, the charge pump is turned on. The charge pump voltage can be measured on the pin Vpp1. This bit cannot be set until after the program data are latched in place by asserting the E1LAT bit. Resetting the E1LAT bit will also reset the E1PGM bit.

E1LAT   Bit 1   EEPROM Data/Address Latch. When this bit is reset to zero, both the E1PGM bit and the E1ERA bit are reset to zero. When the E1LAT bit is reset, data can be read from the EEPROM. The first data write to the EEPROM array after this bit is set is latched until the E1LAT bit is reset. Data can be latched only when the E1PGM bit is reset to zero. This operation allows programming of the EEPROM. E1LAT is automatically reset when the chip is reset or when the STOP instruction is executed.

E1ERA   Bit 2   EEPROM Erase Bit. If the bit E1ERA is reset to zero when E1LAT and E1PGM are set to one, data are programmed into the EEPROM. Otherwise, if E1ERA is set to one and E1LAT and E1PGM are set to one, the specified address in the EEPROM will be erased. E1ERA cannot be set before E1LAT, and resetting E1LAT to zero will cause E1ERA to be reset.

Let us now examine a possible sequence of code that can be used to program and erase locations in EEPROM. First, several macro definitions should be used to define the various parameters used.

```
/* pragmas to identify EEPROM control registers */

#pragma portrw EEPROM_CTL @ 0x07;
#pragma portrw OPTIONS @ 0x100;
    .
    .
    .

/* EEPROM programming specific defines */

#define E1PGM 0
#define E1LAT 1
#define E1ERA 2
#define PROG_TIME 10
    .
    .
/* some function prototypes */

void delay(unsigned long);
void program(int ,int);
void erase(int );
    .
    .
    .
int EEPROM[0xff] @ 0x101; /* Identify the EEPROM */
void program(int address,int value)
{
   EEPROM_CTL.E1LAT=1; /* set the E1LAT bit */
   EEPROM[address]=value;/* put the data and address
        in place */
   EEPROM_CTL.E1PGM=1; /* turn on the charge pump */
   delay(PROG_TIME); /* delay programming time */
   EEPROM_CTL.E1LAT=0; /* reset the E1LAT also
                    resets the E1PGM bit */
}        /* return when done */

void erase(int x)
{
```

```
EEPROM_CTL.E1LAT=1; /* set the E1LAT bit */
EEPROM_CTL.E1ERA=1; /* set the E1ERA erase bit */
EEPROM[x]=0; /* select the address */
EEPROM_CTL.E1PGM=1; /* turn on the charge pump*/
delay(PROG_TIME); /* wait the appropriate time*/
EEPROM_CTL.E1LAT=0; /* reset the E1LAT bit turns
            off both E1PGM and E1ERA
            bits */
}         /* return when done */
```

The above program sequences are compiled and listed below.

The function delay is not included or linked into this program. To handle this type of problem, the registers are set up for the function call, and the instruction JSR $**** is executed. A later linking will replace the unknown function address with the correct value. An appropriate delay() function will be written in the timer section. The instructions for this function call are found at addresses 0x80c to 0x80f in the following listing.

```
0020 0030 #pragma memory ROMPAGE0 [48] @ 32;
0800 1700 #pragma memory ROMPROG [5888] @ 2048;
0050 00B0 #pragma memory RAMPAGE0 [176] @ 80;
0100 0100 #pragma memory RAMPROG [256] @ 256;

   /* pragmas to identify EEPROM control registers */

0007 #pragma portrw EEPROM_CTL @ 0x07;
0100 #pragma portrw OPTIONS @ 0x100;

   /* EEPROM programming specific defines */

0000 #define E1PGM 0
0001 #define E1LAT 1
0002 #define E1ERA 2
000A #define PROG_TIME 10

/* some function prototypes */

void delay(long);
```

```
void program(int,int );
void erase(int );

0101 0101 00FF int EEPROM[0xff] @0x101;

void program(int address, int value)
0050 0051 {
0801 BF 50 STX $50
0803 B7 51 STA $51
0805 12 07 BSET 1,$07 EEPROM_CTL.E1LAT=1;
0807 D7 01 01 STA $0101,X EEPROM[address]=value;
080A 10 07 BSET 0,$07 EEPROM_CTL.E1PGM=1;
080C 5F CLRX delay(PROG_TIME);
080D A6 0A LDA #$0A
080F CD 00 00 JSR $****
0812 13 07 BCLR 1,$07 EEPROM_CTL.E1LAT=0;

0814 81 RTS }

        void erase(int x)
0052 {
0815 B7 52 STA $52
0817 12 07 BSET 1,$07 EEPROM_CTL.E1LAT=1;
0819 14 07 BSET 2,$07 EEPROM_CTL.E1ERA=1;
081B 97 TAX EEPROM[x]=0;
081C 4F CLRA
081D D7 01 01 STA $0101,X
0820 10 07 BSET 0,$07 EEPROM_CTL.E1PGM=1;
0822 5F CLRX delay(PROG_TIME);
0823 A6 0A LDA #$0A
0825 CD 00 00 JSR $****
0828 13 07 BCLR 1,$07 EEPROM_CTL.E1LAT=0;


082A 81 RTS }
```

The function `program()` requires 20 bytes and the function erase requires 22. This is a good point to explore some of the C programming practices that can lead to poor M68HC05 family compiled code. The M68HC05 family is a family of 8-bit machines. There has been no discussion of the programmers' register model of these devices. Programmer models of the larger microcontrollers will be discussed because knowledge of the programmers' model might help in crafting good C code. For these small machines, the watchword is 8-bit. The internal structure of the system is all 8-bit. The width of the single index register is 8 bits, and the width of the accumulator is also 8 bits. The program counter is more than 8 bits in most cases, but it is wide enough to address only the range of the internal computer memory. In fact, the width of the stack pointer in the M68HC05Bx family is only 6 bits. There is no luxury of spare bits in any register.

Therefore, when writing code for the M68HC05 family, keep foremost in your mind that you are dealing with an 8-bit device. If at all possible, avoid 16-bit operations because they will always result in larger memory and/or code usage. The following code demonstrates an example of the careless use of 16-bit implied code in an 8-bit machine.

Consider the `erase()` routine from above. This function could have been written as follows:

```
void erase(int *x)
{
   EEPROM_CTL.E1LAT=1; /* set the E1LAT bit */
   EEPROM_CTL.E1ERA=1; /* set the E1ERA erase bit */
   *x=0; /* select the address */
   EEPROM_CTL.E1PGM=1; /* turn on the charge pump*/
   delay(PROG_TIME); /* wait the appropriate time*/
   EEPROM_CTL.E1LAT=0; /* reset the E1LAT bit turns
            off both E1PGM and E1ERA
            bits */
}        /* return when done */
```

The only change in this version is to pass the integer *x to the function by reference. Remember, since all addresses in the M68HC05 family of parts are greater than 8 bits, the compiler must handle the transfer of the pointer x as a 16-bit number. The statement `*x=0;` compiles into an inline function at the address range 0x81d to 0x82c in the compiled version of the code shown below. This function creates a subroutine that does an indexed store with a 16-bit offset. First, the value to be programmed is placed in the accumulator. Then the op code, 0xd7, to do a store the accumulator indexed with a 16-bit offset is created at the location 0x56 in memory. The 16-bit offset is the address passed to the function in the combination of the x register and the accumulator. This address is placed in the memory locations 0x58 and 0x57, completing the store instruction. At the address 0x59, a return from subroutine instruction, 0x81, is placed to complete the function. The index register is cleared, and this two-instruction subroutine is executed to store the appropriate data prior to the program setting the latch bit.

```
     void erase(int* x)
0052 {
0815 BF 52 STX $52
0817 B7 53 STA $53
0819 12 07 BSET 1,$07 EEPROM_CTL.E1LAT=1;
081B 14 07 BSET 2,$07 EEPROM_CTL.E1ERA=1;
081D B7 58 STA $58 *x=0;
081F 9F    TXA
0820 B7 57 STA $57
0822 4F    CLRA
0823 AE D7 LDX #$D7
0825 BF 56 STX $56
0827 AE 81 LDX #$81
0829 BF 59 STX $59
082B 5F    CLRX
082C BD 56 JSR $56
082E 10 07 BSET 0,$07 EEPROM_CTL.E1PGM=1;
0830 5F    CLRX delay(PROG_TIME);
0831 A6 0A LDA #$0A
0833 CD 00 00 JSR $****
0836 13 07 BCLR 1,$07 EEPROM_CTL.E1LAT=0;
0838 81    RTS }
```

This code sequence requires 36 bytes, plus 4 bytes of uncommitted RAM space, to accomplish what required 22 bytes in the earlier example of the same operation.

You must not avoid the use of pointers because of this one example. There are cases when proper pointer usage will provide the best code that you can generate. When writing code for microcontrollers, use many relatively small functions that you can compile individually and examine if they create outlandish code. Once these small functions are all debugged, you can integrate them into your program as either inline code or as function calls. Hence, the fundamental rule of writing good high-level code for a microcontroller: *the production of good C code for a microcontroller is a joint effort between the programmer and the compiler writer.*

Erasure of EEPROM causes component wear, and most EEPROMs will wear out after a large number of erasures. The nature of the degradation is that the component refuses to erase after many erasures. There has been no evidence that data retention is affected by repeated erasures. The number of erasures that can cause problems is temperature sensitive. Most of these devices are rated for 10,000 write/erase cycles at the maximum rated temperature for the part. At room temperature, the number of write/erase cycles without damage can grow to several hundred thousand. In light of these facts, it is important that programs use care in rewriting the contents of EEPROM.

EXERCISES

1. The EEPROM in microcontrollers erase to the 1 state. Write a function that checks to determine if an erasure cycle erases the contents of a given location in memory.

2. Write a function that compares data to be programmed into EEPROM and determine if it is necessary to erase a byte containing data before it is reprogrammed. This approach will extend the life of the EEPROM.

Timers

The systems of timers placed on microcontrollers are among the most creative engineering efforts most people will ever see. The functions of these timers cover literally dozens of different operations. This set of peripheral components mainly relieve the computer of much work associated with execution of the peripheral function. We will start with the simplest timer and outline different timer capabilities in increasing complexity. The most basic timer in the M68HC05 family is called the 15-bit timer, and probably the most advanced timer system is the 16-bit timer. These different timer systems are literally unrelated in the features they offer. Another timer feature offered in some microcontrollers is the *computer operating properly (COP)* timer. Each of the systems will be examined in some detail in the following paragraphs.

Multifunction Timer—15-Bit Timer

A timer of this nature can be found on the smallest of the M68HC05 components, such as the M68HC05J1 or the M68HC05P8, and it is not found on the M68HC05Bx devices discussed previously. This timer consists of an 8-bit ripple counter followed by an additional 7-bit counter. This counter chain is driven by a signal that is at one-fourth the internal clock frequency of the microcontroller. The internal clock frequency in turn is half the crystal frequency. This portion of the counter is completely uncontrolled. The program can, however, read the value of this counter at any time. A block diagram of this type of timer is shown in Figure 4-1.

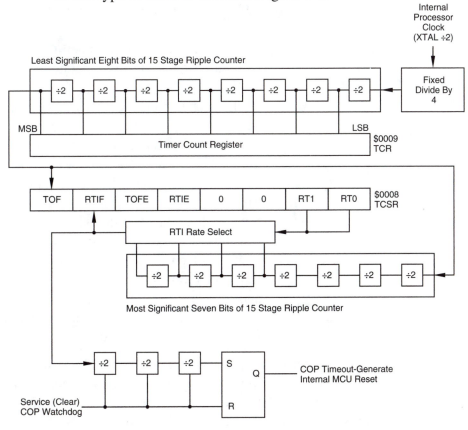

Figure 4-1 15-bit Timer Block Diagram

STATUS REG	Bit 7	Bit 6	Bit 5	Bit 4	Bit 3	Bit 2	Bit 1	Bit 0
0x08	TOF	RTIF	TOFE	RTIE	0	0	RT1	RT0

RT0 Bit 0 Real Time Interrupt Select Rates. These bits control the rate at which the real-time interrupt and the COP reset time will occur. Note that these two

RT1 Bit 1 bits are delivered to the RTI Rate Select register in the diagram. Table 4-3 shows the various RTI and COP rates that can be obtained for different values of RTI and RT0. Reset sets both bits so that the periodic rates will be the slowest possible when the part comes out of reset. It is expected that these bits will not be changed If these bits are altered, the first cycle following the change will be wrong.

RTIE Bit 4 Real Time Interrupt Enable. When this bit is set, a CPU interrupt request is generated whenever the RTIF is set.

TOFE Bit 5 Timer Overflow Enable. When this bit is set, a CPU interrupt request is generated whenever the TOF is set.

RTIF Bit 6 Real Time Interrupt Flag. This bit is set whenever the output from the selected divider stages is set. If the RTIE is also set, setting of this bit will request a CPU interrupt. This bit is cleared by reset or by writing a zero to it. It is not possible to write a 1 to this bit.

TOF Bit 7 Timer Overflow Flag. This bit is set whenever the 8-bit ripple counter overflows from a 0xff to a 0x00. This bit is cleared by reset or by writing a zero to it. It is not possible to write a 1 to this bit. The timing that results from selection of values for RT1 and RT0 are shown in Table 4-1. This table also shows the real time interrupt rate for different values of the real time interrupt select rate bits.

The timer counter register at address 0x09 contains the value found in the 8-bit ripple counter. The contents of this counter can be read at any time, but it cannot be written to. When the part comes out of reset, the timer counter register contains a zero. 4096 clock cycles will follow, during which the TCR will count at the minimum rate.

At the close of this period, the initialization of the part is complete, and the TCR is again reset to zero prior to execution of the code identified by the reset vector. Whenever the RESET line is asserted, the TOF will be loaded with zeros.

Any program can read the TCR, so it is possible to generate asynchronous time events faster than the TOF or the RTIF would indicate.

When the processor enters the WAIT mode after execution of a WAIT instruction, the CPU clock halts, but the timer clock continues to execute. If the interrupts are not masked, a timer interrupt, an external interrupt or a reset will cause the device to exit the WAIT mode.

Table 4-3: RTI And COP Rates for F_{xtal} = 4.0 MHz

RT1	RT0	RTI Rate	Minimum COP Reset
0	0	8.2 ms	57.3 ms
0	1	16.4 ms	114.7 ms
1	0	32.8 ms	229.4 ms
1	1	65.5 ms	458.8 ms

If a STOP instruction is executed, the timer clock is halted along with the CPU clock. The STOP mode is exited when an external interrupt occurs or the RESET line is asserted. In this case, the part performs as described above.

Most microcontrollers are placed in operation with no operator to intervene in the event of a problem. A COP timer will provide one means of recovering if the operation of the microcontroller gets lost. "Gets lost"? The situation that can cause a microcontroller to get lost is usually some type of voltage spike or glitch in the power supply operation. The program counter usually ends up with a value outside of the program, and no one knows what will happen. The COP is simply a timer that counts for a specified amount of time. If the COP timer has not been reset before the specified time elapse, the COP timer overflow causes an internal reset of the microcontroller. If the cause of the problem is a drop in power or other error, in most instances forcing a reset will bring the microcontroller back into normal operation.

The COP control register is located at address 0x7f0 in the M68HC05J1. To service the COP from the program, the program must merely write a zero to bit 0 of this address to reset the COP portion of the timer system.

Good programming practice dictates that microcontroller-specific information be placed in a header file like that shown here:

```
#pragma portrw PORTA @ 0x00;
#pragma portrw PORTB @ 0x01;
#pragma portrw DDRA @ 0x04;
#pragma portrw DDRB @ 0x05;
#pragma portrw TCST @ 0x08;
#pragma portrw TCR @ 0x09;


#pragma portrw __COPSVS @ 0x7f0;
#pragma vector __TIMER @ 0x07f8;
#pragma vector __IRQ @ 0x07fa;
#pragma vector __SWI @ 0x07fc ;
#pragma vector __RESET @ 0x07fe;
#pragma has STOP ;
#pragma has WAIT ;
#pragma has MUL ;


#pragma memory RAMPAGE0 [64] @ 0xc0;
#pragma memory ROMPROG [1024] @ 0x300;


#define RT0 0 /* TSCR Bits */
#define RT1 1
#define RTIE 4
#define TOFE 5
#define RTIF 6
#define TOF 7
```

Listing 4-3: Header File For The M68HC05J1

The #pragma and several important #define commands are microcontroller specific. Therefore, to change the program from one microcontroller to another, the programmer need only change the microcontroller header file. Listing 4-3 is a header file for the M68HC05JJ1 controller. The first six entries identify the locations of the I/O ports, the data direction registers, timer status/control register, and the timer counter Register. The next five entries specify the COP service address and the vector locations for this part. Note that the names associated with the vector locations all start with a double underscore. These names are also listed in all upper-case letters. These entries are the names of the various interrupt service routines. Since C is case sensitive, the names of the functions to be used as interrupt service routines must have the same form.

The three #pragma entries identified as has notifies the compiler that the microcontroller has the STOP, WAIT, and MUL instructions. The next pair of entries defines the memory map for this microcontroller. Finally, the next six entries are #defines that identify the bits in the TCSR. Therefore, mnemonic representations of all registers and bits can be used in the C program.

Several header files for the M68HC05 family are found on the CD-ROM. The conventions in these files are to use bit names and register names that are identical to those used in the technical data books that describe the devices. Therefore, the programmer can safely use register names and bit names found in the books without having to look up the values in the header files. These files include commands to prevent listing of these files in the compiler listing output files.

Listed below is a simple program that shows the use of the 15-bit timer in the M68HC05J1. This program is not aimed at doing more than showing the use of the timer operation. The system will create an inaccurate clock in which the time in hours, minutes, and seconds will be recorded in memory, but no provision to display these values or even set the values will be considered at this time.

Most clocking operations should be interrupt driven. If a periodic interrupt can be generated, the operation of the clock will be transparent to any other operations being conducted in the microcontroller.

```
#include "hc05j1.h"
enum {FALSE,TRUE);
enum {OFF,ON};

#define FOREVER while(TRUE)
#define MAX_SECONDS 59
#define MAX_MINUTES MAX_SECONDS
#define MAX_HOURS 12
#define MAX_COUNT 121

/* define the global variables */
int hrs,mts,sec;
```

```c
int count;

main(void)
{
    count=0; /* start count at zero */
    TCST.RT0=OFF; /* 57.3 ms cop timer */
    TCST.RT1=OFF; /* 8.192 ms RTI */
    TCST.RTIE=ON; /* Turn on the RTI */
    TCST.RTIF=OFF; /* Reset interrupt */
    TCST.TOF=ON; /* flags */
    CLI(); /* turn on interrupt */

    FOREVER
    {
        if(sec>MAX_SECONDS) /* do clock things */
        {
            sec=0;
            if(++mts>MAX_MINUTES)
            {
                mts=0;
                if(++hrs>MAX_HOURS)
                hrs=1;
            }
        }
    /* here is where any applications program should
    be placed. */
    }
}
void __TIMER(void)    /* routine executed every
                        RTI (8.192 ms) */
{
    TCST.RTIF=OFF; /* reset interrupt flag */
    if (++count>MAX_COUNT)
    {
        sec++; /* increment seconds */
        count=0;/* reset the count each second */
    }
}
```

Listing 4-4: A Time-of-Day Program Based On The 15-bit Timer.

A few words about a good programming practice: numbers in a program with no defined meaning are called "magic numbers." You should avoid magic numbers, because a number with no meaning makes life difficult for the program maintenance people. In the program above, several numbers are needed. These numbers are given a name by either `enum` statements or `#define` statements. Then, in the program, you can see every instance of the use of the number does have a meaning relative to the program. Another advantage to avoiding magic numbers is not too evident in the above program, but it is truly an important advantage. If the program is long and complicated, these numbers might be used many times. Then if a maintenance situation requires the change of the value of a number in the program, it can be changed in one place and a recompilation will correct every instance of the number in the program.

Several global variables are used in this program: `hrs`, `mts`, `sec`, and `count`. These variables are all changed in the main program, but they are available in any other part of the program if needed. For example, the `count` variable is initialized to zero in the main program and incremented and reset in the interrupt service routine. One point should be noted in this program: the main program has all of the time calculations based on the current contents of `sec`. The variable `sec` is incremented each second in the interrupt service routine. Some programmers would put the complete time service within the interrupt service routine. That is, they would reset `sec` when it reaches 60, increment `mts`, and so forth within the interrupt service routine. Either approach will provide the same result, and each takes the same total computer time. It is, however, better to keep the time that the program is controlled by the interrupt service routine at a minimum. Interrupts are disabled when a program is in an interrupt service routine. If there are several competing interrupts, execution of an interrupt service routine prevents other interrupts from being processed. Quickest response to all interrupts will be obtained if all of the interrupt service routines are as short as possible.

Program Organization

A compiled version of this program is listed below. Note that the compiler listing routine prints out the contents of the include file. The memory map `#pragmas` puts the RAM in page 0 beginning at

0xc0 and the program memory starts at 0x300. Note that the compiler places the global variables in 0xc0 through 0xc3, and the executable program begins at 0x300 as one would expect.

The first several instructions clear the *count* location and set or reset proper bits in the TSCR. The instruction CLI clears the interrupt bit in the status register of the microcontroller. When this bit is cleared, interrupts will detected and processed.

The beginning of the loop defined by the macro command FOR-EVER is at address 0x30d. This macro causes no code at the beginning of the loop. At the end of the loop, address 0x32b, there is an instruction BRA 0x30d that causes control of the program to start at the beginning of the loop. That is the total code created by the macro FOREVER. Within this loop, the code created by the compiler is straightforward and not very different from code that would be created by a competent assembly language programmer.

```
     #include "hc05j1.h"
0000 #pragma portrw PORTA @ 0x00;
0001 #pragma portrw PORTB @ 0x01;
0003 #pragma portr  PORTD @ 0x03;
0004 #pragma portrw DDRA  @ 0x04;
0005 #pragma portrw DDRB  @ 0x05;
0008 #pragma portrw TCST  @ 0x08;
0009 #pragma portrw TCNT  @ 0x09;


07F0 #pragma portrw __COPSVS @ 0x7f0;
07F8 #pragma vector __TIMER  @ 0x07f8;
07FA #pragma vector __IRQ    @ 0x07fa;
07FC #pragma vector __SWI    @ 0x07fc ;
07FE #pragma vector __RESET  @ 0x07fe;


        #pragma has STOP ;
        #pragma has WAIT ;
        #pragma has MUL ;


00C0 0040 #pragma memory RAMPAGE0 [64]  @ 0xc0;
0300 0400 #pragma memory ROMPROG  [1024] @ 0x300;
```

```
0000 #define RT0  0
0001 #define RT1  1
0004 #define RTIE 4
0005 #define TOFE 5
0006 #define RTIF 6
0007 #define TOF  7

0001 #define TRUE 1
0000 #define FALSE 0
0001 #define FOREVER while(TRUE)

00C0 00C1 00C2 int hrs,mts,sec;
00C3 int count;
     main(void)
       {
0300 3F C3  CLR $C3       count=0;
0302 11 08  BCLR 0,$08    TCST.RT0=0;
0304 13 08  BCLR 1,$08    TCST.RT1=0;
0306 18 08  BSET 4,$08    TCST.RTIE=1;
0308 1D 08  BCLR 6,$08    TCST.RTIF=0;
030A 1F 08  BCLR 7,$08    TCST.TOF=0;
030C 9A     CLI           CLI();
        FOREVER
          {
030D B6 C2  LDA $C2       if(sec==60)
030F A1 3C  CMP #$3C
0311 25 18  BCS $032B
          {
0313 3F C2  CLR $C2       sec=0;
0315 3C C1  INC $C1       if(++mts==60)
0317 B6 C1  LDA $C1
0319 A1 3C  CMP #$3C
031B 26 0E  BNE $032B
          {
031D 3F C1  CLR $C1       mts=0;
031F 3C C0  INC $C0       if(++hrs==13)
0321 B6 C0  LDA $C0
0323 A1 0D  CMP #$0D
```

```
0325 26 04 BNE $032B
0327 A6 01 LDA #$01 hrs=1;
0329 B7 C0 STA $C0  }
                           }
032B 20 E0 BRA $030D }
032D 81 RTS }


     void __TIMER(void)
07F8 03 2E {
032E 1D 08 BCLR 6,$08 TCST.RTIF=0;
0330 3C C3 INC $C3 if (++count==122)
0332 B6 C3 LDA $C3
0334 A1 7A CMP #$7A
0336 26 04 BNE $033C
    {
Timers
0338 3C C2 INC $C2 sec++;
033A 3F C3 CLR $C3 count=0;
    }
033C 80 RTI }

07FE 03 00
```

At the beginning of the interrupt service routine there is an entry
07f8, which has a value of 032e, in the address column. This entry
places the address of the timer interrupt service routine 0x032e
into the timer vector 0x07f8. The code generated in the interrupt
service routine is straightforward and little different from what one
would expect an assembly language programmer to do. Note, however, that the return at the end of the interrupt service routine is an
RTI instruction. This is the instruction that causes the microcontroller
to restore the processor status to the state that existed when the interrupt occurred. The normal return from subroutine RTS does not restore
the processor state. An RTI must be used to return from interrupt
service routines, and any function identified with a vector pragma
will be assumed to be an interrupt service routine by the compiler.

It was noted earlier that this timer routine is inaccurate. It is inaccurate only because 122 periods of 8.192 milliseconds each total
0.999424 seconds. This seemingly small error will cause big problems

if one wants a real clock because the error amounts to 2.1 seconds per hour. One way this error could be corrected is to adjust the crystal frequency of the microcontroller. Suppose we would use a frequency of 3.996354 MHz instead of 4.0 MHz. This number is derived by

```
122*2^15/f=1
```

which yields the above value for f. The 122 periods of 8.196721 milliseconds, which is the real-time interrupt time for this frequency, is exactly 1 second. Another approach involves making small corrections to the time periodically so that on the average the time is correct. An example of an interrupt service routine that makes these corrections is as follows:

```
void __TIMER(void) /*routine executed every RTI
(8.192 ms)*/
{
   static int corr1,corr2,corr3;
   TCST.RTIF=0;         /* flags */

   if (++count==122)       /* increment seconds */
   {               /* To correct for 8.192 */
      sec++;            /* ms per tick. Run 122*/
      if(++corr1==14) /* ticks per second for */
      {          /* 13 seconds, and 123 */
         corr1=0;      /* for the 14th second */
         if(++corr2==80) /* With this algorithm */
         {          /* there are 14.000128 */
            corr2=0;   /* actual seconds per */
            if(++corr3==4)  /* 14 indicated. Then */
            { /* run 79 of these */
               count=1;   /* cycles followed by */
               corr3==0;  /* one cycle of 14 */
            }          /* seconds with 122 ticks */
            else          /* per second. The */
               count=0;   /* elapsed time for this*/
         }          /* cycle is 1120.002048 */
         else    /* seconds for and */
            count=(-1);  /* the count is 1120 */
   } /* seconds. Repeat this */
   else      /* cycle 4 times and on */
```

```
                    count=0; /* the last cycle drop */
         }                   /* one tick makes the */
     }                       /* indicate and elapsed */
                             /* time exactly 4480 sec.*/
```

The three static variables `corr1`, `corr2`, and `corr3` are used to keep track of the number of times the several different loops in the algorithm are executed. C will always initialize these variables to zero and then their value will be retained from call to call of the function.

16-bit Timers

The multifunction timer discussed in the previous section provides for implementation of relatively simple timing functions. Let's assume that the microcontroller clock frequency is 4.0 MHz. The fastest interrupt time with this system is 0.512 milliseconds, and the granularity of the interrupt times is in large, power-of-two blocks for the `RTI` system. Also, the relation of the interrupt times to unity is not "clean"; complicated algorithms or special frequency crystals are needed to get the device to respond accurately in seconds.

Often microcontroller applications must provide more than one time function. The 15-bit timer is set up to provide only one time base. Of course, a programmer can program the timer to control many different functions and at many different times. The limits on the functions and times are difficult to determine. It is clear that the fastest practical time base in the 15-bit timer is 0.512 milliseconds. To obtain any finer time resolution, the programmer would have to compare the `TCR` bits to a specified value on a cycle-by-cycle basis. This type of program completely consumes the microcontroller and leaves no processing time for other functions during the execution of the timing program. If the time base must be other than some multiple of 0.512 milliseconds and not one of the standard RTI times, the processor can probably service only one time function. If the required times can fall on the above values, the processor can execute several time-based functions limited by the total time required to execute the functions and the microcomputer interrupt latency time.

The 16-bit timer addresses these problems. A block diagram of this type of timer is shown in Figure 4-2. This style of timer contains an internal 16-bit counter that is clocked at some fraction of the

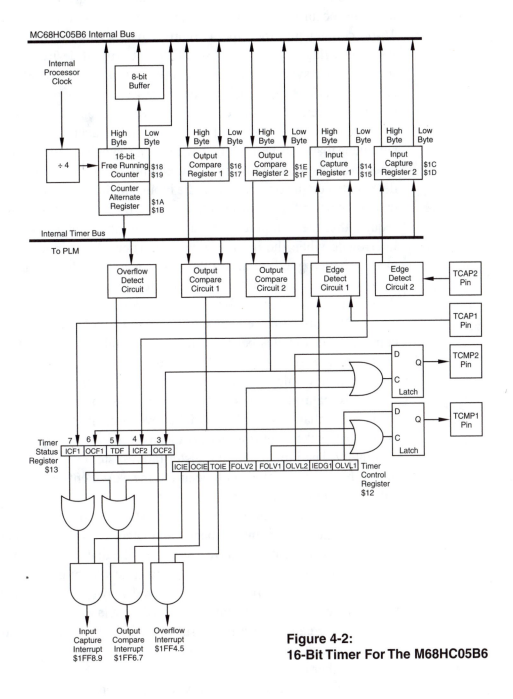

Figure 4-2:
16-Bit Timer For The M68HC05B6

microcontroller crystal frequency. An input capture operation detects the occurrence of an input and transfers the contents of the 16-bit counter into the input capture register. This transfer will always set a flag, and it can cause a CPU interrupt if desired. With an input capture system, precise measurement of time interval is possible. Details such as phase between two waveforms can be determined or slight differences in frequencies between several signals can be detected. The input capture provides far more accurate time measurement than can be obtained with either synchronous polling or asynchronous interrupt time measurements. There is a tiny inherent delay between the occurrence of the input and the setting of the input capture register. Such a measurement made by polling an input would require that the computer have a free running counter available to interrogate when the input is detected. Then, the computer would have to be assigned totally to the job of watching the input for the impending transition. When the transition is detected, the value of the counter would have to be read to determine the time of the transition. Of course, this sequence of operations would require several computer clock cycles per test, and also several cycles would be required to read the counter. Therefore, the accuracy of the time measurement would be compromised by these necessary time delays.

An asynchronous interrupt method to determine the time interval is better than a polled method, but even this method has built-in errors that make it an impractical means to measure time intervals accurately. The input capture register input system resolves most of the problems associated with accurate measurement of time intervals with a microcomputer.

Another type of timing problem exists. Suppose that the time that an event is to occur has been calculated. If the time of occurrence is to be accurate, we have a situation like that discussed above. The processor will have to spend all of its time watching the clock to determine when the correct time has arrived. Any time spent on other tasks during this measurement interval will be a latency during which the processor cannot determine if the specified time has arrived. In this case, the accuracy of the event time will be degraded by the time spent on other tasks.

The 16-bit timer avoids this type of problem nicely. An output compare system is used. The time of occurrence is calculated relative to the internal 16-bit counter. This value is placed into an output compare register. The content of the counter is compared automatically by the microcontroller to the value in the output compare register at each count of the counter. When the two values are equal, a flag is set, an output occurs, and if desired, the CPU is interrupted. The output compare system can be used to generate waveforms, to control phases between different waveforms, to control events based on calculated times.

Different microcontrollers will have differing numbers of input capture and output compare registers. In the discussions that follow, details of a single input capture and output compare register will be discussed. It is assumed that these registers are part of an M68HC05B6, so there are two input captures and two output compares onthe microcontroller. For details on access to the second register set, refer to the appropriate data manual. Later we will see microcontrollers that have many more input capture and output compare systems (up to 16 on one microcontroller).

Timer Control Register

The timer control register (TCR) is located at the address 0x12. This read/write register controls the operation of the 16-bit timer system. Shown below is a diagram of this register, and a listing of the functions of the various register bits.

TCR 0x12	Bit 7	Bit 6	Bit 5	Bit 4	Bit 3	Bit 2	Bit 1	Bit 0
	ICIE	OCIE	TOIE	FOLV1	FOLV2	CLVL2	IEDG1	OLVI1

OLVL1   Bit 0   Output Level 1. The contents of this bit will be copied to the output level latch the next time an output compare occurs. This result will appear at TCMP1. This bit and the output level latch are cleared when the part is reset.

IEDG1   Bit 1   Input Edge 1. This bit determines the transition direction that will cause an input to occur on Input Capture 1:

IEDG1 = 0 Falling Edge

IEDG1 = 1 Rising Edge

The contents of this bit are undetermined and unaffected at reset.

OLVL2	Bit 2	Output Level 2. The contents of this bit will be copied to the output level latch the next time an output compare occurs. This result will appear at TCMP2. This bit and the output level latch are cleared when the part is reset.
FOLV1	Bit 3	Forced Output Compare 1. This bit always reads zero. A one written to this position will force the OLVL1 bit to be copied to the output level latch. This result will appear at TCMP1. A forced output compare does not affect the OCF1 bit in the timer status register. This bit is cleared at reset.
FOLV2	Bit 4	Forced Output Compare 2. This bit always reads zero. A one written to this position will force the OLVL2 bit to be copied to the output level latch. This result will appear at TCMP2. A forced output compare does not affect the OCF2 bit in the timer status register. This bit is cleared at reset.
TOIE	Bit 5	Timer Overflow Interrupt Enable. If the TOIE is set, the timer overflow interrupt is enabled and an interrupt will occur when the TOF flag is set in the timer status register. This bit is cleared at reset and the interrupt is inhibited.
OCIE	Bit 6	Output Compare Interrupt Enable. If the OCIE bit is set, the output compare interrupt is enabled, and an interrupt will occur whenever either the OCF1 or the OCF2 is set in the timer status register. This bit is cleared at reset and the reset is inhibited.
ICIE	Bit 7	Input Capture Interrupt Enable. If the ICIE bit is set, the input compare interrupt is enabled and an interrupt will occur whenever either the ICF1 or the ICF2 is set in the timer status register. This bit is cleared at reset and the reset is inhibited.

Timer Status Register

This register—TSR—is an 8-bit register. The most significant 5 bits of this register contain read only status information. These bits describe the condition of the 16-bit timer system. Their functions are outlined below:

TSR	Bit 7	Bit 6	Bit 5	Bit 4	Bit 3	Bit 2	Bit 1	Bit 0
0x13	ICF1	OCF1	TOV	ICF2	OCF2			

OCF2   Bit 3   Output Compare Flag 2. This bit is set when the content of the free-running counter matches the contents of output compare register 2. OCF2 is cleared by accessing the TSR (specifically the OCF2) followed by an access to the low byte of the output compare register 2, 0x1f. The output compare flag 2 is undetermined at power on and is unaffected by reset.

ICF2   Bit 4   Input Capture Flag. This bit is set when a negative edge is sensed at TCAP2. It is cleared by an access of the timer status register followed by an access of the low byte of the input capture register, 0x1d. The input capture 2 flag is undetermined at power on and is unaffected by reset.

TOF   Bit 5   Timer Overflow Bit. This bit is set by a transition of the free-running counter from a 0xffff to a 0x0000. It is cleared by accessing the TSR with the TOF set followed by an access of the free-running counter low byte, 0x19. The TOF bit is undetermined at power on and is unaffected by reset.

OCF1   Bit 6   Output Compare Flag 1. This bit is set when the content of the free-running counter matches the contents of output compare register 1. OCF2 is cleared by accessing the TSR (specifically the OCF1) followed by an access to the low byte of the output compare register 1, 0x17. The output compare flag 1 is undetermined at power on and is unaffected by reset.

ICF1   Bit 7   Input Capture Flag 1. This bit is set when the proper edge is sensed at TCAP1. The edge is selected by the IEDG1 bit in the TCR. It is cleared by an access of the timer status register followed by an access of the low byte of the input capture register, 0x15. The input capture 1 flag is undetermined at power on and is unaffected by reset.

To clear bits in the TSR, the program must first access the TSR followed by an access of the LSB of the register associated with the bit that must be reset in the TSR. This sequence can lead to problems in dealing with the counter register. Suppose you are attempting to measure an elapsed time and are reading the counter register at random times and you also will read the TSR to service timer requirements. It is possible in these circumstances to accidentally reset the TOF bit when it is undesired. To avoid this problem, an alternate counter register has been designed into the M68HC05 devices. The alternate register always contains the same values as the prime register, but the TOF bit in the TSR is not reset when the alternate register is read.

Counter Register

The counter register is found in the memory locations 0x18 and 0x19. The least significant byte of the counter is in 0x19. An alternate counter register is found in addresses 0x1a and 0x1b with 0x1b being the least significant byte of this register. These registers are clocked at the same time and are incremented from low values to higher values. The counters are clocked at one-fourth of the internal processor clock, which in turn is one-half the oscillator frequency. The clocking frequency is one-eighth the crystal frequency, and the clocking period is 2 microseconds when the crystal frequency is 4 MHz. These ratios are not adjustable in the M68HC05B6.

The free-running counter values can be read at any time. A read sequence that reads only the least significant byte will receive the count value at the time of the read. If the most significant byte of either counter is read, the count value will be received and the contents of the least significant byte will be transferred to a buffer. This value will remain in the buffer until the program reads the contents of the least significant byte of the register. The value received for this read is the buffered value saved when the most significant byte was read. The most significant byte, MSB, can be read several times prior to reading the least significant byte, LSB, and the contents of the buffer will remain unchanged. After the MSB has been read and the LSB has been buffered, the free-running counter continues to be incremented at its normal rate. If the MSB/LSB read sequence is started, it is necessary to read the LSB to complete the sequence.

The counter is 16 bits, and when the register overflows from

0xffff to 0x0000, the timer overflow (TOF) bit is set. This event can cause an interrupt if the TOIE bit is set. Since the register is clocked at 2 microseconds, the interval between TOF is 0.131072 seconds.

Input Capture Registers

There are two input capture registers called ICR1 and ICR2. ICR1 is found at addresses 0x14 and 0x15, and ICR2 is located at addresses 0x1c and 0x1d. The lower address always contains the MSB of a 16-bit number. With the exception of the edge detection system discussed in the TCR section, these two registers operate the same. ICR1 can be set to respond to either a rising edge or a falling edge on the timer compare input pin TCAP1. If IEDG1 is 0, ICR1 responds to a falling edge on TCAP1. Otherwise, if IEDG1 is 1, ICR1 responds to a rising edge on TCAP1. ICR2 responds only to a falling edge on TCAP2. An interrupt will also accompany an input capture if the corresponding ICIE bit is set in the TCR.

The contents of the free-running counter are transferred to the input capture registers each clock cycle. Therefore, the registers contain a value that corresponds to the most recent input capture. After a read of the most significant byte of the input capture register, the transfer of new data to the least significant byte of the input capture register is inhibited until this byte is read. At no time during this sequence is the counter register inhibited.

Output Compare Registers

There are two output compare registers. OCR1 is found at address locations 0x16 and 0x17 while OCR2 is located at 0x1e and 0x1f. Again the lower addresses contain the MSB of these 16-bit numbers. These registers may be read or written at any time regardless of the timer hardware. If the output compare functions are not utilized, these four bytes can be used for data storage. Their contents are not altered at reset. There is only one output compare interrupt bit that is used for both output compares in the system.

The contents of the output compare registers are compared with the contents of the counter register each cycle of the counter register. If a match is found with either output compare register, the corresponding output compare flag—OCF1 or OCF2— bit is set. Also, the value of proper output level bit—OLVL1 or OLVL2—is transferred to the proper output pin, TCMP1 or TCMP2. If the OCIE bit is set in the timer control register, an interrupt will accompany the output compare.

There are times when it is desirable to force an output compare from a program. The FOLV1 and FOLV2 bits can be used for this purpose. These bits will always read 0, but writing a 1 to these bits in the TCR will cause transfer of the corresponding OLVL1 or OLVL2 bit to the specified output compare bit, either TCMP1 or TCMP2. This output does not affect the compare flags, so no interrupt is generated.

Programming the 16-bit Timer

We will examine several different uses of the 16-bit timer system in this section. The first is merely a repeat of the simple timer programmed in the section on the 15-bit system. Here we merely want to keep track of time, hours, minutes, and seconds in memory. No provisions are made yet for reading the time values or to change the values; these problems will be discussed later.

A listing of this program is shown below. In this case, the header file for the M68HC05B6 is used. A listing of this file is found on the CD-ROM. This program will make use of an output compare to generate periodic interrupts to the microcontroller. We will use output compare register 1. It will be set up so that when the first output compare interrupt occurs, the contents of the output compare register will be incremented by 500. Since the clocking time of the counter register is 2 microseconds, 500 2-microsecond periods will allow an output compare every 1 millisecond. This occurrence will be treated in the interrupt service routine.

```
#include "hc05b6.h"

int hrs, mts, sec; /* global variables */
long count=1000;

struct bothbytes /* 16 bit int structure */
{
    int hi;
    int lo;
};
```

```
union both /* and union */
{
   long l;
   struct bothbytes b;
};

union both time_count;
registera ac;

main()
{
   TCR.OCIE=1; /* enable output compare interrupt */
   CLI(); /* enable all interrupts */

   FOREVER
   {
      if(sec>59) /* do clock things each minute */
      {
         sec=0;
         if(++mts>59)
         {
            mts=0;
            if(++hrs>12)
               hrs=1;
         }
      }
      WAIT();
   }
}

void __TIMER_OC(void) /* time interrupt service
    routine */
{
   if(TSR.OCF2==1) /* is this interrupt due to OC2?*/
   {
      ac=OCLO2;    /* Yes. read OCLO2 to disable */
      return;      /* the interrupt and exit */
   }              /* the routine */
```

```
/* the program gets here every millisecond */
   time_count.b.hi = OCHI1;
   ac = TSR; /* Arm OCF1 bit clear */
   time_count.b.lo = OCLO1; /* Clear OCF1 bit */
   time_count.l += 500; /* 500 counts per ms */
   OCHI1 = time_count.b.hi;
   OCLO1 = time_count.b.lo;

if(--count==0)
   return ;
else
{
   sec++; /* here every second */
   count=1000;/* reset count to 1 second */
}
}
```

Listing 4-4: Timer Using Output Compares

The listing shows that microcontroller executing programs are broken into three distinct sections. The first section is referred to as the *initialization* section. In this case, the initialization section is the first two lines following the `main()` invocation. In the initialization section, the code executed sets up the operation of the microcontroller. Generally, this code is executed only once. Therefore, initialization of volatile memory values, setting up of interrupts, establishment of I/O ports and data direction registers are all completed in the initialization of the program. Unless there is a pressing reason, the main system interrupts should not be enabled during initialization.

The second section is the *applications* section. The applications section is usually a loop that contains all of the code to be handled routinely by the microcontroller. All input or output operations should take place within the applications section.

The third section of the program is the *collection of interrupt service routines* (asynchronous service section). These routines are called when appropriate interrupts are generated. In general, interrupt service routines should be short and do as little as possible to service the specified interrupt. When an interrupt is serviced, the status register of the microcontroller is saved and the system interrupt is disabled. Therefore, unless the programmer takes special care to re-enable the interrupts, no other interrupt can be serviced while the microcontroller is executing an interrupt service routine. This operation does not lead to missed interrupts, but it can cause an inordinate delay in service of an interrupt.

In some cases, it is possible that data might be lost if an interrupt is not handled expeditiously. For example, a high-speed serial port might notify the microcontroller that its receiving data buffer is full by an interrupt when the interrupt is disabled. In such a case, if the interrupt service routine being executed is not completed before the next serial data are received, the data in the buffer will be lost.

Sometimes it is necessary to pass data between the applications portion of the program and the interrupt service routine. These data can be stored in global memory, and both routines can access them. Here is a case where you must examine the assembly code generated by the compiler to make certain that no problems will be generated in passing of data between these routines. Problems can be created by passing information this way, but they can be avoided if the program rigorously avoids loading data that is changed in an interrupt service routine into a register in the application section of the program. We will see instances of this problem later.

If you go back to Listing 4-4, you'll see that the program organization discussed above is also used. This organization will be found for every program in this book. We will refer to the initialization, the applications, and the asynchronous service sections of the program. This arrangement works quite well, and is reliable. There should be a strong justification if alternate program forms are to be used.

When programming the 15-bit timer, we found that asking a compiler designed specifically for an 8-bit machine to work in 16-bit quantities often created unwieldy code. A programmer, however, does not always have the freedom to work with 8-bit quantities only. All of the time registers in this system are 16 bits wide, and the time values contained in these registers must be processed. These registers are always located in two adjacent 8-bit memory locations. One method for handling the 8/16-bit dichotomy is to use a *union*. The code sequence

```
struct bothbytes
{
    int hi;
    int lo;
};
```

creates a structure that contains two bytes. This structure is combined with a type `long` in the `union` below. Remember that a `union` provides space to hold its largest member and the different members of the `union` will occupy the same memory space. Therefore, the memory space to store the `long l` is the exact same memory to store the structure `b`. `b.hi` will occupy the most significant byte of `l`, and `b.lo` will occupy the least significant byte of l. It is now easy to deal with the bytes of the 16-bit quantity when moving the data around, and equally easy to invoke 16-bit arithmetic operations on the long combination of the two bytes.

```
union both
{
    long l;
    struct bothbytes b;
};
```

The statement

```
union both time_count;
```

declares that `time_count` is a union of the type `both`. In the interrupt service routine, the members of `time_count` are handled with little difficulty, as shown below.

The type `registera` defined as `ac` above is unique to the M68HC05 compiler. This type specifies that the variable `ac` will be stored in the accumulator. There is also a `registerx` type for the index register.

In the main program, the output compare interrupt is enabled and the system interrupt is enabled with the `CLI()` instruction in the initialization section. The program then enters an endless loop in which the clock is serviced every time `sec` becomes 60. The parameter `sec` will be changed by the interrupt service routine every second so that the system should be a satisfactory clock. This loop is the initialization section.

The last instruction in the `FOREVER` loop is a `WAIT()` instruction. This instruction places the processor into the wait mode. In this mode, processor operations are halted and the microcontroller operation is configured to reduce energy consumption. The operation of the internal timers proceed as usual. The part is removed from the wait mode by either a reset or the occurrence of an interrupt. The interrupt can be either an internal or an external interrupt. Since the output compare timer is set up and its interrupt is enabled, when the internal counter matches the contents of the output compare register, an interrupt will occur and remove the processor from the wait mode.

Sometimes, it is desirable to get a measure of the fraction of the time that the microcontroller is used to execute its program. The use of the wait mode provides an excellent mechanism for measurement of this usage. If there is an extra output port pin available, this pin can be set just prior to entering the wait mode. The pin can then be reset as the first instruction in the interrupt service routine. To make this measurement more accurate, you can set another output pin to the on condition all of the time. Measure these two outputs with an averaging DC voltmeter. One hundred times the ratio of the cycling output to the fixed output is the percentage of time that the microcontroller is available to execute other code.

Examine the following code sequence:

```
time_count.b.hi = OCHI1;
ac = TSR; /* Arm OCF1 bit clear*/
time_count.b.lo = OCLO1; /* Clear OCF1 bit */
time_count.l += 500; /* 500 counts per millisecond */
OCHI1 = time_count.b.hi;
OCLO1 = time_count.b.lo;
```

The first instruction copies the high byte of the output compare register into the high byte of the structure b in the union `time_count`. When the TSR is copied into the a register, the system is set up to clear the OCF1 bit. Then, the low byte of the output compare register is moved into the low byte of the structure b. These operations leave the 16-bit contents of the output compare register in the union `time_count.l`. Note that the high byte is moved from OCHI1, the TSR is accessed, and then the low byte is moved from OCLO1. This sequence, accessing OCLO1 after the TSR has been accessed will clear the output compare flag 1, OCF1, and remove the interrupt source. 500 is added to this 16-bit number, and the result is copied back into the output compare register 1 a byte at a time.

While it is usually best to keep the interrupt service routines short, sometimes other operations can be completed within these routines that can be useful. Recall that in the EEPROM programming routine there was a function called delay. A delay function can be integrated into a time interrupt service routine easily. The interrupt service routine above is entered every millisecond. We specified a function delay that had an unsigned long argument that corresponds to the required delay in milliseconds. Such a function could be implemented as:

```
void delay(unsigned long);

unsigned long time=0;
void delay(unsigned long del)
{
   time = del; /* place the delay time in the global
   memory */

   while(time>0);
}
```

Now we must add one code sequence to the interrupt service routine above:

```
if(--count==0)
{
 if(time>0)
   --time;
 return ;
}
```

The function delay() places the value of the required delay in the unsigned long location time. The program then hangs on the while statement so long as time is greater than 0. Every time the ISR is entered—which is every millisecond— the long unsigned value time will be decremented. When the proper time has passed, the delay() function will return to the calling program.

The function delay() does something that many programmers do not like: the program hangs in a loop until a specified time has passed. This operation may seem to waste the microcontroller resource. Note, however, that the program does not spend all of the time in the loop; interrupts are being serviced during this time. We will see that the timer interrupt is only one of many potential interrupts that will be working on the part at all times. Placing the device into a wait loop for a few milliseconds may stop the main program in its tracks, but the background operations being serviced by the interrupts will continue to be processed unhindered. In this case, a global variable has been used to transfer data between an interrupt service routine and the applications portion of the program. The variable time is set in the applications program, and the completion of the time delay is evaluated in the applications portion of the program, and time itself is decremented in the interrupt service routine.

It is possible to create errors in operation with this type of procedure. One place where a nasty bug can creep into your code is when you are dealing with bit manipulations in both the application portion of the program and the interrupt service routine. Suppose that you want to toggle a bit when an event was detected in the application, and simultaneously you need a periodic bit toggle that is controlled by code in the interrupt service routine. The code sequence might appear as follows:

```
.

.

PORTA.BITAPP = !PORTA.BITAPP;
.

.
```

In the interrupt service routine, the code could be

```
.

.

PORTA.BITINT = !PORTA.BITINT;
```

In each case, this code will compile into

```
lda PORTA
eor 2^BITNUMBER
sta PORTA
```

The same code sequence will appear in both the application and the interrupt service routine with the BITNUMBER appropriately chosen. Suppose that we are in the application, and have just executed the lda PORTA instruction when the interrupt occurs. In the interrupt service routine, the above code will be executed properly, and when control is returned to the program main line, it will continue with the eor instruction. However, the contents of PORTA will have been changed by the interrupt service routine, so the value that was loaded by the `lda` instruction prior to the interrupt is no longer valid. The earlier loaded value of PORTA will be restored by the above sequence in the application code, and the change made in the interrupt service routine will be undone.

This bit of interplay between the applications program and the interrupt service routine is a type of software race. It can be easily avoided. Whenever a variable is changed in both the applications code and in an ISR, the erroneous set/reset can be avoided if the program disables all interrupts before the offending instruction in the applications code. The interrupts must then be re-enabled after the code execution. In this case, the code will be

```
    .
    .
SEI ();
PORTA.BITABB=!PORTA.BITAPP;
CLI();
    .
    .
```

in the applications code. In this case, there can be no interrupt to interfere with the proper handling of PORTA in the application.

In the timer routine above, there is what some might call a glaring oversight. The timer was not initialized in the initializing routine of the program. The first lines of code merely enabled the timer interrupt and then proceeded into the FOREVER loop, which is the applications portion of the program. The initial state of the TCR and the OCR registers is not established at reset. There is no way of knowing when the first output compare will take place. If the initial value of the OCR happens to be one less than that of the TCR, there will be an output compare about 0.13 seconds after the interrupts are enabled, and from then on the interrupt period will be accurate. If it is imperative that the first output compare occur at exactly the specified time, then the code for initialization of the output compare registers must be included in the initialization routine. Otherwise, if this error in the initial timing can be tolerated, it will save bytes of code space that might be desperately needed for another portion of the program.

EXERCISES

1. Write a program that uses an output compare system to generate an accurate waveform with a 1000.0 second period.

2. Devise a convenient method to test the performance of the program in Exercise 1 above.

3. Write a program to generate two waveforms. One output is to be at twice the frequency of the other. The duty factors of the two signals are to be equal to 50%. The frequency of the slowest wave is to be 1000 Hz. The phase of the higher frequency signal is to be such that its rising edge is to occur 260 microseconds following the rising edge of the first signal.

4. Two DC motors are running. Each motor has an optical interrupter on its shaft with 15 interrupts per revolution of the shaft. All but one of the interrupts occupy one-sixteenth of the circumference of the rotation. The fifteenth interrupt occupies one-eighth of the circumference. Using input capture registers, measure the speed of the two motors, and provide a slow down or speed up signal that can be used on either motor to synchronize the rotation of the two motors with the wide interrupter positions on the shafts being in lock-step.

5. What microcontroller characteristics will control the maximum speed at which the motors in Exercise 4 can run? The minimum speed?

Analog-to-Digital Converter Operation

The analog-to-digital converter (ADC) found on the M68HC05 family is moderately simple in its operation. There are a few important items that must be remembered when dealing with the ADC. Most important is that the ADC must be turned on for at least 100 microseconds prior to reading a value. If 100 microseconds has not elapsed, it is guaranteed that the value read will be in error. The ADC is turned on by setting the ADON bit in the ADC control/status register. This register is referred to as AD_CTST. The following code sequence will turn the ADC on:

```
AD_CTST=0;
AD_CTST.ADON=1;
```

The following function will provide a reading of a single channel of the ADC input:

```
unsigned int read_adc(int k)
{
    AD_CTST &=~0X7;
    AD_CTST |=k;
    while(AD_CTST.COCO==0)
        ; /* wait here til COCO is set */
    return AD_DATA;
}
```

The argument k is the channel that is to be read, and k can have a value of 0 to 7 to read the external channels.

The first two lines of code in the above function will place the channel number to be read in the channel bits of AD_CTST. These bits must be cleared by an instruction sequence that will not alter the upper bits of AD_CTST because the ADON bit is in the upper portion of AD_CTST. This bit cannot be reset while the ADC operation is continuing. The first line of code clears the least significant three bits, and the second line places the channel number in these bits.

Writing to AD_CTST will cause the ADC conversion to start. Therefore, all that must be done is to wait until the conversion is completed to read the data into the program. The code

```
while(AD_CTST.COCO==0)
    ; /* wait here til COCO is set */
```

will keep control of the microcontroller in that instruction sequence until the COCO bit, which is the conversion completion bit, is set. At that time the value found in AD_DATA will be the result of the latest conversion.

Often, the ADC results must be subjected to some processing to remove unwanted characteristics of the signal being measured. Here is a case where careful use of assembly language procedures can make a big difference in the execution speed as well as the amount of code needed. An example that is often used is to average the past values of the data. A reasonably simple approach is to allow the latest ADC reading to have a 50% weight and all of the past readings to have a 50% weight. The following example code will accomplish this task in three different ways:

```
#include "hc05b6.h"

unsigned adc_data[8];
unsigned read_adc(int);

main()
{
   unsigned int j;

   AD_CTST=0;
   AD_CTST.ADON=1;

   for(j=0;j<8;j++)
   {
      adc_data[j]=(read_adc(j)+adc_data[j])/2;
      adc_data[j] >>= 1;
      adc_data[j] += read_adc(j)>>1;

   #asm
      ldx j
      lda j
      jsr read_adc
      add adc_data,x
      rora
      sta adc_data,x
   #endasm

   }

}
```

Listing 4-5: Three Different ADC Averaging Routines

The first attempt to read and average the data approaches the problem as simple as practical. The data are read in from the ADC, added to the corresponding stored data in the array adc_data, the result is divided by two, and the final average is put back into the proper location in the array. The following line of code is all that is necessary to accomplish this task:

```
adc_data[j]=(read_adc(j)+adc_data[j])/2;
```

Hidden in this code is the fact that both `read_adc` and `adc_data` are unsigned results. When two unsigned numbers are added together, the most significant bit of each can be 1 so there can be a carry or overflow when the addition takes place. Problems from this carry can be avoided in this case by merely using a `long` or `double` precision add routine in adding the two numbers. Then, when the result is divided by 2, if a bit is carried into the upper byte of the result it will be shifted back into the lower byte. The result of this operation is correct and will fit into a single unsigned `int`.

The compiled version of the above program with `read_adc()` merged into it follows:

```
                  #include "hc05b6.h"

0050 0008     unsigned adc_data[8];
              unsigned read_adc(int);

              void main(void)
              {
0058              unsigned int j;

0100 3F 09 CLR $09          AD_CTST=0;
0102 1A 09 BSET 5,$09        AD_CTST.ADON=1;

0104 3F 58 CLR $58           for(j=0;j<8;j++)
0106 B6 58 LDA $58
0108 A1 08 CMP #$08
010A 24 36 BCC $0142
                       {
010C CD 01 43 JSR $0143  adc_data[j]=(read_adc(j)+
                         adc_data[j])/2;
010F BE 58 LDX $58
0111 EB 50 ADD $50,X
0113 AE 02 LDX #$02
0115 CD 01 57 JSR $0157
0118 9F    TXA
0119 BE 58 LDX $58
011B E7 50 STA $50,X
```

```
011D BE 58 LDX $58    adc_data[j]>>= 1;
011F E6 50 LDA $50,X
0121 44    LSRA
153
0122 BE 58 LDX $58
0124 E7 50 STA $50,X
0126 B6 58 LDA $58    adc_data[j]+=
                          read_adc(j)>>1;
0128 CD 01 43 JSR $0143
012B 44    LSRA
012C BE 58 LDX $58
012E EB 50 ADD $50,X
0130 E7 50 STA $50,X


                   #asm

0132 BE 58 ldx j
0134 B6 58 lda j
0136 CD 01 43 jsr read_adc
0139 EB 50 add adc_data,x
013B 46    rora
013C E7 50 sta adc_data,x
          #endasm


                        }
013E 3C 58 INC $58
0140 20 C4 BRA $0106
0142 81    RTS    }

                   unsigned int read_adc(int k)
0059              {
0143 B7 59 STA $59
0145 B6 09 LDA $09    AD_CTST&=~0X7;
0147 A4 F8 AND #$F8
0149 B7 09 STA $09
014B B6 09 LDA $09    AD_CTST |=k;
014D BA 59 ORA $59
014F B7 09 STA $09
```

```
0151 0F 09 FD BRCLR 7,$09,$0151
while(AD_CTST.COCO==0);
0154 B6 08 LDA $08   return AD_DATA;
0156 81    RTS
                    }


0157 BF 5A STX $5A
0159 B7 5B STA $5B
015B 4F    CLRA
015C 5F    CLRX
015D 5C    INCX
015E 38 5B LSL $5B
0160 49    ROLA
0161 B0 5A SUB $5A
0163 24 03 BCC $0168
0165 BB 5A ADD $5A
0167 99    SEC
0168 59    ROLX
0169 24 F3 BCC $015E
016B 53    COMX
016C 81    RTS


1FFE 01 00
```

The assembly code to execute this single line of C code is found in the address range $10c to $11b, or 17 bytes of code. However, there is a call to the double precision add routine at address $115. This routine occupies the address range $157 to $16c, or 22 additional bytes of code hidden from the main routine. Therefore, this requires 37 bytes of code for its execution.

The next approach is to avoid the overflow problem by dividing by 2 each of the terms to be added prior to the addition. Division by 2 is accomplished by shifting each of the terms to the right by one bit. This approach guarantees that there will be no overflow into the higher byte because the most significant bit of each number will be 0 after the shift, and no binary addition can cause more than a 1-bit overflow. Therefore, the sum will at most have its most significant bit turned on. This number is then stored in the location adc_data[j].

The assembly code created by the compiler for this approach is in the address range $11d to $130. This code uses 21 bytes of memory. Careful examination of this code will show that the instruction at $122 is unneeded, so the code could have been completed in 17 bytes.

The assembly version of the same routine merely adds the two unsigned numbers, and then rotates the result right 1 bit to accomplish the divide by 2. The 1-bit rotate accomplishes the same thing as a right shift by 1 bit with the exception that the carry bit is shifted into the most significant bit of the result. The only way the carry bit could be set is by the most significant bits of both addends being 1. The assembly version of the code resides in $132 to $13c and requires 12 bytes of code.

Here is a case where judicious choice of assembly code will provide a significant improvement in the amount of code needed to execute a specific program. The main reason that the assembly version is shorter is that the carry bit in the condition code register is available to the programmer from assembly. This bit is completely hidden from the programmer in any high-level language. Therefore, tricks like rotating a bit from the carry into a register are not available in the high-level language.

EXERCISE

1. Write a routine to average readings from the ADC on a microcontroller, but weight the current reading three times that of the past average.

Pulse Width Modulator System

In this section on programming of the timers, two approaches to the generation of a pulse width modulation (PWM) signal will be discussed. These approaches both use the output compare system of the 16-bit timer.

What is a PWM system? A PWM signal is a periodic signal where the signal is set high for a calculated duty factor, and then the output goes low for the remainder of the period. If you measure a PWM signal with an averaging voltmeter it will have a value equal to the peak voltage times the duty factor of the PWM. Here the duty factor is defined as the ratio of the on time to repetition period of the signal.

Thus, we have created a simple digital-to-analog converter. Many control applications need analog control signals for parts of the system. PWM systems on microcontrollers provide an excellent means for providing these analog voltages, so long as the limitations we will discuss below are acceptable to the total system design.

The built-in PWMs in the M68HC05Bx family of devices provide you with one of two pulse periods: the programmable timer clock frequency divided by 256 or 4096. These frequencies are fixed, and the interval is divided into 256 parts. Therefore, the analog frequency ranges—and the accuracy of the analog reproduction—is limited to one part in 256 over the range of the output.

When using the output compare system to create a PWM, it is possible to achieve a finer analog resolution at slower frequencies, or perhaps a faster signal with poorer accuracy. A PWM created from an output compare system is somewhat more flexible than the usual built-in PWM. With the PWM systems built around the output compare system, the computer is responding to frequent interrupts. When an interrupt occurs, the system status is pushed onto the stack and the address of the interrupt service routine is placed into the program counter. It takes approximately 10 clock cycles to respond to an interrupt after the completion of the executing instruction. Return from an interrupt requires 9 clock cycles. Therefore, it is not possible to accomplish the full goal of a PWM with an output compare system. The goal in this case is to achieve full on at the one extreme input level and full off at the other. The times required to process an interrupt will limit the performance at one extreme, the other or perhaps both when a PWM is implemented with an output compare system.

A first example is shown in the following listing. This example shows the interrupt service routine along with the necessary defines only. It is assumed that the output compare interrupt bit along with the interrupt bit are properly set to allow the output compare 2 to interrupt the processor. Also shown in this example is code to control a second output compare—1—to provide the system with a fixed time base.

In operation, the applications portion of the program must place a pair of complementary numbers, pwm_number and off_count, in place. The sum of these two numbers must equal a constant that establishes the period for the PWM signal. For the M68HC05B4, the period will be 2 microseconds times the total count value. Prior to starting use of the PWM, the bit `flag.ON` should be reset and the value for `count`, which is the main time base period count, should be set.

```c
/* bits in flag */

    #define ON 0

    union both time_cnt;
    unsigned long off_count,pwm_number,count;
    bits flag;

void __TIMER_OC(void)
{
    if( TSR.OCF2==1) /* accesses the TSR */
    {
        if(flag.ON==0)
        {
            flag.ON=1;
            TCR.OLVL2=0; /* Timer Compare 2 off*/
            time_cnt.b.hi=OCHI2;/* get start time */
            time_cnt.b.lo=OCLO2; /* reset the OCF */
            time_cnt.l +=pwm_number;
            OCHI2=time_cnt.b.hi;
            OCLO2=time_cnt.b.lo;
            return;
        }
        else
        {
            flag.ON=0;
            TCR.OLVL2=1;/* Timer Compare 2 to on */
            time_cnt.b.hi=OCHI2;/* get start time */
            time_cnt.b.lo=OCLO2; /* reset the OCF */
            time_cnt.l +=off_count;
            OCHI2=time_cnt.b.hi;
            OCLO2=time_cnt.b.lo;
            return;
```

```
        }
    }
    else /* must be that OCF1==1 rather than OCF2
*/
    {
        time_cnt.b.hi=OCHI1; /* output compare */
        time_cnt.b.lo=OCLO1; /* get lo byte */
        time_cnt.l +=count; /* time for interrupt */
        OCHI1=time_cnt.b.hi;
        OCLO1=time_cnt.b.lo;/* reset the OCF */
        ms_10++;
        return;
    }
}
```

Listing 4-6: ISR For PWM System

Since the interrupt bits are set to allow output compare interrupts, an interrupt will occur. The interrupt service routine will be entered, and the instruction sequence

```
if( TSR.OCF2==1) /* accesses the TSR */
{
    .

    .

    .
```

accomplishes two things. First, it determines if the output compare 2 caused the interrupt, and, secondly, it accesses the TSR and arms the reset of the OCFX when the proper lower byte of the output compare register is read later.

For the moment, assume that OCF2 bit is set: the interrupt was caused by a compare in output compare register 2. In this case a test is made to determine if flag.ON is reset. Since the initialization routine reset this bit, it will be 0. Therefore, the code sequence

```
flag.ON=1;
TCR.OLVL2=0; /* Timer Compare 2 to turn off*/
time_cnt.b.hi=OCHI2; /* get the start time */
time_cnt.b.lo=OCLO2; /* reset the OCF */
time_cnt.l +=pwm_number;
```

```
OCHI2=time_cnt.b.hi;
OCLO2=time_cnt.b.lo;
return;
```

will be executed. This code is the "on" code for the PWM system. The first instruction in this sequence will set the bit flag.ON, so that this code will not be executed in the next execution of the interrupt service routine. The bit TCR.OLVL2 is the state that will be transferred to TCMP2 pin when an output compare occurs. TCR.OLVL2 is then set to 0 so that TCMP2 will be set low when the next output compare occurs. The code sequence that follows gets the contents of the output compare register 2, increments this value by pwm_number, and places this new value back into output compare register 2. Thus, the time of the output compare is established and the interrupt service routine is exited.

Eventually, the next output compare 2 interrupt will occur. Since the bit flag.ON has been set, the following code will be executed. This code is the "off" code for the PWM system. The first business to take care of is to reset the bit flag.ON so that the on portion of PWM code will be executed the next time into the ISR. TCR.OLVO2 is set so that TCMP2 will go on when the next output compare occurs.

```
flag.ON=0;
TCR.OLVL2=1; /* Timer Compare 2 to turn on */
time.cnt.b.hi=OCHI2; /* get the start time */
time_cnt.b.lo=OCLO2; /* reset the OCF */
time_cnt.1 +=off_count; /*complete the cycle*/
OCHI2=time_cnt.b.hi;
OCLO2=time_cnt.b.lo;
return;
```

The contents of OCR2 are handled as above, but this time their value is incremented by the contents of off_count. The main line code is then re-entered by the use of the return instruction.

There is another approach to generation of PWM signals with output compare systems that will materially aid this problem. However, this approach requires two output compare systems. One output compare system is used to create the time base and the other is used to generate the on or off time. Let's examine the following interrupt service routine. Output compare 1 is used here to generate the PWM period. The way that this routine works, it is possible for an interrupt to occur as a result of a compare on output compare 2. This interrupt is to be ignored, and the first five lines of the code below tests for this condition and clears interrupt when it occurs. This interrupt occurs at the end of the on time. Control of the program must be out of this portion of the ISR and in the main line code before the next interrupt caused by OCF1 occurs. The time required for the first five lines of code here is about 18 microseconds, which is the minimum on time.

```
void __TIMER_OC(void)
{
   if(TSR.OCF2==1)
   {
     ac=OCLO2; /* if interrupt is caused by out-
put*/
     return; /* compare 2 reset it and return */
   }
   time_cnt.b.hi=TCHI; /* get in the TCR */
   time_cnt.b.lo=TCLO; /* get lo byte */
   time_cnt.l +=count; /* bump the time counter
                           for next*/
   OCHI1=time_cnt.b.hi;/* interrupt */
   OCLO1=time_cnt.b.lo;/* reset the OCF */
   if(flag.PWM==1)
   {
     TCR.OLVL2=1;/* Timer Compare 2 to set output */
     TCR.FOLV2=1;/* turn on Timer compare 2 output */
     TCR.OLVL2=0;/* Timer Compare 2 to reset output*/
     time_cnt.b.hi=TCHI; /* get the start time */
     time_cnt.b.lo=TCLO;
     time_cnt.l +=pwm_number;
     OCHI2=time_cnt.b.hi;
     OCLO2=time_cnt.b.lo;
   }
}
```

Listing 4-7: Alternate PWM ISR

The next sequence is the code required to establish the time base. This time base operates on output compare 1. This routine is similar to those discussed above. A bit in the bit field flag is tested to determine if the PWM signal is to be turned on. If `flag.PWM` is set, the PWM code will be executed. When this routine is entered, `TCMP2` will be reset to zero. The first instruction sets `TCR.OLVL2` so that `TCMP2` will be on at the next output compare. The next instruction, `TCR.FOVL2=1`, sets the force overflow bit for `TCMP2` which will cause this bit to be turned on since `TCR.OLVL2` is set. This time is the beginning of the on period of the PWM signal.

The next instruction

```
TCR.OLVO2=0;
```

is so that `TCMP2` will go off or be reset at the next output compare 2. Control is then passed to the main line of code. When `TCMP2` occurs, the output will return to 0 and an interrupt will occur. This interrupt is detected by the first lines of the interrupt service routine.

One could write a code sequence that would generate the mirror image signal to that above. That is, the off time would be controlled by `TCMP2` and the on time would be the difference between the time of `TCMP2` and the main time base.

Other Program Items

Several other small programs come to mind that are very useful in programming microcontrollers. With these small parts, it is usually desirable to avoid library functions and, if possible, use a bag of tricks to arrive at the desired results. For example, dealing with numbers for either input or output offers a good place to exercise some experience over expedience. One case where it is often important to minimize code space is in generating binary coded decimal (BCD) numbers from integer numbers to output from a computer. Perhaps the most direct approach to accomplish this conversion is as follows:

```
/* convert a binary number less than 100 to BCD */

unsigned char convert_bcd(unsigned char n)
{
    unsigned int result;
    if(n>99) return 0xff;
    result=n/10<<4;
    result += n%10;
```

```
    return result;
}
```

Listing 4-8: First BCD Conversion

This small function takes an 8-bit character n that is less than 99 and converts the number into one 8-bit result. The upper nibble is the number of tens in the number and the lower nibble is the number of units. A hexadecimal number 0xff is returned if the number is greater than 99. Note that the calculation requires an integer divide operation and a modulus operation which is equivalent to a divide operation. The code to execute the divide and modulus operations must be associated with this function to complete its task.

Another approach is shown below. This approach avoids any external function calls and actually requires less total code than the version in Listing 4-8, even though the C code is somewhat longer. In this case, the while loop essentially divides the input number by 10 and places the result in the most significant nibble of the result. After the tens have been removed from the number, all that is left is the units. This value—the number of units—is ORed on the result before it is returned.

```
/* convert a binary number less than 100 to BCD */

unsigned char convert_bcd(unsigned char n)
{
    unsigned int result;

    if(n>99) return 0xff;
    result=0;
    while((n-10)>=0)
        result+=0x10;
    n+=10;
    result += n;
    return result;
}
```

Listing 4-9: Second BCD Conversion

The function shown below was originally written for use on a M68HC05, but was later used on a M68HC11 as well. A two-digit number must be sent to a seven-segment LED display. The number to be shown is contained in the memory locations for tens and units.

The numeric display takes four inputs that are merely the BCD value of the number to be shown. With this particular display, a 4-bit number between the values of 0 and 9 will be displayed. If each of the four input lines to the display is turned on, the display will be turned off. The parameters passed to the function—high and low—are flags to indicate whether the corresponding output is to be turned on.

```c
void display(int high, int low)
/* Display the contents of units and tens on the
appropriate LED displays. High corresponds to
tens, and low corresponds to units. If the proper
argument is TRUE, the corresponding LED will be
turned on. If the argument is FALSE, the LED will
be turned off. */

{
    unsigned int save;
    save = tens<<4;
    save |= units70x0f;
    PORTA=save;
    if(!high)
        PORTA |= 0xf0;
    if(!low)
        PORTA |= 0xf;
}
```

Summary

In this chapter, we have discussed programming techniques for a few of the more important peripheral components found on microcontrollers. Timers and ADC applications will be reconsidered in later chapters. In later chapters, serial communications, attendant programming of look-up tables, interpolation between data points in look-up tables, and synchronous communications from standard digital I/O pins rather than an SPI will be covered. Some small 8-bit microcontrollers have pulse width modulation (PWM) outputs that can be used as a digital-to-analog converter (DAC). Often the ranges available from these fixed PWM systems are not satisfactory for the required application. Other methods of accomplishing the PWM operation will be discussed in later chapters.

Many of the peripheral components found on the small 8-bit devices are found on larger microcontrollers. The next chapter introduces a group of larger microcontrollers, the M68HC11 family. The compiler used for the development of the M68HC05 code does not extend to the M68HC11 family. Therefore, the source code will look a little different in the following chapters, but it will still be all C. To generate code for the M68HC05, the compiler has had to "bend" the concepts of ANSI C to create code that would work with that family. Larger microcontrollers accommodate more of the large computer features so use of ANSI C is possible.

Chapter 5

Programming Large 8-Bit Systems

This chapter on the programming of large 8-bit systems will make use of the MC68HC11 microcontroller. It is absolutely necessary that any programmer understand the device when writing code for a microcontroller application. If you are not familiar with the MC68HC11 family, then read the M68HC11 Reference Manual and the M68HC11 E Series Technical Data Manual on the accompanying CD-ROM to get the needed background to be able to continue the work in this chapter.

Header File

The CD-ROM contains the C header file HC11E9.H. This file should be included with any program that is going to be used on the MC68HC11E9 or the MC68HC711E9. An abbreviated version of this header file is listed below. The header file has about 400 lines of source code, and most of that code is repeats of the portions of code shown in the following listing.

```
#ifndef HC11e9
#define HC11e9

unsigned int Register_Set = 0x1000;

typedef struct {

        unsigned  char bit0 : 1;
        unsigned  char bit1 : 1;
        unsigned  char bit2 : 1;
        unsigned  char bit3 : 1;
```

```
        unsigned  char bit4 : 1;
        unsigned  char bit5 : 1;
        unsigned  char bit6 : 1;
        unsigned  char bit7 : 1;
} Register;

#define PORTA (*(volatile Register*)(Register_Set+0))

typedef struct {
   unsigned char STAF :1;
   unsigned char STAI :1;
   unsigned char CWOM :1;
   unsigned char HNDS :1;
   unsigned char OIN  :1;
   unsigned char PLS  :1;
   unsigned char EGA  :1;
   unsigned char INVB :1;
}Pioc;

#define PIOC  (*(volatile Pioc*)(Register_Set+2))

#define PORTC (*(volatile Register*)(Register_Set+3))
#define PORTB (*(volatile Register*)(Register_Set+4))
     .

     .

     .

#define TCNT  (*(unsigned int *)(Register_Set+0xE))
#define TIC1  (*(unsigned int *)(Register_Set+0x10))
     .

     .

     .

/* To clear bits in the flag registers use the form

   TFLG1 = OC1

  to clear OC1F in TFLG1.  Use this form only in the
    two flag registers TFLG1 and TFLG2 below.
*/
```

```
#define OC1F 0x80
#define OC2F 0x40
#define OC3F 0x20
#define OC4F 0x10
#define I4O5F 0x08
#define IC1F 0x04
#define IC2F 0x02
#define IC3F 0x01

#define TFLG1 (*(unsigned char*)(Register_Set+0x23))
```

.
.
.

```
typedef struct {
   unsigned char TCLR :1;
   unsigned char SCP  :2;
   unsigned char RCKB :1;
   unsigned char SCR  :3;
}Baud;

#define BAUD  (*(volatile
Baud*)(Register_Set+0x2B))
```

.
.
.

```
/*    Macros and function to permit interrupt service
      routine programming from C.

      To use the vector call, do vector(isr,
      vector_address)where isr is a pointer to the
      interrupt service routine, and vector_address
      is the vector address where the isr pointer
      must be stored.    */

#define vector(isr,vector_address) (*(void **)(vector_address)=(isr))
#define cli() _asm("cli\n")
#define sei() _asm("sei\n")
```

```
#ifndef NULL
#define NULL (void *)0
enum {FALSE,TRUE};
enum {STOP,START};
enum {OFF,ON};

#define FOREVER while(TRUE)
typedef unsigned int WORD;
typedef unsigned char BYTE;

#endif
#endif
```

The first instruction in the file is

```
#ifndef HC11E9
#define HC11E9
```

and the last entry in the file is

```
#endif
```

These lines of code are useful to prevent multiple definitions of the items defined within the file. If the header file has not been previously compiled as a part of the program when the above statement is seen, HC11E9 will not be defined. The first instruction determines if HC11E9 is not defined, and if it is not, the second line defines it. Then all of the code until the matching #endif will be compiled. If HC11E9 is already defined, then the compiler will skip the code lines until the matching #endif is found. Therefore, the programmer can put an

```
#include <hc11E9.h>
```

at the beginning of each program module, and it will be used in the program only once even if several modules are combined into one and the above statement is included several times in a single file. This approach is convenient because it allows each module to be compiled and tested and then the several modules can be merged and compiled as a unit without worry about multiply defined variables and values found in header file.

The file next entry is a simple typedef and declaration of a structure called Register. This structure contains eight 1-bit entries. Each of these bits corresponds to a specific bit in a register field, and the bits are given the names `bit0`, `bit1`, . . . `bit7`. Therefore, when dealing with specific bits within a type `Register`, the programmer should use `bitx`, where x is the number of the bit being referenced.

In the second portion of the file that follows, all of the I/O registers are declared.

The external variable `Register_Set` is given a value 0x1000. This value is the initial location of the I/O register map in the system. In the MC68HC11, bit manipulation assembly instructions have a one-byte address that can be an offset from an index register. With the indexed version of the instruction, any single byte or bit in the entire memory map can be accessed or tested with a single instruction. However, the address operation must be indexed. If you use a single address with offsets to each of the registers as is done in the header file, the compiler will automatically place the value of `Register_Set` into an index register and use the offsets specified to allow indexed access to the data in the registers from anywhere in the program.

The value of `Register_Set` is not fixed by the microcontroller. It can be changed within the first 64 clock cycles following reset. To make this change, the programmer must assign the correct value to the INIT register in the I/O memory space. This value should be changed in the initialization routine for the program, which is usually written in assembly language. After the INIT register is changed, a new proper value assigned to `Register_Set` will allow the desired access to all registers and bits in the I/O memory map.

The definition of `Register` shows that any instance of this variable type is a collection of eight bits. Any location that is defined as a type `Register` is truly a collection of bits and each bit must be processed individually. For example, `PORTA` is declared to be of the type `Register`. Therefore, an expression like

```
PORTA = 0x3f;
```

will result in an illegal assignment error because `PORTA` is of the type `Register`, not `char`.

It is possible to make assignments to ports defined in the above manner as either a byte-wide field or as bit fields. Return to the initial declaration of a register in the header file:

```
typedef struct
```

```
{
signed char bit 0 :1;
. . .
unsigned char bit 7 :1;
} Register;
```

We can now declare:

```
typedef union
{
char byte;
Register bits;
} Mix_Register;
```

and thus

```
#define PORTA(*volatile
Mix_Register*)(Register_Set +0)
```

will allow the programmer to use

```
PORTA.byte = 0x2E
```

to set the value of all bits in the port with one instruction and in the same program to use

```
PORTA.bits.bit3 = 1;
```

to set, reset, or test the individual bits inside of the port. In this book, we will use the bit fields only as shown in header file.

Since PORTA is of the type Register, it is possible to deal with the individual bits within this location by normal C constructs such as

```
PORTA.bit3 = 1;
if(PORTA.bit6 == 0)
. . .
```

Of course, it is much better to define practical names to the various bits within the port to achieve even clearer code:

```
#define ON TRUE
#define MOTOR bit3
#define PUSH_BUTTON bit6
```

```
.
.
PORTA.MOTOR = ON; /* turn the motor on */
.
.
.
if(PORTA.PUSH_BUTTON==ON)
{
        do push button things
}
```

With this compiler, an `int` is a 16-bit value. Therefore, the registers that are two bytes are cast onto the type `int`. Usually these registers are accessed as `int`s only and there is no need to have the individual bit access afforded by the use of the `Register` type. The timer counter register (TCNT) is one such register that is accessed as an `int` only. There are also a few one byte, or 8-bit, registers that are accessed as bytes only. No bit accesses within these registers are needed. In these cases, the register is cast onto the type `char`. The several ADRx registers are examples of this type. The ADRx registers contain the result of an analog-to-digital conversion that is usually handled as an 8-bit unit only.

In most instances, register locations should be `unsigned`. The ADRx registers each contain the result of analog-to-digital conversions. These results are all unsigned. Therefore, these registers should be cast as `unsigned char`. Also note that all of the timer count and input capture or output compare registers are also declared as `unsigned`.

In the listing above, you will note that there are two parts to the declaration of each register. The first identifies all of the bits in the register through a structure `typedef`. Then a macro definition of the port name causes each instance of the port name in the program to be replaced by the dereferenced value of the register address cast onto a pointer of the correct type. The port name is the name of the port found in the data manual, and the bit names given the bits in the structure are the names found in the data manual. Therefore, if you wish to set the bit named HNDS found in the register PIOC, you need to use

```
PIOC.HDNS=ON;
```

There are two register locations that work differently from the rest. These are the two flag registers whose bits are set by the occurrence of an interrupt. The bits in these registers are turned off when the code instructs the bit to be set. Since these two registers are so different from the remainder of the registers in the system, I handle them differently. Rather than declare a structure for these registers, the individual bits are #defined as power-of-two values. Therefore, the bit OC2F is defined as 0x40 and the bit IC2F is #defined as 0x02. The flag1 register TFLG1 is forced to the address Register_Set+0x23. Now to clear the bit OC2F, you need to set

```
TFLG1=OC2F;
```

The final section of the file contains several macros that are helpful in handling interrupt service routines. The first macro is

```
#define vector(a,b) ((*(void **)b) = (a))
```

This macro is used to place the address of an interrupt service routine into a vector address. The argument a is a pointer to the interrupt service routine, and b is the vector address. If one asks what b is, you must say that b is a pointer to a location that contains the address of the interrupt service routine. The interrupt service routine address is also a pointer to function that has no (void) return. Therefore, the vector address is a pointer to a pointer to the type void and must be cast as such before it can be used. This value must be dereferenced to be able to place the address of the interrupt service routine into it. You can use the macro vector(a,b) to place the address of each interrupt service routine used into the proper vector location.

This macro will create code that copies the address of an interrupt service routine to the specified memory location. This operation is needed whenever the vector table is stored in RAM, so that the vector table must be rebuilt each time the microcontroller is powered up. Another approach must be used to place interrupt service routine addresses in the vector table when this table is contained in ROM. This latter case is probably more common than the former. In this case, we are trying to fill a memory array with the values of the addresses of the several interrupt service routines that the program might use. One way to do this operation is to build an array that contains these addresses, compile this array, and then at link time force the array to be linked to the memory location corresponding to the beginning of the vector table. Consider the following code sequence.

```
extern void IC1_Isr(), OC3_Isr(),_stext();
void (* const vector[])()={0,0,0,0,0,0,0,OC3_Isr,0,
0,0,0,IC1_Isr, 0,0,0,0,0,0,0,_stext};
```

This two-line sequence identifies three functions, each of which returns nothing. The first two are interrupt service routines that will process interrupts from input capture 1 and output compare 3, respectively. The third entry is the name of the entry point in the start-up routine that is linked to the C program. The second line of code indicates that vector is an array of const pointers to functions of the type void. There is one entry in this array for each interrupt vector and the reset vector for the micro-controller. This array is initialized with either zeros or the addresses of the interrupt service routines in the proper locations. The address of the start-up routine is placed in the last location in the array. This little program will be compiled and linked to the final program. The name of the file that contains this file is interrup.c, and it must be modified for each program in which it is used. At link time, the address of vector[ ] will be forced to 0xffd6 which is the beginning of the vector table in the MC68HC11 family.

Some programmers will initialize the vector table with a known address rather than 0. In the event that a spurious interrupt occurs and takes the processor to an unused vector location, the processor would certainly get lost if the vector table were filled with zeros. Placing a known program into all unused vector locations will prevent this problem.

Another problem can occur in the operation of unattended microcontrollers. It is possible that control of the microcontroller could be diverted to unused ROM locations by serious noise spikes. Such a loss of control will not be devastating if the system uses a computer operating properly (COP) system. However, another safety back-up that the programmer can incorporate into the program is to fill all unused memory with the one-byte instruction swi (software interrupt). This instruction causes the program to save the machine status and pass control to the function addressed in the SWI vector location. If this value contains the address of the start-up program, the system will be restarted immediately if control is accidentally moved to unused ROM.

There are several small routines associated with the processing of interrupts that are not in the C library. These routines are written as functions to make it easier to access these important operations. The `cli()` and `sei()` functions allow the program to clear or set the interrupt bit in the condition code register (CCR). The programmer can with these two functions and the interrupt masks in the I/O memory map control all interrupt operations of the part.

The compiler must be notified that a function is to be an interrupt service routine. An interrupt service routine can have no arguments, and it must return nothing. Therefore, the function prototype of an interrupt service routine might look like

```
void isr_clock( void );
```

However, the compiler would still have no way of knowing that this function is an interrupt service routine. The reason that the compiler must know an `isr` is that the MC68HC11 family stacks the complete machine status when an interrupt is accepted by the device. This status must be restored when the program control is returned to the interrupted program. The machine status is restored when an `rti` instruction is executed. Therefore, any return from an interrupt must be the assembly instruction `rti` rather than the instruction `rts` usually used to return from a function. An interrupt service routine is identified to the compiler by the sequence @port ahead of the definition of the function in the function prototype. The function prototype of an interrupt service routine should be

```
@port void isr_clock( void );
```

This flag will cause all returns from the function to be `rti` instructions.

There is no guarantee that there will be macro definitions for useful numbers and functions. A series of enumerations that define TRUE, FALSE, ON, OFF, START, STOP, etc., are included to provide mnemonics for these often-used values. The pointer constant NULL is also defined by a macro here. These constants are often defined in other header files, so the protection against including these constants more than one time is also included. Also useful is the macro FOREVER which is included.

EXERCISES

1. Write a header file that contains definitions of all vectors shown on page 3 of the HC11 E Series Programming Reference Guide, M68HC11ERG/AD found on the CD-ROM. This file should be included with any program that is intended to make use of interrupts.

2. Create a small program that uses the function `vector(a,b)`. Compile the function and observe the assembly language code generated to accomplish this operation.

The Cosmic Compiler

The Cosmic compiler was originally known as the Whitesmiths compiler. Cosmic retained the maintenance contract on this compiler through a couple of corporate owners, and now has the rights to the compiler. This compiler used for the MC68HC11 family is somewhat more complicated to use than the ByteCraft compiler used with the C68HC05. The compiler should be installed using the install program that comes with it. You should allow the compiler to modify your autoexec.bat and config.sys files unless you plan to do it yourself. The system path should contain the path to the bin directory in the directory that contains the compiler. There are also two set commands that should be inserted into the autoexec.bat file. These changes, once completed, will make compilation of programs rather easy. This compiler creates an intermediate assembly language program that must be assembled. The result of the assembly is a relocatable object module. This module must be linked with other modules of the program and basic library modules to comprise the final program. The linking phase of the compilation places all of the parts of the program in their proper memory location.

Fortunately, the compiler can make use of command files to control the compilation sequence. These files relieve the programmer of the need to remember the dozens of little details that must go into each compilation. We will start with the highest level of the command files and work down into the lower levels. The basic command file to compile a program and create an S Record version of the program is as follows:

```
c -dlistcs +o %1.c
lnkh11 < %1.lnk
hexh11 -s -o %1.hex %1.h11
```

This file is called `comp.bat` and it is invoked by

```
comp <filename>
```

The `%1` in the command file will be replaced by `<filename>` when the command file is run. The first line of the command file tells the compiler, named `c`, to execute with the options `-dlistcs` and `+o`. The name of the file will be the filename entered on the command line with a `.c` extension. The `-dlistcs` option causes a listing file to be generated and saved in a file of the same name filename but with an extension `.ls`. The `+o` option informs the compiler to create a relocatable object module of the program.

The basic invocation syntax of the compiler is

```
c [options] file.[c | s | o] [filen. [c | s | o]]
```

Any portion of the command line in the above sequence that is enclosed in square brackets [ ] is optional. Therefore, the only required command line entry following the `c` call is the file name. (Refer to the compiler manual for the variety of options that can be used on this command line.)

Several files can be included on the command line. Each of these files can have one of three extensions, `c`, `s`, or `o`. If the extension is `.c`, the compiler expects a C program. If it is `.s`, the compiler will process an assembly language program. When the extension is `.o`, the compiler invokes the linker.

The compiler creates a relocatable object module that is linked by the next line in the command file.

```
lnkh11 < %1.lnk
```

The direct call to the linker `lnkh11` is handed a command file to control the linking. Each program must have its own link command file, and the extension of this file is `.lnk`. An example linker command file is shown below. This command file is for a program developed later in this chapter.

```
#
# link command file for motorfr program
#
+h # multi-segment output
-o motorfr.h11 # output file name
+text -b 0xd000 # program start address
+data -b 0x0000 # data start address
crtsmot.o # start-up routine
motorfr.o # application program
libi.h11 # integer library
libm.h11 # machine library
+text -b 0xffd6 # vectors start address
interrup.o # interrupt vectors
+def __memory=__bss__ # symbol used by library
```

Comment lines are preceded with a # in this command file. The first executable line of code in the file contains a command +h which notifies the linker that the program will be in multiple memory segments rather than in a single block in memory. The second line

```
-o motorfr.h11
```

tells the linker to write the output file to motorfr.h11. This particular linker file is for a program named motorfr. The +text option places the beginning of the code portion of the next module to be linked at the offset 0xd000. The +data option places the data memory for the next module at the offset 0x0000. Finally, the names of the modules to be linked are next. The module crtsmot.o is a relocatable object module version of a start-up program. This routine will be discussed below. The next module to be linked is motorfr.o which is the output of the compiler. This line is followed by the invocation of two library calls. The first is to the integer library, and the second is to the machine library. These libraries will provide necessary code for function calls that are not contained directly in the program. Another library named libf.h11 contains floating point operations that might be needed in your program.

Another +text option will force the beginning of the code linked next to the address 0xffd6. This address is the beginning of the MC68HC11 vector table, and the code linked at this point is the vector function discussed earlier. This entry will create the vector table at the correct location in memory, and it will also place the addresses of the designated vectors in the proper locations in the table.

The start-up routine mentioned earlier is shown below. This routine is patterned after one provided with the compiler. This assembly language program contains all of the code necessary to begin the operation of the C program to which it is linked. But note this point on syntax: within C, all function and memory names generated by the code are modified by the addition of an underscore _ at the beginning of the name. Therefore, if the function main() in the C program were to be referred to by an assembly program, the assembly program would be required to use _main. You will notice that there are many names beginning with both single and double underscores. For example _main has a single underscore, and __memory has a double underscore in the external statement that follows. Whenever an assembly language memory location or function name has underscores preceding the name, one of the underscores must be discarded when this same memory or function name is used in a C program to be linked to the module.

Here you see an example of why the programmer should not use an underscore for the first character of a name. The compiler writer assumes complete freedom to use the single underscore to begin any name needed by the operation of the compiler, and the programmer should concede this freedom to the compiler writer by avoiding the use of the underscore for the first character of a name anywhere in his program.

An example of this linkage has already been seen. In a previous section where the vector table was programmed, the entry point into the program was the function _stext(). Note in the code that follows that the function name in the assembly language part of the program is __stext with a double underscore. This start-up routine has been modified for use with a program that we will use later in the text. The most significant modification is the first two executable instructions which set the y register to the beginning of the I/O register memory area and then sets bits 0 and 1 of the location offset 36 (0x24) from the y register content. These instructions set the prescaler of the timer counter to 4 so that the timer will increment at the slowest possible rate. These two instructions modify the contents of TFLG2 and they must be changed within 64 bus cycles after the microcontroller exits reset. These changes and others needed in the INIT register, the

OPTION register, and the CONFIG register must be placed in this location in the program. These registers become read only memory locations after the first 64 bus cycles following reset.

The value __memory is calculated in the command file motorfr.lnk and it is the number of bytes of memory that the program uses for volatile memory. The start of the base memory section is _sbss and it is zero for this program. The instruction sequence starting with ldx #__sbss and ending with bne zbcl will clear all of the volatile memory used by the program. Finally the stack pointer is set to a value of 0xff and the C program is executed by the jsr _main call.

```
.processor m68hc11
;
; C START-UP FOR MC68HC11
;
.external _main, __memory
.public _exit, __stext, __return
;
.psect _bss
__sbss:
.psect _text
__stext:
ldy #4096
bset 1,36,y ; set the prescaler to /4
clra ; reset the bss
ldx #__sbss ; start of bss
bra loop ; start loop
zbcl:
staa 0,x ; clear byte
inx ; next byte
loop:
cpx #__memory ; up to the end
bne zbcl ; and loop
prog:
sts __sdata ; save sp for monitor return
xgdx ; initialize stack pointer
ldd #00ffH ; put stack pointer at ff for HC11E9
xgdx
```

```
txs
jsr _main ; execute main
_exit:
lds __sdata ; restore stack pointer
rts ; and return to calling program
__return:
rti ; used by default interrupt vector
;
.psect _data
__sdata:
.word 0 ; avoid any pointer to null
.end
```

The start-up program must be written to account for any special problems that might be needed for your program. Through this routine, memory can be initialized with data stored in ROM, the special registers that must be processed within the first 64 bus cycles can be set, the stack can be placed where the programmer desires, and so forth. The code at the end of the routine restores the stack pointer to the value it contained when the start-up routine was entered and executes an `rts`, return to subroutine, to return control of the processor to a monitor that started the program if one exists.

The location `__sdata` is the first byte following the volatile memory used by the program. This value will be used by the memory allocation functions to provide memory on the system heap.

Finally, the last line of the compiler command file

```
hexh11 -s -o %1.hex %1.h11
```

causes execution of the program `hexh11`, which converts the object module from machine code to an S record format that can be written into an evaluation module or used to burn an EPROM version of the microcontroller. This S record file can also be delivered to the factory and used to define a mask for a mask ROM version of the microcontroller. The file created by this line of code will have an extension `.hex`.

EXERCISES

1. Write and compile a small program that will show some bit manipulations in PORTA. Make the program test, set, and clear bits in this

port. Does the assembly language generated by the compiler appear to be efficient? If not, what can be done to improve the code?

2. Write and compile a program that uses Output Compare 3 to generate a periodic interrupt. This interrupt should occur each millisecond.

3. With the time tick generated by the results of Example 2 above, create a clock that will keep track of the time of day accurate to the nearest one-hundredth of a second.

The Compiler Optimizer and Volatile

The compiler optimizer that is a part of the Cosmic C compiler does an excellent job of reducing code size. One of the steps involved is to remove code required to change values that are not changed by the program. Recall that the volatile qualifier identifies varibles that should not be optimized. The following code sequence will show this problem:

```
volatile char able;
char baker

test(void)
{
   char dog;

   dog = able;
   dog = able;
   able = 0;
   able = 0;
   dog = baker;
   dog = baker;
   baker = 0;
   baker = 0;
}
```

Note that this function makes multiple assignments of dog, able, and baker. The assignments involving able show how the compiler responds to a volatile variable, and baker to a nonvolatile variable. The following listing is the compilation of the above program:

```
1 ; Compilateur C pour MC68HC11 (COSMIC-France)
32 .include"macro.h11"
```

```
3 .list +
4 .psect _text
5 ; 1 volatile char able;
6 ; 2 char baker ;
7 ; 3
8 ; 4 test(void)
9 ; 5 {
10 _test:
11 0000 3C pshx
12 0001 34 des
13 0002 30 tsx
14 .set OFST=1
15 ; 6 char dog;
16 ; 7
17 ; 8 dog = able;
18 0003 F60000 ldab _able
19 LL4:
20 0006 E700 stab OFST-1,x
21 ; 9 dog = able;
22 0008 F60000 ldab _able
23 LL6:
24 000B E700 stab OFST-1,x
25 ; 10 able = 0;
26 000D 7F0000 clr _able
27 LL01:
28 ; 11 able = 0;
29 0010 7F0000 clr _able
30 LL21:
31 ; 12 dog = baker;
32 0013 F60001 ldab _baker
33 ; 13 dog = baker;
34 0016 E700 stab OFST-1,x
35 ; 14 baker = 0;
36 ; 15 baker = 0;
37 0018 7F0001 clr _baker
38 ; 16 }
39 001B 31 ins
40 001C 38 pulx
41 001D 39 rts
```

```
42 ; 17
43 .public _test
44 .psect _bss
45 _baker:
46 0000 .byte [1]
47 .public _baker
48 .psect _data
49 _able:
50 .byte [1]
51 .public _able
52 .end
```

Lines 17 through 24 correspond to the two lines

```
dog = able;
dog = able;
```

The memory location of able in the program is _able, and the memory location for dog is at OFST-1,x. Note that the value in _able is read and placed into dog twice as the program stated. Also, the instruction clr _able is executed twice to account for the two lines

```
able = 0;
able = 0;
```

This code is what you should expect when the volatile keyword is used when able is declared. On the other hand, lines 31 through 34 show the compilation of the two C lines

```
dog = baker;
dog = baker;
```

In this case, you will note that the value for baker, stored at _baker, is placed into the address of dog only once even though the code line is repeated. Similarly, the two lines

```
baker = 0;
baker = 0;
```

result in assembly code that clears the location _baker only once. This type of optimization will save you much code space. It is certainly a problem whenever there is a memory location that can be changed from outside the program. The volatile keyword will cause the optimization operation to skip over all variables that can be changed by the hardware portion of the system without knowledge of the program.

Sorting Programs

You might not expect that sorting routines would be important when programming microcontrollers, but there have been several instances in my experience where sorting of a list of data was needed in a microcontroller application. Therefore, let's examine some sorting routines that can be used with a microcontroller and determine which of several popular approaches is best for use in a microcontroller program.

There are several sorting routines that are popular with programmers. The three most-used routines are named the *entry* sort, the *Shell* sort, and the *quick* sort. The entry sort is a routine that is thoroughly discredited because it is extremely slow when compared with the other routines. An entry sort orders the contents of an array in place. The first entry in the array is compared with the remaining entries and, if any array entry is smaller than the first, these values are swapped. This operation is then repeated for the second entry in the array followed by the third entry in the array until all entries in the array are compared with all others. When this sequence is completed, the array is sorted. If there are *n* entries in the array, the first scan involves *n* compares, the second involves *n*-1 compares, the third *n*-2 and so forth. Therefore, the total number of compares is

```
n+(n-1)+(n-2)+...1
```

The value of this sum is $n*(n-1)/2$. In other words, the entry sort requires on the order of n^2 compares.

The Shell sort was discovered in 1959 by D. L. Shell. Like the entry sort, the Shell sort orders the contents of an array in place. The Shell sort splits the *n*-element array into two halves. The contents of these two halves are compared entry-by-entry and, if the value in the one half is larger than that in the other, the entries are swapped. These arrays are then split again, and the contents of the four arrays are scanned and ordered two at a time as above. Then the four arrays are split into eight arrays and the scan and ordering operation is repeated again. This operation of splitting the arrays and ordering the contents is repeated until there are n/2 arrays of 2 elements each. After this last scan and sort operation, the array is ordered.

There are n/2 compares completed each scan of the array. However, the sort will be completed after the logarithm based-2 of *n* times through the array, because $\log_2 n$ is the number of times that an array that contains *n* entries can be cut in half. Therefore, the time to execute a Shell sort of an array of *n* items will be nearly proportional to `n*log (n)`. For even medium-sized arrays, this time is significantly^2 shorter than for the entry sort. For example, an array of 1000 entries requires time on the order of 1000000 to execute an entry sort and 10000 to execute a Shell sort.

Another sort is the quick sort developed by C. A. R. Hoare in 1962. This sort is a recursive routine. It sorts an array in place. Assume that the array runs from left to right and we want to sort the array in ascending values from left to right. An array scan will place at least one of the array values in the proper final array location. This entry is called the pivot entry. As the scan is completed, all entries larger than the pivot value are swapped to the right end of the array and smaller values are placed in the left end of the array. When this operation is completed, the two portions of the array—excluding the pivot entry—are again quick sorted and the result of this recursive operation is that the final result is sorted. This sorting operation also requires time proportional to `n*ln(n)/2`.

The question now arises whether we should use a Shell sort or a quick sort when writing code for a microcontroller. Let's examine these functions. The code for a Shell sort is as follows:

```
/* shell_sort(int v[ ], int i): sort the array v
of i things into ascending order */
void swap ( int *, int *);
void shell_sort(int v[ ], int i)
{
   int split,m,n;
   for( split = i/2 ; split > 0; split /=2 )
     for( m = split ; m < i ; m++)
        for ( n = m-split; n >=0 &&
           v[n] > v[n+split]; n -= split)
           swap(v+n,v+n+split);
}
```

The outer loop in the preceding routine controls the way that the array is split. It first splits the array in half and then shrinks the arrays by halves until the sub arrays become zero in width. The second loop steps along the elements of the arrays, and the third loop compares the elements that are separated by split and swaps them when necessary. The quick sort is:

```
/* quick_sort( int v[ ], int left, int right) :
   sort the array v into ascending order */
void quick_sort(int v[ ], int left, int right)
{
   int i,pivot;
   if(right > left)
   {
      swap(v+left, v+(right+left)/2); /* chose the
               value at center to pivot on */
      pivot = left;
      for(i=left+1; i<=right; i++)
          if(v[i] < v[left])
              swap(v+(++pivot), v+i);
      swap(v+left, v+pivot); /* put the partition
               element in place*/
      quick_sort(v,left, pivot-1);
      quick_sort(v,pivot+1,right);
   }
}
```

The swap routine for both of these sorts is the same:

```
/* swap(int *, int *) : swap two integers in an
array */
void swap(int *i, int *j)
{
   int temp;
   temp = *i;
   *i=*j;
   *j=temp;
}
```

One of the nice things about both of these programs is that the swap operation is not built into the program. Therefore, you can swap values as was done above, but also, you can swap pointers from an array of pointers to order a set of data that would be difficult to swap. For example, if you had a collection of words, you could use a `srtcmp()` routine to determine whether one word was lexically larger than the other and then write a swap routine that merely swapped the pointers to these words rather than swapping the words themselves. This approach to sorting a list of words is relatively simple. On the other hand, it would be extremely difficult—nearly impossible, in fact—to sort a list of words that are packed into memory.

The following program is used to test the performance of the above sort routines. The program simply makes an array of 11 entries and sorts the array by both the Shell sort and the quick sort. The result of this program is printed out below.

```
#include <stdio.h>
main()
{
    int v[]={81,99,23,808,3,77,18,27,128,360,550};
    int i,k[11];
    for (i=0;i<11;i++)
        k[i]=v[i];
    shell_sort(k,11);
    for( i=0;i<11;i++)
        printf ("%d ",k[i]);
    printf("\n");
    quick_sort(v, 0,11 -1);
    for( i=0; i<11; i++)
        printf("%d ",v[i]);
    printf("\n");
}
```

The original array is not printed out by the program, but the order of the array is seen in the above listing. The two sort programs sort the values correctly.

```
3 18 23 27 77 81 99 128 360 505 808
3 18 23 27 77 81 99 128 360 505 808
```

The quick sort routine is not the best example of recursive programming. Usually, the use of recursion results in a program that is significantly shorter than would be expected with conventional linear programming. Here, the Shell sort requires fewer lines of C code and, as we will see, the Shell sort has a smaller assembly language program. You would not expect the linear program to be shorter in either C or assembly language. These two programs were compiled and the results analyzed.

The Shell sort requires 0x78 (120) bytes of code while the quick sort needs 0xaa (170) bytes of code. This amount of extra code would not be a serious cost if that were the end of the costs. However, each time a recursive routine calls itself it must establish an argument set that contains all of the variables to be passed to the function. When the quick_sort routine calls itself, it must pass three parameters. These parameters are usually passed in either the stack or in computer registers. When quick_sort returns control to the calling program, the stack pointer must be corrected. quick_sort will continue calling itself repeatedly while the value of the parameter right is greater than the value of the parameter left when the function is entered. Each time the function is called, the program will create a new stack frame which must be made available from RAM. Unfortunately, RAM is usually a more precious commodity than ROM to a microcontroller.

It is possible to get an idea about how much RAM is needed to do a quick sort. The stack frame consisting of six bytes—four bytes for parameters plus two bytes for the return address—must be created each time quick sort is entered. You can count the maximum number of times recursive routine calls itself. The maximum depth is the number of times the function is called before a normal return to the calling function is executed. The depth is found by incrementing a count each time the function is entered. Whenever control is passed back to the calling function through the normal return sequence at the end of the function, the depth will be evaluated to determine if it is the largest value seen so far, and then the depth will be returned to zero. The two listings below incorporate these modifications.

```
/* quick_sort( int v[ ], int left, int right) :
sort     the array v into ascending order */
void quick_sort(int v[ ], int left, int right)
{
   int i,value;
   extern int calls, depth, max_depth;
```

```
    calls++; /* count the number of times called */
    depth++;
    if(right > left)
    {
        swap(v+left, v+(right+left)/2); /* chose the
                value at center to partition on */
        value = left;
        for(i=left+1; i<=right; i++)
            if(v[i] < v[left])
                swap(v+(++value), v+i);
        swap(v+left, v+value); /* put the partition
                    element in place*/
        quick_sort(v,left, value-1);
        quick_sort(v,value+1,right);
    }
    if(depth>max_depth)
    max_depth=depth;
    depth=0;
}

void swap ( int *, int *);
void quick_sort(int v[ ], int left, int right) ;
#include <stdio.h>
int calls, depth, max_depth;
main()
{
    int v[11]={81,99,23,808,3,77,18,27,128,360,550};
    int i;
    quick_sort(v, 0,11 -1);
    for( i=0; i<11; i++)
        printf("%d ",v[i]);
    printf("\n");
    printf("calls=%d\nmax_depth=%d\n",calls,max_depth);
}
```

Note that the global variables calls and depth are incremented each time quick_sort() is entered. When quick_sort() is exited through its normal return procedure at the end of the routine, depth is examined to determine if it is greater than max_depth.

If it is, `max_depth` is replaced by `depth`. `depth` is reset to zero prior to returning through the end of the `quick_sort()` routine. The result of this program is as follows:

```
3 18 23 27 77 81 99 128 360 550 808
calls=15
max_depth=4
```

To sort the above 11 element array, the program called `quick_sort()` 15 times and the maximum number of stack frames that were created at one time was four. Each stack frame consists of four bytes for data and two bytes for the return address. Therefore, at one time in the execution of this program, the sort routine required 24 bytes of stack space in addition to the memory area required to hold the array being sorted. This is distressing when you remember that the array is 22 bytes long!

Careful analysis of the program will show that the expected maximum number of stack frames for the quick sort is proportional to $\ln_2 n$. Therefore, the stack frame performance will get better when the size of the array is larger. The main problem with the use of any recursive routine is that you do not know exactly the number of stack frames needed and therefore you must always analyze the maximum stack frame depth and allow that amount of stack for execution of the function.

The quick sort is faster than the Shell sort under most—but not all—circumstances. Let's take the example of having a quick sort and a Shell sort compiled to run under the same host computer. If larger arrays are used to do the measurement and random arrays are used, we will find that the sort time for both quick sort and Shell sort depends on the array. The reason for this difference is the number of swap routines that must be executed during the sort operation. Therefore, there will be some arrays that the Shell sort will sort quicker that the quick sort. However, the quick sort is usually faster than the Shell sort.

The Shell sort is every bit as flexible as the quick sort and can be used to sort different types, character strings, and swap pointers rather than swap the objects being sorted. Therefore, in a microcontroller application where memory is always limited, you should use a Shell sort rather than a quick sort because the Shell sort has no funny memory requirements that will always occur with a recursive routine like the quick sort.

The purpose of the above discussion is to demonstrate that the programmer must exercise much more care in the details of program design for a microcontroller than is necessary for a larger machine. The above sorting routine would be of no concern for a typical desktop machine because RAM is usually of no consideration with such machines. It might not be a problem for some microcontroller applications, but in a case where RAM is limited and needed, it might make the difference between being able to get the program to run or not.

In most instances, the elegance of recursive code has a cost of RAM. Usually a recursive program will be slower than the equivalent linear program. The microcontroller programmer must examine each instance of code to determine if the elegant recursive program has hidden side effects that end up costing more in computer resources than is gained in small code size.

EXERCISES

1. Write a program to run on a host computer that will test the time required to execute both a quick sort and a Shell sort on random large arrays. Make the array sizes selectable from the keyboard.

2. Test both the quick sort and Shell sort for various array sizes. Examine the time required to execute the sorts for several random arrays. Explain what is happening when arrays of 4090, 4095, and 4096 entries are sorted.

Data Compression

Another program problem that can arise with microcontroller applications is handling displays in which there are many phrases and names each consisting of several letters. The question that must be considered is the most efficient method of storing these phrases and names, in memory so that they can be recalled for display. The oldest, and yet very effective, method of compressing the data for storage is to use a Huffman code. This code is developed based on the statistical occurrence of each letter in the alphabet. A letter that occurs frequently will be given a short code sequence. A letter that is seen infrequently can be assigned a long code, which will not appreciably affect the number of bits per letter in the message.

Huffman codes are self-synchronizing. A long string of characters can be represented as a series of bits, and the decoding program can find each character in the sequence without any special flags to designate the end of each character. Messages stored as ASCII characters require at least 7 bits per character, and because of the nature of storage in a computer, 8 bits are usually assigned to each character. With a Huffman code, it is possible to reduce the average number of bits per character for the whole English language to about 4.5 bits per character. Sometimes even fewer bits are needed with limited messages.

The programmer does all of the Huffman encoding and the computer must decode the messages for display. Huffman decoding requires extensive bit manipulation, so C is an ideal language to decode the messages.

A Huffman decoding scheme can be represented by a strictly binary tree. A binary tree is comprised of a collection of nodes. Each node has two descendents, often called the right and left descendents (thus the term binary tree). The terminating node in a tree is called a leaf. In a strictly binary tree, every nonleaf node has both a right and left descendent. If a strictly binary tree has n leafs, it has a total of $2n - 1$ nodes.

The Huffman code is decoded in this manner. The sequence always starts at the root node. The code sequence to be decoded is examined a bit at a time. If the bit is a 0, move to the left node. If the bit is a 1, move to the right node. The tree is traversed until a leaf is found which contains the desired character. The sequence is then restarted at the root node. If the tree is not known, the sequence of bits is bewildering. Usually it is found that the occurrence of ones and zeros is about equally likely, so no statistical analysis can be applied to decode the sequence easily. A disadvantage is that if a single bit in the sequence is lost, the remainder of the code stream is undecipherable. Of course, the advantage to the Huffman code is that it takes somewhere between one-quarter and one-half the number of bits to represent a message as is required by the ASCII code. A Huffman Code tree is shown in Figure 5-1. The bit sequence 11111110010101111010100000 as decoded by this tree is seen to spell the word "committee." This sequence requires 23 bits, and the ASCII code needed to represent committee is 72 bits long. Of course, this tree was designed specifically for the word "committee," but try to encode other words that can be made of these same letters such as "tic," "me," "come," "tom," "mite," etc., and you will find that every word that can be derived from this tree requires fewer bits than the corresponding ASCII sequence would need.

If you look at Figure 5-1, you will note that the letters e, m, and t require only 2 bits each to determine the value. These letters were selected to be represented by 2-bit values because they are the most frequently occurring letters in the message. The remainder of the letters require more bits but they do not occur so often in the message.

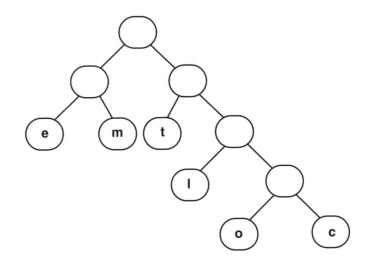

Figure 5-1: Huffman Code Tree

The code must be designed specifically for the messages to be represented. First, all of the characters in all of the messages are collected into a histogram of the occurrence of each character. Remember to include spaces and other such characters in the list. Once you have determined the number of different characters in the list, you can design a tree that will compress the code. If the list contains n different characters, there will be $2n - 1$ nodes in the tree. Each level in the tree, starting from the root node at the zeroth level, can contain 2^k nodes or leafs. This tree cannot be balanced like a binary sort tree because we want some of the leafs to be determined by as few as 2 bits, and we do not care about the number of bits needed to determine a character that occurs infrequently. In Figure 5-1, levels 0 and 1 contain only nodes with no leaves. Level 0 contains three leaves and one node. It is through this single node that all of the remainder of the characters must be found.

A series of avionics-related items were put together and an analysis of the frequency of characters in these messages was completed. A Huffman code that will encode these data effectively was created. This code is shown in Table 5-1. Frequently occurring letters like e, i, or a are encoded with only two or three bits. On the other hand, letters that occur very infrequently—like c or v—require 9 bits to encode. Even though some letters require more than 8 bits to encode, the average number of bits per character for this particular message set is only 4.032.

Decoding a Huffman code with a program is relatively easy. The program must contain the Huffman tree with all of its nodes and leaves. If we want the entire alphabet and a space, a period, a comma and an end of message, there will be 59 entries in the tree. This or an equivalent tree is an overhead that must be carried for every Huffman program. Assume that the statistical analysis of the messages along with the construction of the Huffman tree is complete. The approach taken to build this tree in memory is to allow one byte for each node. The tree is searched from the root node, which is the lowest address. The message to be decoded is examined. If the bit in the message is a zero, the node pointer is incremented by one and the next bit is examined. If the bit in the message is a one, the node pointer is incremented by the content of the node. All printable characters are entered into the tree as negative numbers. The contents of the node are examined. If the content of the node is positive, the next bit from the message is examined and so forth. Whenever the content of a node is negative, its most significant bit is removed, and the character is sent to the output. At that time, the node pointer is returned to the root node, and the sequence is repeated until an end of message is found.

Character	Code	Character	Code
E	00	D	110000
I	01	' '	110001
A	100	G	111100
'\n'	101	U	111101
N	11001	L	1111100
R	11010	O	1111101
T	11011	F	11111100
M	11100	H	11111101
S	11101	P	11111110
		C	111111110
		V	111111111

Table 5-1: Huffman Code

The listing shown below is a simple Huffman-encoded routine to print out to the screen several aircraft-oriented terms that might be used in an on-board avionics system. The first part of the function is a list of the several words or phrases.

```c
#include <stdio.h>
/* The messages to send out */
static unsigned char M1[]={0x9f,0x36,0xef,0xdc,0x0a};
static unsigned char M2[]={0xfd,0x26,0x0e,0x7c,0xa0};
static unsigned char M3[]={0xc1,0xee,0xe6,0x7f,0xc6,
            0x3a,0x39,0x1c,0xb9,0xf2,0x80 };
static unsigned char M4[]={0xfc,0xf4,0xf9,0x8e,0x8e,
             0x47,0x2e,0x7c,0xa0};
static unsigned char M5[]={0x8e,0xb1,0xef,0xf0,0x61,
              0x40};
static unsigned char M6[]={0x73,0x8f,0xef,0xdf,0x75,
              0xda};
static unsigned char M7[]={0xdb,0xc2,0x80};
static unsigned char M8[]={0x3b,0xb7,0x93,0x66,0x18,
             0xed,0xe1,0x8f,0xdf,0xcc,0x66,
             0xb4,0xff,0xe7,0xca};
static unsigned char M9[]={0xf3,0x5f,0x7d,0xce,0x18,
             0xf7,0xf8,0x30,0xa0};
/* The Huffman tree */
const static char
Node[]={4, 2, 'E'|0X80, 'I'|0X80, 4, 2,'A'|0X80,
'\n'|0X80, 10, 6, 4, 2,'D'|0X80, ' '|0X80,
'N'|0X80, 2, 'R'|0X80, 'T'|0X80, 4, 2, 'M'|0X80,
'S'|0X80, 4, 2, 'G'|0X80, 'U'|0X80, 4, 2,
'L'|0X80, 'O'|0X80, 4, 2, 'F'|0X80, 'H'|0X80, 2,
'P'|0X80, 2, 'C'|0X80, 'V'|0X80};

void decode(unsigned char *M)
{
   unsigned char mask = 0x80;
   char i=0,j=0,k=0,l=0; /* j is the node pointer,
                           i is the byte pointer, M
                           is the message pointer */

   while(k !='\n')
```

```
{
    if((mask & M[i])==0)
        j++; /* use next node entry if bit is zero */
    else
        j+=Node[j];/*jump to designated node when
                        one */
    if(Node[j]<0)
    { /* if a printable, send it out */
        putchar(k=Node[1]&0x7f);
        j=0; /* also go to the root node */
    }
    if((mask>>=1)==0)  /* if the mask is zero,turn
                        on MSB */
    {
        mask = 0x80;
        i++; /* and get the next byte from message */
    }
}

    }
```

Listing 5-1: Huffman Decode Function

The next part of the program is the Huffman tree needed to decode the data. This tree was constructed to decode data specifically for the nine messages of this problem. You will note that each byte in the table is a node. Intermixed in the table are numbers that dictate jumps to the next node depending on whether the incoming bit is a zero or a one. The leaves each contain a negative number that is generated by the inclusive OR of the character to be printed and the number 0x80.

The messages are sent to the function decode as a pointer to a bit array. The data must be examined one bit at a time to implement the decoding. The bit selection is accomplished by ANDing the data in the incoming message with the contents of the variable mask. Mask is initially set to 0x80, so the most significant bit of the first byte of the incoming data is selected. Later in the routine, the value of mask will be shifted right to select other bits in the incoming sequence. The variable k is loaded with the character to be output during the program. The linefeed character '\n' was used to determine the end of message, so the decoding sequence will be executed until the character that was output from the program is a '\n'. Recall the way that the tree was built. If an incoming bit is 0, the next node in the tree will be taken. If the incoming bit is 1, the content of the current node will be added to the node pointer to determine the next node in the tree. If at any time a negative number is found in the tree, a leaf has been found. The data portion of this byte, bits 0 through 6, will be sent out to the screen.

After each bit is processed, the next bit in the sequence must be processed. The next bit is selected by shifting the mask byte to the right one bit. If the result of this shift creates a zero mask, a new mask with a value of 0x80 is created and the next byte from the incoming message code is selected when i is incremented.

The simple program below causes the nine messages to be printed on the screen:

```
main()
{
decode(M1);
decode(M2);
decode(M3);
decode(M4);
decode(M5);
decode(M6);
decode(M7);
decode(M8);
decode(M9);
}
```

Listing 5-2: Message Printing Program

The result of execution of this program is shown below. The nine messages contain 124 characters that would require 992 bits of memory to store the messages in standard ASCII format. The encoded sequence requires 500 bits or 63 bytes of storage.

ALTITUDE
HEADING
DISTANCE REMAINING
FUEL REMAINING
AIR SPEED

IN HOURS
TIME
ESTIMATED TIME OF ARRIVAL
GROUND SPEED

The function below was written to show the difference in memory required for the Huffman code and conventional ASCII coding. This program was compiled as a function to be run on the MC68HC11. The function decode given in Listing 5-1 was compiled to MC68HC11 code. The result of these two compilations showed that Listing 5-1 created an object module that was 255 bytes long, and Listing 5-2 created an object module that was 277 bytes long. The Huffman code for even the small message list in this case provided nearly 10% reduction in code size. A larger message list would provide an even greater memory savings because the code required to decode the messages would be allocated over more message characters to be sent out.

```c
#include <stdio.h>
char M1[]="ALTITUDE";
char M2[]="HEADING";
char M3[]="DISTANCE REMAINING";
char M4[]="FUEL REMAINING";
char M5[]="AIR SPEED";
char M6[]="IN HOURS";
char M7[]="TIME";
char M8[]="ESTIMATED TIME OF ARRIVAL";
char M9[]="GROUND SPEED";
void decode(char *M)
{
   while(*M !=0)
     putchar(*M++);
   putchar('\n');
}
```

Listing 5-3: Nonencoded Output Function

EXERCISES

1. How could the tree in Listing 5-1 be altered to reduce its size by 8 bytes? What effect would this change have on the routine decode? Would there be a net savings in overall code?

2. Do an analysis of a substantial piece of writing and create a Huffman code that will encode the data efficiently. It is recommended that the first three pages of a novel be used for this analysis. Compute the average number of bits per character that this code generates. Hint: you might want to write a program in C for the host computer to calculate the histogram and help create the Huffman codes.

3. Write a program that will implement the above code so that an operator can type the data into a computer and the Huffman code sequences will be generated. You should break the message sequences into moderate size bit strings, 500 to 1000 bits, and restart.

4. Create a Huffman tree table as was used in Listing 5-1 to decode the Huffman code developed in Exercise 2.

5. How can you double the number of entries in a Huffman tree by adding only one bit to the code strings?

Timer Operations

The programs written in the earlier sections on sorting and data compression were more computer programs than microcontroller programs. The code written would work on a desktop system or a mainframe computer if needed. In this section, we graduate to true microcontroller programming. The set up of the MC68HC11 family requires that the programmer have a detailed knowledge of the operation of the device. Even though the program is written in a high-level language, it is the responsibility of the programmer to properly set all of the necessary control bits to make the device work as desired. Unfortunately, there are few helpful tools that can guide you through this portion of the program. You must first understand what you want the device to do and then dig through its specifications to find the necessary bits to be set to make it perform as desired. It is highly recommended that prior to an attempt at programming this device that you familiarize yourself with the technical data manual for it as well as the reference manual for the family that is found on the CD-ROM.

The timer subsystem in the MC68HC11 family contains both input capture operations and output compare functions. In this section, we will explore these subsystems associated with the MC68HC11Ex series. The main difference between the Ex series and the other devices lies in the number of output compare and input capture systems on each device. The Ex series has four output compares, three input captures, and one timer that can be programmed as either output compare or input capture. The other devices have three input captures and five output compares.

Output Compare Subsystems

What does an output compare do? The explanation of the timer subsystem must always begin with the 16-bit timer counter called TCNT that is always counting at some fraction of the system bus frequency. The bus frequency is always one-fourth of the system crystal frequency. The prescaler frequency is set to the bus frequency divided by either 1, 4, 8, or 16 depending on the bits PR1 and PR0 in the register TMSK2. The timer counter register can be read at any time but it cannot be written to. The Output Compare 1 is special. All of the discussion that follows is valid for all output compares, but—as will be shown later— Output Compare 1 has capability beyond the others.

Associated with each output compare system there is a single 16-bit output compare register containing the time that the program needs an event to occur. When TCNT contents matches that of OCx, an OCx event occurs. Note that I say an event, not output. What happens when the contents of the TCNT matches the contents of OCx is up to the program. Associated with each output compare is a pin that can be set, reset, or toggled when an output compare occurs. When an output compare occurs, the corresponding OCxF flag bit is set in the TFLG1 register. This bit can be examined asynchronously by the program to determine whether an output compare has occurred. If the corresponding bit in the TMSK1 register is set when the OCxF flag bit is set, an interrupt will be requested by the part and the event will be processed asynchronously. These are the operations of all output compare subsystems in the part.

There are three input captures, four output compares, and one timer that can be programmed as either an output or an input. This timer is controlled by the I405 bit in the PACTL register. When this bit is a zero, the programmable timer is an output compare. Otherwise when the bit is set, the programmable timer is an input capture. Often it is desirable to have an output compare operation be initiated by the completion of another output compare. Output Compare 1 is set up in just this manner. The bits in the OC1M and the OC1D registers control the coupling between the Output Compare 1 and the other output compare subsystems. When one of the mask bits, say OC1M5, is set in the OC1M and a compare occurs on OC1, in addition to the normal operation that usually occurs when OC1 happens, the contents of the corresponding bit, OC1D5, in the OC1D register is sent to the output compare pin, which in this case is OC3. Completely independent of the operation of OC1, OC3 could be set to toggle at some time. Output Compare 3 could be programmed to be a PWM output where Output Compare 1 establishes the period of the output, and Output Compare 3 establishes the on time.

The applications for use of the coupled output compares are unlimited. An accurate PWM is but one of several. These outputs could be set up to establish an output sequence to drive a stepper motor. The control of acceleration or deceleration of the motor is easily controlled by merely selecting the base time used with OC1.

If you recall, the output compare-based PWM system discussed for the MC68HC05 had several limitations. For example, that system required that the system operate from interrupt service routines. The latency time of interrupt service was so long that the minimum pulse width was considerably longer than one would expect. Also, the interrupt service timing dictated the maximum on time for the pulse, which again was not nearly 100%. Let us look at how we would do a PWM system with the coupled output compare systems.

```
#include "hc11e9.h"

/* This program will provide a PWM output to OC3, or
   PA5. The period will be the integer value
   found in period, and the on time will be the
   integer value found in time_on. Keep time_on
   less than period. */

#define PERIOD 0X1000
#define TIME_ON 0x0800

WORD period=PERIOD, time_on=TIME_ON;

main()
{
    OC1M.OC1M7=ON; /* sent OC1 out to PA7 */
```

```
OC1M.OC1M5=ON; /* couple OC1 to OC3 */
OC1D.OC1D5=ON; /* turn on OC3 when OC1 occurs */
TCTL1.OL3=ON; /* toggle OC3 when OC3 occurs */
PACTL.DDRA7=ON; /* make OC1 an output to PA7 */
TOC1=TCNT+period; /* set OC1 to the period */
TOC3=TOC1+time_on; /* set OC3 to the time on */
FOREVER
{
  if(TFLG1&OC1F)
  {
    TFLG1=OC1F; /* reset OC1 interrupt flag */
    TOC1+=period;
      OC1D.OC1D7 ^=ON; /* toggle the output
                          Compare 1 bit */
  }
  if(TFLG1&OC3F)
  {
    TFLG1=OC3F;/*reset OC3 interrupt flag */
    TOC3=TOC1+time_on;
  }
}
}
```

Listing 5-4: Pulse Width Modulation Routine PWM.C

This routine starts with the inclusion of the header file hc11e9.h. It is created as a main program to demonstrate how it will work, but should be changed to a subroutine later. The period and time_on interval are declared as global variables. They are declared to be the type WORD. WORD is a typedef synonym for the type unsigned int . The main program begins with a series of five initialization instructions. The first two instructions

```
OC1M.OC1M7=ON;
OC1M.OC1M5=ON;
```

declare that the output of OC1 should be sent to the outside, which will be to pin PA7, and that whenever OC1 occurs, OC3 through pin PA5 should be activated.

The instruction

```
OC1D.OC1D5=ON;
```

indicates that when `OC1` occurs, `OC3` should be turned on. The next instruction

```
TCTL1.OL3=ON;
```

causes `OC3` to toggle when its time expires, and the instruction

```
PSCTL1.DDRA7=ON;
```

sets the `DDRA7` bit so that signals from `OC1` will be sent to the output pin `PA7`. `TOC1` and `TOC3` are initialized by the next two instructions. `TOC1` is given a value of the contents of period greater than the contents of the timer counter register `TCNT`. `TOC3` is set equal to the contents of `TOC1` plus the `time_on`.

The routine then enters an endless loop. Within this loop, two flags are examined and if they are set, they are reset. Also, new times are calculated for when the next output compares should occur. The `OC1F` flag in `TFLG1` is set whenever an Output Compare 1 occurs. This flag must then be reset. The instruction sequence

```
if(TFLG1 & OC1F)
{
    TFLG1 = OC1F;
    .
    .
    .
```

will accomplish the required assembly instructions to test the bit and reset it if the bit is turned on.

The time of the next `OC1` is calculated as `TOC1+period`. In this particular instance, the bit `OC1D.OC1D7` is complemented so that the output observed on `PA7` will toggle with the occurrence of each `OC1`. `TFLG1.OC3F` is tested to determine if an Output Compare 3 has happened. If it has, this bit is reset, and `TOC3` is set to a new value of `TOC1+time_on`. With this setup, `OC3`—`PA5`—will go on when each `OC1` occurs, and will be reset an amount corresponding to `time_on` after it goes on. Therefore, the period of this system can be set and the on time can be any value less than the period.

Normally, the operation of a PWM is to generate an analog signal that is present continuously. The above program does perform this task, but it also assumes that the computer does not have much else to do. Other tasks could be built into the above program, but it is absolutely necessary that the program have a cycle time that is shorter than the period of the PWM signal. If the loop time of the FOREVER loop is less than the PWM period, then you can use the above synchronous approach. It makes no difference when in the cycle that the two if statements in the FOREVER loop are executed. It is necessary that they be executed prior to the occurrence of the OC1 that designates the end of the period. If it becomes necessary to include so much code in the FOREVER loop that it is impossible to guarantee that the loop time will always be less than the period, then an asynchronous approach should be used. This program was compiled and executed on an evaluation module. The value of time_on was adjusted to determine the range of outputs that could be created with this program. The minimum time on must be greater than the time required to execute the code

```
if(TFLG1&OC3F)
{
    TFLG1=OC3F; /* reset OC3 interrupt flag */
    TOC3=TOC1+time_on;
}
```

when OC3F is found on. The reason for this timing is that TOC3 must be updated after TOC1 is given a new value and the update must be complete prior to the passing of time_on . Otherwise, the period of TOC1 will pass before the OC3 will occur. This time was measured, and it was found to be 6 clock cycles. Therefore, reliable performance is obtained with a minimum time on of 6 clock cycles.

The maximum time on was found to be 0xffe with the above code. The output signal at 0xfff was on all of the time with no single cycle off period as the program would indicate.

An interrupt can be requested whenever an output compare occurs. We are servicing two output compares in this case. One's initial thought might be to have two interrupts, in this case one for OC1 and one for OC3. Is this approach really necessary? If it is guaranteed that the time_on parameter is always less than period, then an approach that can be used is to delay the reset of the OC1 flag until the OC3 has occurred. OC3 will always happen after OC1. The problem with this approach is that OC3 occurs very near the end of the period, and there might not be enough time to reset the OC1 interrupt flag prior to the expiration of the period. The following program can be used to demonstrate this approach:

```c
#include "hc11e9.h"
/* This program will provide a PWM output to OC3, or
   PA5.  The period will be the integer value
   found in period, and the on time will be the
   integer value found in time_on. Keep time_on
   less than period. This program uses asynchro-
   nous service of output compare occurrences. */

WORD period=0x1000, time_on=0x0800;
@port void OC3_Isr(void); /* need a prototype for the
                                           ISR */
main()
{
   OC1M.OC1M7=ON;  /* sent OC1 tout to PA7 */
   OC1M.OC1M5=ON;  /* couple OC1 to OC3 */
   TMSK1.OC3I=ON;  /* enable the OC3 interrupt */
   OC1D.OC1D5=ON;  /* turn on OC3 when OC1 occurs */
   TCTL1.OL3=ON;  /* toggle OC3 when OC3 occurs */
   PACTL.DDRA7=ON;  /* make OC1 an output to PA7 */
   TOC1=TCNT+period;  /* set OC1 to the period */
   TOC3=TOC1+time_on;  /* set OC3 to the time on */
   cli();  /* enable the system interrupts */
   FOREVER
   {     /* wait here forever */
   }
}


@port void OC3_Isr( void)
{
   TFLG1=OC1F;  /* reset OC1 interrupt flag */
   TOC1+=period;
   OC1D.OC1D7 ^=ON;  /* toggle the output */
   TFLG1=OC3F;  /* reset OC3 interrupt flag */
   TOC3=TOC1+time_on;
}
```

Listing 5-5: System Using Asynchronous Time Service PWM1.C

Notice that this code is very similar to the earlier program shown in Listing 5-4. The interrupt service handling is set up by use of the `@port` construct and the two added instructions which first enable the `OC3` interrupt and then enable the system interrupts. The code that was contained within the `FOREVER` loop in Listing 5-4 has been moved into the interrupt service routine in this program.

This program was run, and it was experimentally determined that the maximum value that can be allowed for `time_on` is `0xff0` when period is `0x1000`. This maximum value indicates that 16 clock cycles or 8 microseconds at an 8-MHz crystal is needed to service the interrupt prior to the occurrence of an `OC1`. The minimum on time found here is slightly better than found with the code in Listing 5-4. In this case, the minimum on time is 1 clock cycle, which is the expected minimum value.

We have already seen that a significant time is required to process an interrupt, and it is usually impossible to have an output event occur during the interrupt service routine. With the above code it is easy to have a minimum on time of one clock cycle. This small time is accomplished by having the setting of the output signal not be attended by an interrupt. The interrupt will occur when the output signal is reset. Therefore, when the coupled output signal with `OC1` goes high, the output compare on the coupled channel will have its event even if the time is as short as one clock cycle beyond the occurrence of `OC1`. Assume that the coupled channel is `OC3`. When the event occurs on `OC3`, an interrupt will occur. In this interrupt service routine, both `OC1` and `OC3` will have to be set up to operate in the correct manner. This operation has a problem with long on times. If the event associated with `OC3` occurs a few clock cycles prior to the occurrence of the `OC1` event, and the interrupt is caused by `OC3`, then the `OC1` set-up might not be completed when the next `OC1` time arrives. In this case, the whole base period would go out of kilter, and there would be at least one cycle of the output that would be based on the timer overflow cycle rather than the desired time base.

As is often found in engineering operations, there is a choice that can be made. If the interrupt is based on the reset time of the PWM cycle, then a minimum on period of one cycle can be achieved. With this choice of operation, the maximum on time will be several clock cycles—perhaps 20 to 30—short of 100% on. On the other hand, if the interrupt is based on the set time of the PWM cycle, the minimum on time is poor and the maximum on time can be up to one clock cycle shy of the base period. If you wish to have good performance at both ends, minimum on time at the same time as maximum on time, you can examine the on time in your program and if it is less than 50% of the base time period, control the PWM by an interrupt on reset. Otherwise, control the PWM by an interrupt on set. Implementation of the code for this approach is left to the exercises.

Output compare operations OC2 through OC4 and the programmable timer subsystem I4O5 are all identical. These versatile timers can be used to generate complex timing waveforms that are useful in keeping time, running stepper motors, etc. The examples shown above demonstrate how these outputs can be converted into a PWM digital to analog output. The coupling between OC1 and the other outputs provides an excellent mechanism for synchronizing two time events for the outside circuitry. Of course, these subsystems can create events in time, but they are completely unable to measure time. We will later examine the analysis of timed events that can be measured with the help of the input capture subsystems.

EXERCISES

1. Write the code that will determine from the on time for the PWM the interrupt operation that will allow the minimum off and maximum on time simultaneously.

2. Modify the code in Listing 5.4 to eliminate the need for a conditional test if (OC1D.OC1D7 = = 1).

Input Capture Subsystems

On the MC68HC11EX family, there are three input capture timers and one timer that can be programmed as either an input capture or an output compare. These sub-systems are used to measure time interval. The same 16-bit timer used in the output compare systems is used to support the input captures. When an input occurs on one of the input capture pins, the value contained in the 16-bit timer is saved in a register designated for that pin, a flag is set, and the processor can request an interrupt. With input captures, we can measure period and hence frequency (or speed). One of the leading applications where the input capture is used is in automatic braking systems. We shall choose a somewhat simpler system for an example, but the general approach used here is much the same as would be used in the design of an automatic braking system.

Suppose that we have a DC motor that is being controlled by an MC68HC11E9 and wish to be able to set the speed of the motor. The motor has a magnetic sensor whose output will cycle once each rotation of the motor. To determine the speed of the motor, we must measure the cycle time of the output from the sensor. Let us examine the minimum speed of the motor. The TCNT register is being clocked at a frequency that is either one-fourth of the crystal frequency of the system or it can be altered by a prescaler value of either 4, 8, or 16. The 16-bit TCNT register will overflow every 65,536 clocks at its input. Therefore, if we use an 8-MHz clock, the timer overflow period will be 32.768 ms with no prescaler, 131.072 ms with a prescaler value of 4, 262.144 ms with a prescaler value of 8, or 524.288 ms with a prescaler value of 16. The minimum speed at which the motor can move and still be detected unambiguously must be greater than one revolution in the timer overflow time. If the motor rotates any slower than this value, the differences in times between two inputs will be such that the program cannot tell if the period is longer than one TCNT overflow time or a very short time. It is possible to extend the minimum unambiguous time that can be measured by the system through the use of a timer overflow interrupt, but let's examine an approach without these steps and then look at the time extended approach later.

The maximum practical time to be measured with the divide-by-16 prescaler is 500 milliseconds. The minimum rotation speed must be greater than 1 revolution per 500 milliseconds or 120 revolutions per minute. The motor that we will use in this example has a minimum speed of 1000 rpm, so that the prescaler need not be as great as 16. We will use a prescaler value of four, which will provide about a minimum speed of 500 rpm so that there is a safety factor in our measurement.

With the divide-by-four prescaler, the minimum time that can be measured is (4 [crystal to bus frequency division]* 4 [prescaler divide ratio]/ 8 MHz [clock frequency]), which is 2 microseconds. At this time interval, the maximum speed of the motor shaft that can be measured is one revolution in 2 microseconds, or 500000 * 60 revolutions per minute. It is quite clear that such a system is better suited to handle high-speed operations than low speed. If there is no clear reason otherwise, it is usually best to operate a control system at the maximum possible speed. One reason that is important is the operation of the PWM DAC that accompanies the operation of the control system. It is usually best to have a PWM run at the highest practical speed. If the PWM is slow, then conversion of the pulse output from the system to a DC voltage is difficult and not very accurate on an instantaneous basis.

The following code segment might be used as an interrupt service routine to handle the input capture operation.

```
@port void IC1_Isr( void)
{
   TFLG1=IC1F; /* reset IC1 interrupt flag */
   measured_period=TIC1-time1;
   time1=TIC1;
}
```

Here it is assumed that the maximum time between input captures is less than the time of a timer overflow. In this case, the time is merely the difference between the current value and the preceding value which is stored in `time1`.

This approach has only one major problem. Most inputs such as will be obtained from reed switches, push buttons, and even optical interrupt type devices will be noisy when the contact is closed. This noise is called switch bounce, and it will always be present with a contact closure. Therefore, the switch must be debounced in some manner before its data are reliable.

The most common way of debouncing a contact closure is to observe the closure, and then wait a time and observe if the contact is still closed. If closed, it is assumed that the contact is good; otherwise, another wait period in allowed to elapse and the contact is observed again. This procedure is repeated until the contact is closed on a pair of successive observations. Only then is it assumed that the contact is closed. The time between observations is the subject of much engineering debate. Often the designer can place an oscilloscope on the contact and repeatedly close and open it. With proper synchronization of the instrument, it is possible to see the signal caused by the bouncing contacts. If this time can be measured, then having a debounce time of perhaps twice the bounce time will probably give a safe time for the debounce. However, you should not assume that this value will be correct over the lifetime of the contact. As the mechanism ages, it is possible that the bounce time will change, and it will always change in the worst possible way. Therefore, be safe, and keep the debounce time at least twice the bounce time, and perhaps longer for the sake of safety. The range of values used for debounce times ranges from 2 to 30 milliseconds.

This system is measuring the time of an input. How can we stick an arbitrary time for debounce into our measurement equation and be reliable? It is possible to do the debounce and not have the time as part of the measured interval. The contact closure will cause an observable input. Our real concern with debounce is to guarantee that no additional inputs occur after the initial input has been processed. Therefore, we can implement a debounce by merely not re-enabling the input capture interrupt until the debounce time has passed. This way, the time of the first edge seen in the input sequence is processed, and all other inputs caused by bounce will not be processed because the input capture interrupt is disabled. The time that the interrupt is disabled can also be adjusted to meet the particular needs of the program. For example, in the system we will construct below, the speed is measured with a reed switch and a magnet on the motor shaft. The circuit will be built such that the closing of the switch will cause a voltage to fall from +12 volts to 0 volts. There is a resistor between the 12-volt supply and the switch. The bottom side of the switch is grounded, and as the switch is closed and opened, the voltage measured at the top of the switch will change from 12 volts to ground. When the switch closes, it is expected that it will bounce and the voltage will "chatter" between 0 and 12 volts. Surprisingly, the opening of the switch, while it will exhibit less bounce, will also generate these unwanted signals. If we want to block out all of the error signals that can be introduced by the bounce, it is necessary that the time following the initial signal drop be protected, as well as a time around the signal rise that occurs midway during the shaft rotation. In the first example, we will provide a fixed time for the debounce. Later when the program is measuring the speed of the motor, we will provide a debounce time that is somewhat longer than one half the measured period to block out all bounce signals that occur on both the rising and falling edge of the input signal.

The input capture subsystem will capture any specified input when enabled. These inputs are captured whether the interrupts are enabled or not. In other words, if an input occurs after the one that was captured, it will be captured and the original saved value will be lost. Therefore, the very first operation in the interrupt service routine for an input capture should be to save the contents of the input capture register. Then subsequent inputs from bounces will not affect the measured time. Another problem can occur, however. Whenever an input occurs on an input capture line, not only are the time data captured, but also, the appropriate input capture interrupt flag is set. If this flag is set, an interrupt will be requested whenever the input capture interrupt is enabled. Therefore, in the debounce interrupt routine, the input capture interrupt flag for the channel being used should also be reset.

We still must control the time that the input capture is disabled. How this control is implemented depends on the system. If the system is not busy, one might calculate a value that equals the contents of the TCR plus the number of timer counter ticks in the debounce time. This value could then be compared with the contents of the TCR, and when the TCR equals the calculated value, the input capture interrupt would be reenabled, and control returned to the interrupted program. This approach provides an accurate debounce time, but the processor is devoted entirely to the measurement of this time during the debounce time. It is not wise to tie up a microcontroller for milliseconds at a time and lock out other important actions that might take place during that time.

If all other events that are occurring within the microcontroller are interrupt driven, the programmer could re-enable the system interrupts prior to entering the delay time. This approach would at least keep the processor available for other asynchronous events that might occur during the debounce period. Yet another approach would be to disable the input capture interrupt and within the applications program, provide a time measurement that would have to expire before the input capture interrupt is reenabled. Both of these methods can be implemented without the use of an output compare. If an output compare is available, it could be used to execute the debounce timeout. This output compare would be set up in the input capture interrupt service routine. Also within the input capture interrupt service routine, the input capture interrupt would be disabled. The output compare interrupt service routine would disable the output compare operation and reenable the input capture. This approach is by far the best, and will be used here. The code for this method of debounce is included in the listing shown in Listing 5-6. In this case, a fixed debounce time is used. We will see the variable debounce time in a later program.

A beginning program or framework from which to build this application is shown in Listing 5-6. All this program contains is the setup of the interrupts and the interrupt service routines. Input capture 1 will serve as input from the motor shaft encoder. The motor will be controlled by a PWM signal. This signal will be filtered to provide a DC signal that can drive the motor. As with the MC68HC05 programs, this code is broken into three basic parts: 1) the initialization section, 2) the applications section, and 3) the asynchronous service section. In this case, there is no applications section, so it is designated by a FOREVER command followed by an empty block.

The program begins with the inclusion of the HC11E9.H header file. Immediately following this entry are the prototype entries for all of the interrupt service routines. The declarations of the global variables follows. The first set of variables are all associated with the input capture, and the second line of variables controls the output compare portion of the program.

Notice that the bit set-up instructions are all grouped so that the bits from each register are set in the same location. It is not necessary to group these instructions; however, if they are grouped by register as is done here, the compiler will use a single bit manipulation instruction to set all of the bits in each register. The first 8-bit manipulation instructions in the following code will require only six assembly instructions as they are grouped.

```
#include "hc11e9.h"

#define MS3_DEBOUNCE 1500
#define PERIOD 0X1000
#define TIME_ON 0x0800

@port void IC1_Isr(void);
@port void OC2_Isr(void);
@port void OC3_Isr(void);

WORD measured_period, time1, delpc;
WORD PWM_period=PERIOD, time_on=TIME_ON;

main()
{
    TCTL2.EDG1B=ON;/* capture falling edge only */
```

```
   OC1M.OC1M7=ON; /* sent OC1 tout to PA7 */
   OC1M.OC1M5=ON; /* couple OC1 to OC3 */
   TMSK1.OC3I=ON; /* enable the OC3 interrupt */
   TMSK1.IC1I=ON; /* enable the IC1 interrupt */
   OC1D.OC1D5=ON; /* turn on OC3 when OC1 occurs */
   TCTL1.OL3=ON;  /* toggle OC3 when OC3 occurs */
   PACTL.DDRA7=ON; /* make OC1 an output to PA7 */
   TOC1=TCNT+PWM_period; /* set OC1 to the period */
   TOC3=TOC1+time_on; /* set OC3 time on */
   cli();          /* enable the system interrupts */

   FOREVER
   {
           /* put application code here */
   }
}

@port void IC1_Isr( void)
{
   time1=TIC1;
   TFLG1=IC1F; /* reset IC1 interrupt flag */
   measured_period=TIC1-time1;
   TOC2=TCNT+MS3_DEBOUNCE; /* 3 ms debounce time */
   TMSK1.IC1I=OFF; /* disable IC1 interrupt */
   TMSK1.OC2I=ON; /* enable OC2 interrupt */
}

@port void OC2_Isr(void)
{
   TFLG1=IC1F|OC2F; /*reset IC1 & OC2 interrupt flag */
   TMSK1.OC2I=OFF; /* disable OC2 interrupt */
   TMSK1.IC1I=ON; /* enable IC1 interrupt */
}

@port void OC3_Isr( void)
{
   TFLG1=OC1F; /* reset OC1 interrupt flag */
   TOC1+=PWM_period;
   OC1D.OC1D7 ^=ON;
```

```
TFLG1=OC3F; /* reset OC3 interrupt flag */
TOC3=TOC1+time_on;
}
```

Listing 5-6: Motor Control Program Framework

The interrupt service routine for IC1 resets the IC1F bit in the TFLG1 register. Then the elapsed time from the last input capture time is calculated. The result of this calculation is saved in period1 which is the current elapsed time required for the motor to rotate one revolution. At the end of this calculation, the contents of the input capture register is saved in time1 to be used as the old time in the next calculation.

The PWM output from OC3 will be used to drive motor 1 whose shaft sensor is fed into IC1. The value in speed will be the desired speed of the motor in revolutions per minute. The range of this number will be 1000<=RPM<=10000, and it will have to be put in place through a debugger for this test. This range of motor speed was chosen to satisfy the speed of a small DC motor used in the final system. Let us plan a servo type operation where the motor is driven by a feedback loop that is controlled by the speed error. Therefore, within the applications loop, there must be code to check and control the PWM signal to force the speed1 parameter to match the input speed parameter.

The input speed is measured in RPM, and the values measured by the microcontroller are all times. A question that should be considered is whether the number placed in speed1 will actually be times or should this number be RPM. To calculate the time, note that there is one interrupt per revolution of the shaft and one minute is 60 seconds, so RPM/60 will be revolutions per second. What we need is the number of bus cycles in the time corresponding to this period. With a prescaler count of four, the clocking speed of the TCNT register is one count every two microseconds, or 500,000 counts per second. Therefore, the conversion from RPM to time for a revolution of the shaft in bus cycle counts is

$$Counts = \frac{500000*60}{RPM} = \frac{30000000}{RPM}$$

At 3000 RPM, the count value will be 10000 while at 16000 RPM the count value will be 1875. These numeric values can be handled by the timer system in the MC68HC11.

The above expression can be used to calculate RPM from the count value as

$$RPM = \frac{30000000}{Counts}$$

Neither calculation is convenient in a microcontroller. The numerator is larger than an `int` but smaller than a `long`. Both `Counts` and RPM have a range that can be contained in an `int`. One would expect that the conversion to time should be done on the input speed. That way, the division would be done only once for each speed setting. If the measured times were converted to RPM each time a time were measured, the computer program would be loaded with complicated divisions in its real time application portion. But let's examine the nature of the error with the different types of measures. Suppose the input RPM were converted to time and the measurements were based on time. Then the error signal, which is the difference between the measured value and the desired value, would have a wrong sense. That is if the motor is moving too slowly, then the error signal would be negative, which would cause the motor to go even slower. If the calculation were done based on RPM and the motor were going too slow, the error signal would be positive. In this case, the control would drive the motor faster which is the needed correction.

The difficulty with the time-based system could be solved by using the negative value of the error signal for the feedback control. A simple analysis will show the potential problem with this approach. Suppose that we work with two counts, C_d and C_m. Let us use the value K for the constant in the above equation. Therefore,

$$e = C_d - C_m = K\left(\frac{1}{RPM_d} - \frac{1}{RPM_m} \right)$$

The error signal is seen to be

$$e = K\left(\frac{RPM_m - RPM_d}{RPM_d * RPM_m} \right)$$

When the two speeds are nearly the same, this expression is very nearly proporstional to the difference between the two speeds. However, when one of the speeds deviates significantly from the other, it will cause the resultant error signal to be less sensitive to difference than would be expected. Also, there is the problem that occurs if the motor is stopped and the error signal in that case is undefined.

If the counts are converted to RPM prior to calculation of the error signal, the reduction of sensitivity for large errors would not be a problem. The problem that occurs when the motor is stopped still exists. In this case, no input capture would occur when the motor is not rotating, and the count value would be undefined. There seems to be no clear-cut reason to choose the time or the velocity measurement. Each has advantages and each has drawbacks. The drawbacks are about the same for each, so we will choose the case that has the simplest program. Therefore, the time-based or count-based system will be used.

Now comes the interesting problem of the calculation of 3000000/RPM. Each time the desired speed of the motor is input to the system, this calculation must be completed. The straightforward calculation of this value will yield code as follows:

```
#include "hc11e9.h"

WORD count(WORD RPM)
{
   unsigned long num = 30000000;

   return num/RPM;
}
```

The listing file of the compiled version of this function is

```
1 ; Compilateur C pour MC68HC11 (COSMIC-France)
2        .include"macro.h11"
3  .list +
4  .psect _text
5 ; 1 #include "hc11e9.h"
6  .psect _data
7 _Register_Set:
8 0000 1000      .word 4096
9 ;  2
10 ;  3 WORD count(WORD RPM)
11 ;  4 {
12     .psect _text
13 _count:
14 0000 BD0000          jsr    c_kents
```

```
15 0003 0C .byte 12
16 .set    OFST=12
17 ;  5 unsigned long num = 120000000;
18 0004 CC0E00 ldd #3584
19 0007 ED0A std OFST-2,x
20 0009 CC0727 ldd #1831
21 000C ED08 std OFST-4,x
22 ; 6
23 ; 7 return num/RPM;
24 000E EC0C ldd OFST+0,x
25 0010 6F02 clr 2,x
26 0012 6F03 clr 3,x
27 0014 ED06 std OFST-6,x
28 0016 EC02   ldd 2,x
29 0018 ED04   std OFST-8,x
30 001A EC08   ldd OFST-4,x
31 001C ED02   std 2,x
32 001E EC00   ldd 0,x
33 0020 C3FFF7     addd #-9
34 0023 188F   xgdy
35 0025 EC0A   ldd OFST-2,x
36 0027 BD0000     jsr c_ludv
37 002A AE00   lds 0,x
38 002C 38   pulx
39 002D 39   rts
40 ; 8 }
41 ; 9
42 .public _count
43 .public _Register_Set
44 .external    c_kents
45 .external    c_ludv
46 .end
```

This function requires 0x2d bytes (45 bytes) of code and that does not count the functions c_ludv and c_kents that must be linked to this function. Our innocuous little one-line piece of code creates a rather formidable piece of assembly code. If at all possible, it would be desirable to shorten this function. A slight modification of the above function code is

```
#include "hc11e9.h"

WORD count(WORD RPM)
{
    return 300000001u/RPM;
}
```

The compiled version of this program is

```
1 ; Compilateur C pour MC68HC11 (COSMIC-France)
2       .include"macro.h11"
3     .list +
4     .psect _text
5 ; 1 #include "hc11e9.h"
6     .psect   _data
7 _Register_Set:
8 0000 1000     .word 4096
9 ; 2
10 ; 3 WORD count(WORD RPM)
11 ; 4 {
12       .psect _text
13   _count:
14 0000 BD0000       jsr   c_kents
15 0003 08   .byte 8
16       .set OFST=8
17 ; 5 return 1200000000lu/RPM;
18 0004 EC08     ldd OFST+0,x
19 0006 6F02     clr 2,x
20 0008 6F03     clr 3,x
21 000A ED06     std OFST-2,x
22 000C EC02     ldd 2,x
23 000E ED04     std OFST-4,x
24 0010 CC0727   ldd     #1831
25 0013 ED02     std 2,x
26 0015 EC00     ldd 0,x
27 0017 C3FFFB       addd #-5
28 001A 188F     xgdy
29 001C CC0E00       ldd #3584
30 001F BD0000       jsr   c_ludv
```

```
31 0022 AE00    lds 0,x
32 0024 38      pulx
33 0025 39      rts
34 ; 6 }
35 ; 7
36 .public _count
37 .public _Register_Set
38 .external    c_kents
39 .external    c_ludv
40 .end
```

Note that this version differs from the first only in that the number 30000000 is not created in a declaration, but rather is created in line when it is needed. This version requires 37 bytes of code.

The purpose of this exercise is to demonstrate that there are usually many different ways that any part of a program can be approached. If the programmer grabs the first idea and plods through without any careful examination of the code being generated by the compiler, the program will almost always suffer. The assembly listings of programs should be examined, and where it seems as if the compiler is creating clumsy code, a new approach should be considered. Writing C code for a microcontroller is a joint exercise by the programmer and the compiler to create the most efficient overall program.

Let's return to the problem: we wish to set the speed of a motor and the measurable control signal is the time of a revolution. Most servo type devices work to position an output. Ours must set a speed which is a different problem from most. The input will be an RPM which will be converted to a time. Our system must compare the desired time with the measured time and correct the speed to make the two times match. A signal to drive the motor will be generated based on the desired time, and this signal will be adjusted by the error signal calculated as the difference between the measured time and the desired time. The following pseudocode sequence will accomplish the desired operation.

```
FOREVER
{
   do
      convert to time;
      calculate the required PWM output count;
      apply calculated time error to PWM count;
```

```
read in the motor speed;
while ( motor speed does not change);
}
```

For the moment, let us defer the problem of getting the desired motor speed. This value will be converted to time by use of the count() routine and will be saved in motor_period. Next, we need to calculate the required motor voltage. For this calculation, let us revert to an experimental technique that can be used usefully. An open loop system was set up to test the operation of the motor system. Several input values were entered, and the performance of the motor was recorded. Table 5-2 shows the result of these measurements.

PWM_Count	PWM_Count	icap period	period ms	voltage	RPM
Hex	decimal	counts			
0x700	1792	15812	31.62	2.186	1897
0x780	1920	11405	22.81	2.341	2630
0x800	2048	8581	17.17	2.498	3494
0880	2176	6841	13.68	2.653	4385
0x900	2304	5616	11.23	2.809	5342
0x980	2432	4759	9.52	2.964	6304
0xa00	2560	4059	8.12	3.118	7391
0xa80	2688	3530	7.06	3.273	8499
0xb00	2816	3085	6.17	3.424	9724
0xb80	2944	2736	5.47	3.578	10965

Table 5-2: Motor Performance Measurements

The data for this table were gathered by the use of an M68HC11EVM and a simple motor driver. This driver is a demonstration board provided by Motorola to show the use of the MC33033 pulse width motor driver and the MPM3002 FET motor driver H bridge. A small DC motor with a speed range of 1000 to 11000 rpm was mounted on the board, and the appropriate circuitry was added to interface the board to the computer. This interface consisted of a simple RC integrator to receive the PWM signal from the computer board and a circuit to measure the rotation of the motor. This latter circuit was quite crude. A magnetic reed switch was mounted near the motor shaft and a magnet cemented to the shaft. For each rotation of the shaft, the reed switch would close and open one time. A resistor was connected between the 12 volts of the motor driver and one side of the switch; the other side of the switch was grounded. Therefore, during each rotation the top side of the switch would jump from 12 volts to 0 volts and back. This voltage is larger than the maximum input for the microcontroller, so the 12 volt square wave was clamped to a maximum high value with a resistor and a diode connected to the +5 volt power supply of the EVM11 board. The schematic diagram for this test setup is shown in Figure 5-2 and a photograph of the complete system is shown in Figure 5-3.

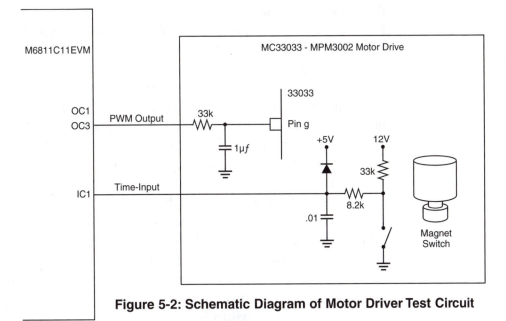

Figure 5-2: Schematic Diagram of Motor Driver Test Circuit

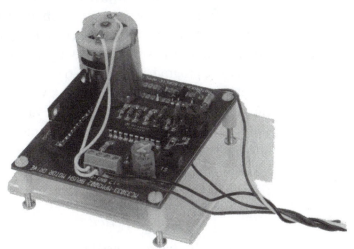

**Figure 5-3: Photograph Of
Assembled Motor Driver Test Circuit.**

The program used to obtain the data in the above table was a slight modification of the motor control program listing given in Listing 5-6. The only change in this program was that the function `@port void IC1_Isr(void)` from Listing 5-7 was used. This routine stores in the location `measured_period` a value equal to the average of the last eight measured periods.

For several reasons, a reed switch is probably the worst example of an input sensor that you can find for a microcontroller. I chose this input because it is the poorest, and the result has been satisfactory. With reed switches, there are serious bounce problems on both the rising and falling edges of the signals. The life expectancy for a reed switch is relatively small, about ten million closures. During the course of preparing this text, I broke one reed switch and had probably a total of several hours of service out of the two used for these experiments. In practice, it is recommended that rotational measurements be made with optical interrupters. Do not, however, think that an optical interrupter is free from bounce problems. An optical interrupter will exhibit both bounce and also a relatively slow rise time. In such a case, it is necessary to use a software debounce technique and also an external Schmidt trigger circuit to create a quick and stable rise time that can be captured by the microcontroller.

The data for the above table were measured on the circuit shown in Figure 5-2. It was found that the PWM to voltage conversion was very accurate and repeatable with many changes in the system. For example, the prescaler value was changed from the value of 4 discussed above to 2 and up to 16. No difference could be noted in the PWM to voltage conversion over this range. Also, the input capture values were quite stable. The program in Listing 5-6 was compiled and used to complete the measurements. The `PWM_count` value was changed manually over the range of values shown above. The value for icap period is the value found in `measured_period` in the program. The voltage measured was the integrated value that was sent into the input of the MC33033 motor driver. The RPM was calculated as

$$RPM = \frac{60000}{measured\_period}$$

Data from this table are used to create an equation that relates RPM to the `PWM_count`. This equation will be used in the program to close the loop and regulate the motor speed. The relation that can be derived from the above data is

$$\text{PWM\_count} = \frac{RPM + 12528}{7.875}$$

This expression is accurate to about 1% across the range of interest. The measured data is the input capture count, which we will call p. What we must do is to calculate a new value for the PWM_count from a measured p. An error in p is called p and this expression means a change in p. The PWM_count will be designated as pc and also means a change in pc. We shall now attempt to arrive at an equation that expresses the necessary change in pc to correct for an error in p. The error in p is the difference between the desired cycle time and the measured cycle time. RPM in the above expression is converted to cycle time p by noting that the RPM is 60000 / shaft cycle time in milliseconds, and that the clock is ticking 500 times per millisecond. Therefore, we have

$$p_c = \frac{1}{7.875}\left(\frac{12528 + 60000 * 500}{p}\right)$$

If there is a error in the cycle time say, Δp, then the change in pc, Δpc, is seen to be

$$p_c + \Delta p_c = 1591 + \frac{3809500}{p + \Delta p}$$

This expression, after some simplification and approximation reduces to

$$\Delta p_c = -3809500 + \frac{\Delta p}{p^2} \qquad\qquad (5\text{-}1)$$

This value calculated by the above expression will be used to change the PWM_count that drives the motor. The data contained in measured_period are placed there by the input capture interrupt service routine. It is assumed that the data there are always current.

The calculation to determine the PWM_count correction is substantial. We shall use this approach here to determine the correction, but recognize that there is another approach that probably requires less total calculation. This approach is to use a lookup table.

We will see the look-up table in some detail in Chapter 6.

We will probably run into overflow problems if care is not exercised in the calculation of the correction Δp_c. In Equation 5-1 above, p is the time of a shaft rotation in milliseconds. This number can vary from 1000 to 20000 depending on the motor speed. The best approach we could use is to divide the constant -3809500 by the value of p. Then multiply the result by Δp and finally divide the result by p a second time. This approach will minimize the overflow problems. In the expression below, p is the `measured_period`, pc is the `PWM_count`, and Δp is the calculated difference `motor_period - measured_period`.

```
FOREVER
{
    if(old_motor_speed!=motor_speed)
    {
    motor_period=300000001u/motor_speed;
    old_motor_speed=motor_speed;
    PWM_count= ((motor_speed+12528)/63)*8;
    PWM_count=limit(PWM_count);
    delpc=(3809500/motor_period);
    delpc=delpc*(motor_period-measured_period)/p;
    PWM_count -= delpc;
    /* read in the motor speed; */
    }
}
```

Recall that division by 7.875 is needed in the calculation of `PWM_Count`. This floating-point operation is avoided by dividing the expression by 63/8. Most often, floating point operations can be approximated by integer operations with sufficient accuracy.

It was mentioned earlier that it is possible to extend the maximum measurable time to something greater than the time required to clock the timer counter register 65535 times. If the longest time exceeds this value, the timer overflow interrupt can be used to an advantage. Suppose that you want to measure a long time with input capture 1. When this measurement is started, the value in the `TCNT` will be saved, and the timer overflow interrupt will be enabled. Also a counter will be reset to zero. In the timer overflow interrupt service routine, the counter will be incremented. Eventually, an input capture will occur, and the time between inputs is calculated as the time remaining prior to the first TOI, plus 65535 times the number of TOIs that have occurred, plus the time indicated in the input capture register. With a scheme of this nature, there is essentially no limit to the length of time that can be measured accurately. The resolution of this measurement is the time increment of the prescaler output.

A velocity servo is a particularly difficult application. Remember that velocity or motor speed is the derivative of position. Any derivative operation is prone to noise, and the successful servo must be carefully put together. For example, the equation calculated above to determine the feedback term is essential to the proper operation of the system. An interesting challenge is to try to duplicate the operation of the system described here with a normal gain type feedback. Even though the program seems to be messy, this approach will work, and most other approaches give you a wildly hunting or oscillating system. The calculation of the proper feedback value will avoid many of these problems.

Another problem area will be found. If the above FOREVER loop is allowed to run without any time control, the computer will be calculating a new feedback value many times before the effect of an earlier calculation will be seen in the system performance. This operation again causes the system to become unstable, and usually the system never finds the final value but hunts over some wide range around it. This operation can be corrected by slowing the rate at which feedback corrections are calculated. In particular, several rates have been tested, and it was found that a stable, accurate system would be obtained with a new feedback value calculated each quarter of a second. This time control is shown in the listing below. The main application program is broken into two parts: the first part is the code that will be executed whenever the motor speed is changed and the second part is the code that will alter the PWM_count which is the feedback portion of the program. The first portion is executed under the control of an if statement

```
if(old_motor_speed!=motor_speed)
```

There is no means to change the motor speed in this program, and this change will be introduced in the next section. Therefore, it is expected that the portion of the code controlled by this if statement will execute the first time the program is executed, and it should never be executed again until the system is reset. The second if statement

```
if(tick)
```

will execute its control code each time tick is TRUE. `tick` is initialized to TRUE when it is created, and it is set to FALSE each time the `if (tick)` routine is entered in the application program. If you go to the end of Listing 5-7, you will find that `tick` is reset to TRUE about each quarter of a second in the PWM timer routine. This sequence will cause the feedback calculation to be executed about each quarter of a second.

The initialization portion of the program shown in Listing 5-7 is little changed from that of the one shown in Listing 5-6. As mentioned above, we have added an application section that contains the calculations to control the motor speed. Also, there is one simple function that is used by the application section.

```
#include "hc11e9.h"

#define DIVIDE_8_SHIFT        3
#define COUNT_8               8
#define COUNT_MAX             3300
#define COUNT_MIN             1600
#define COUNT_ONE_QUARTER     32
#define PERIOD                0X1000
#define TIME_ON               0x0800
#define IMPOSSIBLE            3500
#define TOO_LOW               100

@port void IC1_Isr(void);
@port void OC2_Isr(void);
@port void OC3_Isr(void);

long limit (long);
long measured_period,delpc;
WORD time1,time2, motor_period,
motor_speed=IMPOSSIBLE;
WORD old_motor_speed=TOO_LOW,rpm,mparray[COUNT_8];
long PWM_period=PERIOD, PWM_count=TIME_ON;
int tick=TRUE,count=0;

main()
{
```

```
/* The initialization portion of the program */
TCTL2.EDG1B=ON; /* capture falling edge only */
OC1M.OC1M7=ON; /* sent OC1 out to PA7 */
OC1M.OC1M5=ON; /* couple OC1 to OC3 */
TMSK1.OC3I=ON; /* enable the OC3 interrupt */
TMSK1.IC1I=ON; /* enable the IC1 interrupt */
OC1D.OC1D5=ON; /* turn on OC3 with OC1 */
TCTL1.OL3=ON; /* toggle OC3 when OC3 occurs */
PACTL.DDRA7=ON; /* make OC1 an output to PA7 */
TOC1=TCNT+PWM_period; /* set OC1 */
TOC3=TOC1+PWM_count; /* set OC3 time on */
cli(); /* enable the system interrupts */
/* the applications portion of the program */
FOREVER
{  /* All of the numbers used below are derived
        in the text */
   if(old_motor_speed!=motor_speed)
   {
      motor_period=30000000Lu/motor_speed;
      old_motor_speed=motor_speed;
      PWM_count= ((motor_speed+12528)/63)*8;
      PWM_count=limit(PWM_count);
   }
   if(tick)
   {
      tick=FALSE;
      delpc=(3809500Lu/motor_period);
      delpc=delpc*
         (motor_period-measured_period)/
         motor_period;
      PWM_count-=delpc;
      PWM_count=limit(PWM_count);
      rpm=30000000L/measured_period;
   }
   /* input the new motor speed */
   }
}

/* functions of the main program */
```

```
/* range of acceptable PWM_count is 1600 to 3300 */
long limit(long x)
{
   if(x<COUNT_MIN)
      return COUNT_MIN;
   else if(x>COUNT_MAX)
      return COUNT_MAX;
   else
      return x;
}


/* The asynchronous service portion of the program */
@port void IC1_Isr( void) /*the motor speed
                                  measurement*/
{
   static int i;
   int j;
   time2=TIC1;
   TFLG1=IC1F; /* reset IC1 interrupt flag */
   mparray[i]=time2-time1;
   if (++i==COUNT_8)
   {
      i=0;
      measured_period=0;
      for(j=0;j<COUNT_8;j++)
         measured_period += mparray[j];
      measured_period >>= DIVIDE_8_SHIFT;
   }
   time1=time2;
   TOC2=time2+5*measured_period/8; /*debounce time
                                  5/8 revolution */
   TMSK1.IC1I=OFF; /* disable IC1 interrupt */
   TMSK1.OC2I=ON; /* enable OC2 interrupt */
}


@port void OC2_Isr(void) /* the debounce isr */
{
   TFLG1=IC1F|OC2F; /* reset interrupt flags */
```

```
   TMSK1.OC2I=OFF; /* disable OC2 interrupt */
   TMSK1.IC1I=ON; /* enable IC1 interrupt */
}


@port void OC3_Isr( void) /* the PWM isr */
{
   TFLG1=OC1F; /* reset OC1 interrupt flag */
   TOC1+=PWM_period;
   OC1D.OC1D7 ^=ON;
   TFLG1=OC3F; /* reset OC3 interrupt flag */
   TOC3=TOC1+PWM_count;
   if(++count==COUNT_ONE_QUARTER)
   {
      count=0; /* enter the speed control */
      tick = TRUE; /* about each quarter of
                          of a second */

   }
}
```

Listing 5-7: Motor Control Program

In this program, dynamic debounce is used. In the function IC1_Isr() the debounce time is a calculated value rather than a mere 3 ms as was done in Listing 5- 6. Here the debounce lock-out time is five-eighths of the current measured period. This value goes beyond the half-way point in the input waveform enough to avoid any bounce that will occur during the rise time of the input signal. The remainder of this program is quite similar to that in Listing 5-6.

We have completed most of the application. The remaining parts of the program are to find a means to read into the microcontroller the speed that is desired for the motor, and to integrate and debug the whole program. Usually, the speed is established by some measurement being done with the microcontroller. Let us approach the problem a little differently. Instead of making the speed a result of a measurement, let's enter the speed into the part through the asynchronous serial port.

Serial Communications Interface (SCI)

The SCI is another of the very useful peripheral functions found on the MC68HC11 family of microcontrollers. In fact, the SCI is found on parts from all families of Motorola microcontrollers. The SCI performs the operation of a full function universal asynchronous receiver transmitter (UART). It can operate with most standard baud rates, and several nonstandard rates. Control of this peripheral is through the BAUD register, the SCSR (serial communications status register), and both SCCR1 and SCCR2 (serial communications control registers 1 and 2). There is one additional register called the serial communications data register. This register when written to is the transmit data register, and when read the receiver data register. Both the transmit and receive data registers are double buffered.

The BAUD register contains two test bits and two divide control registers that set the baud rate for the SCI system. The bits SCP control the baud rate prescaler. This prescaler has the unlikely prescale value of divide-by-13 when both bits are set. This value sets the maximum baud rate to 9600 when the system is driven by an 8 MHz clock. With this prescaler value, all of the standard baud rates from 9600 down to 150 are available to the SCI. The second divider is comprised of SCR. The bit patterns in these bits will select any power of two divisors between 1—2^0—and 128—2^7. For our system, we will choose 9600 baud, which is sufficiently slow that there should be few communications problems. The code to set up this baud rate is as follows:

```
BAUD.SCP=3;
BAUD.SCR=0;
```

These lines of code must be added to the initialization portion of the program to set up the baud rate to 9600 baud for the SCI system.

We will use the SCI system to read in the required motor speed for the program started in the preceding section. Such a system must be able to read in data through the SCI, but it must also echo the same data back to the originating source to provide full duplex operation. Therefore, both serial transmit and receive must be implemented. With the above system, there is no serious need to implement an interrupt driven system. The computer is clearly not being used to its capacity, so we will add the SCI input and output sequence to the applications portion of the code and feel safe that no input data will ever be lost because of an overrun error. (An overrun error occurs when a new input overwrites an old input before the old input is processed.) Therefore, none of the bits in SCCR1 need be changed from their reset value. In the SCCR2 register, there are two bits, transmit enable and receive enable, that must be set to enable the SCI. The following two lines of code will enable these bits.

```
SCCR2.TE=1;
SCCR2.RE=1;
```

These two lines of code must also be added to the initialization portion of the program.

The code to read the data in is as follows:

```
if (SCCR.RDRF==1)  /* read in data if it is there */
{
    new_character=SCDR;/*get new byte and reset RDRF*/
    while(SCCR.TDRE==0);  /* wait until transmit
                          buffer empty */
    SCDR=new_character;   /* send out byte and reset
                          TDRE */
}
```

This sequence of code does quite a bit more than you might expect. Before this code sequence can be executed, both the RDRF and TDRE flags must be set and reset. The RDRF is set when a character has been received by the SCI, and the TDRE is set when the SCDR is empty and can receive a character to send. Both of these bits are reset by the sequential read of SCSR with the bit set followed by a read for the RDRF or a write for the TDRE to the SCDR register. The above sequence accomplishes all of the proper bit resets, transmits the newly received character back to the sender, and leaves the new character in the location new_character . After receiving the data, the following sequence converts it into a binary number to be processed by the remainder of the program.

```
if (SCSR.RDRF==1) /* read in data if there */
{
    new_character=SCDR;/*get new byte and reset RDRF*/
    while(SCSR.TDRE==0); /* wait until transmit
                          buffer empty */
    SCDR=new_character;   /* send out byte and reset
                          TDRE */
    /* if a number, process it */
    if(new_character>='0' && new_character <='9')
```

```
    new_speed = 10*new_speed + new_character-'0';
else if(new_character=='\r')
{
/* reject any number out of range */
/* and start over again */
  if(new_speed>=1000 && new_speed<=10000)
     motor_speed=new_speed;
  new_speed=0;
}
else
  new_speed=0;  /* reject everything else */
}
```

The variable new_speed is initialized to 0 outside of this portion of the program. If the input character is a number (a number is any character that lies between '0' and '9'), it is converted from a character to a number when the character '0' is subtracted from it. This value is added to ten times the value stored in new_speed, and the result is saved in new_speed. Repeated executions of this code sequence will convert a series of ASCII character numbers to an appropriate unsigned integer value. The entry sequence is terminated when a nonnumber character is received. If this character is a line terminator (a carriage return escape character in this case), the new speed value is put into motor_speed to cause the motor to change to the new value, and new_speed is reset to zero to await for the next input. If any other character is received, the data saved in new_speed is lost, and the whole entry of a new speed into the system must be repeated from the beginning.

In Listing 5-7 there was a lonely line of code

```
/* input the new motor speed */
```

That line has been replaced by the above and entered into Listing 5-8. The new motor speed input is placed in the main loop of the applications program, so a test is made each time through the loop to determine if there is a new input from the serial port. This routine also sends out the current speed of the motor once each half second. This output is controlled by the parameters tick1 and in_process. tick1 is set to TRUE each half second, and if in_process is FALSE the motor speed is calculated and sent to the serial port output.

This sequence repeats each half second until `in_process` is set to TRUE. Whenever a character is read in through the serial port, `in_process` is set to TRUE to disable the continual output from the system while a new input speed value is being entered. If there is a new character in the input buffer, this character is read in and the RDRF flag is reset. The new character is immediately echoed back to the sending device as part of full duplex operation. If the new character is a digit it is processed, if it is a line feed it is also processed, and any number string is converted to an integer so that it can be used by the computer. The size of the number is tested to be certain that it is within the acceptable speed limits for the motor, and if it passes all of the tests, it is placed in the variable location `motor_speed` to indicate that a new motor speed is here and should be processed.

```
#include "hc11e9.h"

#define DIVIDE_8_SHIFT        3
#define COUNT_8               8
#define COUNT_MAX             3300
#define COUNT_MIN             1600
#define COUNT_ONE_QUARTER     32
#define COUNT_ONE_SECOND      128
#define PERIOD                0X1000
#define TIME_ON               0x0800
#define IMPOSSIBLE            3500
#define TOO_LOW               100
#define RPM_MIN               1000
#define RPM_MAX               11000
#define CR                    0x0d
#define LF                    0x0a


/*function prototypes */
@port void IC1_Isr(void);
@port void OC2_Isr(void);
@port void OC3_Isr(void);
int putchar(char new_character);
void dprint (unsigned int c);
void do_crlf(void);
long limit(long );
```

```
/* external variable definitions */
long measured_period,delpc;
WORD time1,time2,
motor_period,motor_speed=IMPOSSIBLE;
WORD new_speed;
WORD old_motor_speed=TOO_LOW,rpm,mparray[COUNT_8];
long PWM_period=PERIOD, PWM_count=TIME_ON;
int new_character,tick=TRUE,count=0,in_process;
int tick1=TRUE,count1=0;
WORD got_new=TRUE;
main()
{
   /* The initialization portion of the program */

   TCTL2.EDG1B=ON;/* capture falling edge only */
   OC1M.OC1M7=ON;  /* sent OC1 out to PA7 */
   OC1M.OC1M5=ON;  /* couple OC1 to OC3 */
   TMSK1.OC3I=ON;  /* enable the OC3 interrupt */
   TMSK1.IC1I=ON;  /* enable the IC1 interrupt */
   OC1D.OC1D5=ON;  /* turn on OC3 when OC1 occurs */
   TCTL1.OL3=ON;   /* toggle OC3 when OC3 occurs */
   PACTL.DDRA7=ON;/* make OC1 an output to PA7 */
   TOC1=TCNT+PWM_period; /* set OC1 to the period*/
   TOC3=TOC1+PWM_count; /* set OC3 time on */

   BAUD.SCP=3;     /* set up the SCI */
   BAUD.SCR=0;     /* 9600 baud */
   SCCR2.TE=ON;
   SCCR2.RE=ON;
   cli();          /* enable the system interrupts */
   /* the applications portion of the program */

   FOREVER
   {
      if(old_motor_speed!=motor_speed)
         {  /* All of the numbers used below are
                 derived in the text */
         motor_period=300000001u/motor_speed;
         old_motor_speed=motor_speed;
```

```
      PWM_count= ((motor_speed+12528)/63)*8;
      PWM_count=limit(PWM_count);
   }
   if(tick)
   {
      tick=FALSE;
      delpc=(38095001u/motor_period);
      delpc=delpc*(motor_period-
         measured_period)/motor_period;
      PWM_count -=delpc;
      PWM_count=limit(PWM_count);
      rpm=30000000L/measured_period;
   }
/* input the new motor speed */
/* Send out the measured RPM to the terminal
   periodically */
   if(tick1&&!in_process)
   {
      tick1=FALSE;
      rpm=30000000L/measured_period;
      dprint(rpm);
      do_crlf();
   }

   if (SCSR.RDRF==ON)  /* read in data if it is
                          there */
   {
      in_process=TRUE;
      new_character=SCDR;  /* get new byte and
                              save it */
      while(SCSR.TDRE==OFF);  /* wait until transmit
                                 buffer is empty */
      SCDR=new_character;  /* send out byte and
                              reset TDRE */
      /* got a number, process it */
      if(new_character>='0' && new_character <='9')
      new_speed = 10*new_speed + new_character-'0';
      else if(new_character=='\r')
```

```
        {
          do_crlf();
          /*reject any number out of range */
          /* and start over again */
          if(new_speed>=RPM_MIN && new_speed <=RPM_MAX)
          {
            got_new=TRUE;
            in_process=FALSE;
            motor_speed=new_speed;
            new_speed=0;
          }
        }
    else
          new_speed=0; /* reject everything else */
      }
    }
}


/* functions of the main program */
/* the range of acceptable PWM_count is 1600 to
                   3300 */

long limit(long x)
{
   if(x<COUNT_MIN)
   return COUNT_MIN;
   else if(x>COUNT_MAX)
   return COUNT_MAX;
   else
   return x;
}


/* send out a single carriage return-line feed
                  sequence */

void do_crlf(void)
{
```

```
      while(SCSR.TDRE==OFF);
      SCDR=CR;
      while(SCSR.TDRE==OFF);
      SCDR=LF;
}


/* send a single character to the serial port */

int putchar(char new_character)
{
   while(SCSR.TDRE==OFF);
/* wait until transmit buffer empty */
   SCDR=new_character;
/* send out byte and reset TDRE */
}


/* convert an integer and send it to the serial
   port as a string */

void dprint (unsigned int c)
{
if ( c/10) /* recursively determines if */
 dprint(c/10); /* a zero has been reached and then */
putchar(c%10+'0'); /* sends out the characters */
}


/* The asynchronous service portion of the program */

@port void IC1_Isr( void)/*the motor speed
                              measurement*/
{
   static int i;
   int j;
   time2=TIC1;
   TFLG1=IC1F; /* reset IC1 interrupt flag */
   mparray[i]=time2-time1;
   if (++i==COUNT_8)
   {
      i=0;
```

```
      measured_period=0;
      for(j=0;j<COUNT_8;j++)
         measured_period += mparray[j];
      measured_period >>= DIVIDE_8_SHIFT;
   }
   time1=time2;
   TOC2=time2+5*measured_period/8; /*debounce time
               Is 5/8 revolution */
   TMSK1.IC1I=OFF; /* disable IC1 interrupt */
   TMSK1.OC2I=ON; /* enable OC2 interrupt */
}

@port void OC2_Isr(void) /* the debounce isr */
{
   TFLG1=IC1F|OC2F; /* reset interrupt flags */
   TMSK1.OC2I=OFF;  /* disable OC2 interrupt */
   TMSK1.IC1I=ON;   /* enable IC1 interrupt */
}
@port void OC3_Isr( void) /* the PWM isr */

{
   TFLG1=OC1F;         /* reset OC1 interrupt flag */
   TOC1+=PWM_period;
   OC1D.OC1D7 ^=ON;
   TFLG1=OC3F;         /* reset OC3 interrupt flag */
   TOC3=TOC1+PWM_count;
   if(++count>=COUNT_ONE_QUARTER)
   {
      count=0;         /*enter speed control about */
      tick=TRUE;       /*each quarter of a second */
   }
   if(++count1==COUNT_ONE_SECOND)
   {
      count1=0;
         tick1=TRUE;
   }
}
```

Listing 5-8: The Completed Application

There are three input/output routines that have been written for this program. These routines, `putchar()`, `dprint()`, and `do_crlf()`, can be used with other systems with a serial input/output system. The Cosmic compiler does provide the usual I/O routines like `printf()`, `gets()`, `puts()`, etc. It does not provide a basic `putchar()` and `getchar()` which is used by all of these library routines. The reason that these routines are not provided by the compiler is the wide variety of what the programmer will want to implement with the built-in SCI ports on the MC68HC11 family. The `putchar()` shown above will work in most instances. The `dprint()` routine is a recursive routine that converts an integer into an ASCII string and sends it to the SCI port.

There is one final modification to the program. In the last lines of the PWM timer routine, count1 and tick1 are processed to set tick1 to be TRUE each second. This flag is then used to control the writing of the motor speeds to the terminal screen.

Summary

There has been no attempt to work all of the peripherals on the MC68HC11. The various peripherals are similar to those on the other parts that we have discussed in other chapters or will discuss later. We have seen several timer applications both in the MC68HC11 and in the MC68HC05. We will see other timer applications in the following chapters.

We have seen detailed use of the output compare timer subsystem to make a pulse width modulation digital-to-analog converter system. Depending on the program, the system allowed excellent performance in either short on times or maximum on times, but not both without the addition of a significant amount of code. We will see a system in the next chapter that provides excellent performance for both minimum and maximum on times. This performance is not a limitation of the MC68HC11, merely a limitation of the programs presented so far.

The input capture subsystem has been used to measure motor speed in a simple DC motor controller. This system used a primitive reed switch to measure the rotation of the motor shaft, and the performance of the switch was poor. A debouncing system was developed that prevented input captures to occur for a specified time after the first input was detected. This approach uses an output compare channel, but it does not tie up the microcontroller to wait out any delay times during the debounce period. This program is not too removed from many of those encountered in the real world.

The organization of the program is similar to how most applications can be programmed, and the way in which the program was developed showed how most problems should be approached. The problem was broken down into a set of small operations that could each be handled easily. These different parts of the program were developed, tested, and debugged separately. This approach keeps the development of the individual parts of the program manageable, and debugging is not too difficult. If the whole program were written and then debugging started, it would have been nearly impossible to separate out the effects of one part of the program on the others. The main interrupt service routines were written first and tested as well as possible by themselves. With these important functions behind us, it was easy to attack the applications portion of the program in which the closed loop system was implemented along with the management of the input/output through the serial port of the device.

The MC68HC11 is a powerful enough computer that it is possible to make an ANSI compliant compiler. Parameters can be passed to functions on the stack, and re-entrant or recursive functions can be written for this part as was demonstrated with the `dprint()` function. Remember, it is the microcontroller that limited the ANSI compliance with the MC68HC05—not the compiler.

Chapter 6

Large Microcontrollers

In this chapter we'll examine the programming of systems employing large microcontrollers. The realm of the large-microcontroller system is not all that different from the 8-bit systems. One of the main advantages of the use of a high-level language is that it keeps the nasty details of the underlying computer hidden from the programmer. Usually, the programmer will not see much difference between the code for different types of computers. The fallacy to this idea is that when programming any microcontroller, the programmer must know about and use all of the on-board peripherals found on the microcontroller. These peripherals will vary from machine to machine, and how they are accessed will differ from device to device. In this chapter, however, we are going to see an application where a substantial amount of assembly language is required. The chip that we are going to use here has a Digital Signal Processor section. This processor is accessed through special core chip registers and the core condition code register. The abstract machine that the compiler creates code for contains no registers. Therefore, the only access to these features is through assembly language. We will show how to create assembly language functions that can be accessed from your C program.

The part that we will use for the 16-bit discussions is the Motorola MC68HC16 family of components. These are similar to the MC68HC11 in many ways, but there are important differences. First, there are some new registers that must be programmed directly to make the MC68HC16 work as desired. (The only register in the MC68HC11 that must receive special assembly instructions is the condition code register.) Also, the MC68HC16 does not have automatic stacking of its registers when an exception occurs, unlike the MC68HC11. Therefore, all interrupt service routines must begin with code that saves the status of the machine before the normal interrupt operations can proceed. We will see a few other differences between these devices, but the main difference is in the way peripherals are handled.

On the MC68HC16, the core processor is called the CPU16. This central computer is interfaced to an internal bus called the inter-modual bus (IMB). The IMB is very much like the bus a hardware designer would put onto an external circuit board. It has address and data busses and all of the necessary control signals to control any peripheral that the microcontroller might have applied. In the standard device, the MC68HC16Z1, there are five built-in peripherals: the system integration module (SIM), the analog-to-digital converter (ADC), the queued serial peripheral interface module (QSPI), the general-purpose timer module (GPT), and the static random access memory module (SRAM). These modules and others can be added or deleted in future components as the customer needs dictate.

The several internal peripheral modules are each interfaced to the IMB. Each module has a specific set of registers that are located at an address relative to the base address of the module. These base addresses are set at design time. This is interesting because the same modules are used on the MC68300 series of microcontrollers, which are 32-bit microcontrollers. Therefore, the material presented in this chapter is directly applicable to the MC68300 series of microcontrollers. The only difference that the programmer will see is that the base memory locations of the various peripheral modules are different, but even these differences disappear because of the careful design of the microcontrollers. We therefore can consider this chapter to be on large microcontrollers rather than on the 16-bit systems alone.

The MC68HC16

The MC68HC16 is a truly complicated device. However, for our purposes, it can be divided into its several components, and each component is somewhat as one would expect from a programming standpoint. Therefore, we will examine this family of parts as a collection of modules. Each module is rather straightforward. The core processor, which is known as the CPU16, is indeed a complete and competent microcomputer. The bulk of its complexity is hidden by the fact that the program is written in C. This section contains a brief description of the MC68HC16Z1 microcontroller that can be used to aid in writing programs for the component in C. There are several Motorola manuals that are useful adjuncts to this chapter. [1][2][3][4][5][6]

Copies of these manuals are all to be found on the attached CD-ROM. It is recommended that these manuals be reviewed prior to any attempt to write code for this family.

CPU16 Core Processor

The CPU16 represents an attempt to bridge the difference between the large 8-bit microcontrollers and the high-end components embodied in the MC68300 family. This processor is a 16-bit processor. As such, its instructions are each 16 bits wide, instructions are read from memory 16 bits at time, and the instructions are processed 16 bits at a time. On this particular part, a 20-bit address bus with additional control allows the program to access two individual one-megabyte address maps. The controls available distinguish between program memory and data memory.

It has been stated several times that the programmer's model for a microcontroller is not too important when writing code in C. To be able to access all of the features of a CPU16, you will have to use assembly language-based functions. There are certain features that are simply outside the concept of a high-level language. In these cases, it is recommended that functions which access and control these features be written and then you can call these functions from the C program. The main properties of the CPU16 that are not available to the C programmer are the DSP type registers found in the part. Programming of these registers is the subject of a later section of this chapter.

The 20-bit address space of the CPU16 is a significant deviation from the normal 16-bit address space of the MC68HC11. This change has been handled by the addition of several 4-bit extension registers for those registers that deal specifically with addresses. These registers are the program counter (PC), the stack pointer (SP), the three

---

[1] M68HC16 Family MC68HC16Z1 Users Manual MC68HC16Z1UM/AD
[2] M68HC16 Family CPU16 Central Processor Unit Reference Manual CPU16RM/AD
[3] Modular Microcontroller Family GPT General Purpose Timer Reference Manual GPTRM/AD
[4] Modular Microcontroller Family ADC Analog-to-Digital Converter Reference Manual ADCRM/AD
[5] Modular Microcontroller Family QSM Queued Serial Module Reference Manual QSMRM/AD
[6] Modular Microcontroller Family SIM System Integration Module Reference Manual SIMRM/AD

index registers (IX, IY, and IZ), and the EK register. All these registers except for the EK are 16 bits wide. The EK register is an extension register that is used with the extended addressing mode. Any extended address calculation will result in a 16-bit number. The result will be concatenated with the EK register to create a 20-bit address. (The EK register has nothing to do with the E accumulator found in the device.) The extension registers are named PK, SK, XK, etc. The extension registers are usually set during initialization of the device, and there is no need to change them during program operation. The content of an extension register is a page pointer. The pages in this case are each 65536 bytes long. Transition from one page to another is automatic. For example, if an address is calculated for a jmp which will pass program control to code in another page, the proper address will be calculated for both base and extension register automatically. The calculation will alter the contents of both the base and the extension register without programmer concern.

The MC68HC16 family currently has the several internal modules mentioned above, and it is planned that future versions of the part will have more modules. This complexity has suggested that the arrangement of the header file for the part be broken into several different files: One file for the main processor, and several different header files, one for each of the peripheral modules in the individual part. This approach is shown in the HEADER/HC16HEADERS directory on the CD-ROM. There you will find a header file named hc16.h along with files named adc.h, gpt.h, sim.h, sram.h, and qsm.h. When writing code for any specific module, you should include the hc16.h file along with the proper files for the peripheral portions needed. The hc16.h file must be included first because there are items defined in this file needed by the other headers.

One major difference between the large and small devices is the way the exception vector table is handled. Recall that, in the MC68HC11 family, all interrupt vectors are placed at the top of memory. With the MC68HC16 family, the vector table is contained within the first 512 bytes of memory. Upon reset, the CPU16 core processor reads the first four words of memory where it must find certain data for the operation of the program. The address 0 must contain a word whose least significant 12 bits are the contents of the ZK, the SK and the PK registers when the part comes out of reset. The address 2 contains the initial program counter, and the address 4 must contain the initial stack pointer. Finally the address 6 must contain the initial value of the IZ register. Note that the word addresses in the MC68HC16 are always even. In this machine you will find words on even boundaries, bytes anywhere, and long words on even boundaries. These data are loaded into this memory area by the use of a vector routine similar to that seen in Chapter 5 for the MC68HC11. With this approach, all of the registers needed to begin operation of the basic computer are loaded from memory at reset time. Of course, this operation does not eliminate the need for program initialization; it merely provides a mechanism by which the processor will start accessing memory at the correct addresses when the device comes out of reset.

Table 6-1 contains a listing of the uses of each entry in the vector table. Note that vectors 0x0 through 0x37 have assigned functions. The vectors 0x38 through 0xff are available for user-defined operations. Note the relationship between the vector number and the vector address. The vector address is always twice the vec-

Table 6-1: Exception Vector Table

Vector Number	Vector Address	Type of Exception
0	0x0000	Reset—Initial ZK, SK, PK
1	0x0002	Reset—Initial PC
2	0X0004	Reset—Initial SP
3	0X0006	Reset—Initial IZ
4	0X0008	Breakpoint
5	0X000A	Bus Error
6	0X000C	Software Interrupt
7	0X000E	Illegal Instruction
8	0X0010	Division by Zero
9-E	0X0012-0X001C	Unassigned, Reserved
F	0X001E	Uninitialized Interrupt
10	0X0020	Level 0 Autovector
11	0X0022	Level 1 Autovector
12	0X0024	Level 2 Autovector
13	0X0026	Level 3 Autovector
14	0X0028	Level 4 Autovector
15	0X002A	Level 5 Autovector
16	0X002C	Level 6 Autovector
17	0X002E	Level 7 Autovector
18	0X0030	Spurious Interrupt
19-37	0X0032-0X006E	Unassigned, Reserved
38-FF	0X0070-0X01FE	User-defined Interrupts

tor number. When programming the interrupt vectors in the several peripheral modules of the MC68HC16, the programmer will select the interrupt vector. When it comes time to place the interrupt service routine address in the proper memory location, it is to the vector address—NOT the vector number—that this value must be assigned.

When setting up the vector table, a wise programmer will fill all of the possible unused vector addresses with the address of a dummy function that provides an orderly return to the program in the event of an unexpected interrupt. Usually the first 0x18 or 24 vectors should be filled with this address. An example function that can be used for this type of operation is

```
static @port void _init_vector(void)
{ }
```

This program will compile to a single RTI (return from interrupt) that will return the program control to the location when the interrupt occurred. A static function is not used often. When a function is declared static, it can be seen only in the file in which it is defined.

Following is a listing of the routine vector.c. This program is modeled closely after that provided with the Cosmic MC68HC16 C compiler.

```
extern @far @port void _stext(void); /*startup routine */
extern @port void OC3_Isr(void); /* ISR address */
static @port void _init_vector(void);

static const struct reset {
@far @port void (*rst)(void); /* reset + code
extension */
unsigned short isp; /* initial stack pointer */
unsigned short dpp; /* direct page pointer */
@port void (*vector[252])(void); /* interrupt vectors */
} _reset = {
  _stext,       /* 1-start address */
  0x03fe,       /* 2-stack pointer */
  0x0000,       /* 3-page pointer */
  _init_vector, /* 4-Breakpoint */
  _init_vector, /* 5-Bus Error */
```

```
_init_vector, * 6-Software Interrupt */
_init_vector, /* 7-Illegal Instruction */
_init_vector, /* 8-Divide by Zero */
_init_vector, /* 9-Reserved */
_init_vector, /* a-Reserved */
_init_vector, /* b-Reserved */
_init_vector, /* c-Reserved */
_init_vector, /* d-Reserved */
_init_vector, /* e-Reserved */
_init_vector, /* f-Uninitialized Interrupt */
_init_vector, /* 10-Reserved */
_init_vector, /* 11-Level 1 Interrupt Autovector */
_init_vector, /* 12-Level 2 Interrupt Autovector */
_init_vector, /* 13-Level 3 Interrupt Autovector */
_init_vector, /* 14-Level 4 Interrupt Autovector */
_init_vector, /* 15-Level 5 Interrupt Autovector */
_init_vector, /* 16-Level 6 Interrupt Autovector */
_init_vector, /* 17-Level 7 Interrupt Autovector */
_init_vector, /* 18-Spurious Interrupt */
    /* vectors 0x19-0x37 unassigned, reserved */
0,0,0,0,0,0,0,
0,0,0,0,0,0,0,0,0,0,0,0,0,0,0,0,
0,0,0,0,0,0,0,0,0,0,0,0,0,0,0,0,
/* put timer at vector 0x46 0x40 from ICR and 6 for OC3 */
0,0,0,0,0,0,OC3_Isr /* OC3_Isr vector at 0x46 address at 0x8c */
};

static @port void _init_vector(void)
{ }
```

Listing 6-1: The Vector Initialization Routine `vector.c`

We have already seen the `@port` command in Chapter 5. The `@far` command is unique to the Cosmic compiler. This command notifies the compiler that the pointer associated with the command is not the usual 16-bit pointer. An `@far` pointer is an extended pointer. This 20-bit pointer will be placed into the first two memory words by the `vector.c` routine. The rightmost 4 bits in word 0 is the program counter extension PK, so that the placement of the 20-bit program counter in this location will provide proper initialization of the program counter.

In the example above, the interrupt service routine OC3_Isr is placed in the address 0x46. Unfortunately, there is no easy way to accomplish this placement. The structure address must be counted by hand and the pointer placed at the correct location. The structure members are initialized only up to the vector with the highest address. Note that the structure is global, so that its members will all be initialized to zero unless otherwise assigned. Notice also that the vectors up to vector number 0x18 are initialized. These vectors include all of the program-generated exceptions. These areas are where one would expect most of the problems to occur in debugging a program. The vector numbers 0x19 through 0xff are left initialized to 0. These vectors are all accessed by either the internal modules or from external interrupts. If there are no hardware problems with the system, it is unlikely that external devices will cause uncalled-for interrupts, and if one of the internal modules causes an interrupt with an uninitialized vector, the CPU16 will access the uninitialized interrupt vector.

One other problem can arise when debugging programs that involve user-specified interrupts. In the event that the isr address is improperly placed in the vector table, the program will become lost whenever the interrupt occurs. If such a program exhibits bizarre behavior, a good trick is to place a break point at the address 0. If the vector is wrong, an interrupt will take the proper vector which will contain a 0 and attempt to execute the code at the address 0. The break point at this location will stop execution and give you a clue as to the program error. A break point at _init_vector can also be useful to determine where the program is when an unexplained exception occurs.

If a compiler provides a mechanism that complies to the ANSI standard and one that does not, it is better to choose the ANSI standard mechanism rather than the nonstandard approach. For example, ANSI states nothing about how to establish a vector table. The above approach is but one of several that can be used to handle the placement of the vectors in the vector table. Another approach is to use the vector macro that is found in hc16.h . The disadvantage to the vector macro is that it can place a vector into RAM only. Often with embedded controls, it is necessary to place the vectors into ROM. In that case, the use of the routine vector.c is the best approach.

The use of the @port command is certainly not found in the ANSI standard. It is possible to create a complete interrupt service routine without the use of the @port. However, if C is used to the maximum, the isr will have some built-in inefficiencies. For example, the language has no direct register commands so it is necessary to create a function that saves the status of the computer on entry to the routine, and another to restore the machine status prior to the return from the interrupt. Also, a special function must be created to execute the RTI instruction at the end of the interrupt service routine. These assembly routines can be created as function calls and saved in the header file. When they are used, it is wise to study carefully the code generated by the compiler to make certain that there are no errors in the code. For example, if a function should happen to clear some space on the stack for local storage, the placement of an RTI instruction in the C code sequence would cause the return operation to be executed before the stack is restored. Such an error will cause serious problems in system performance if not corrected.

Both the @far and the @port commands are not in compliance with the ANSI standard. It is recommended that you use these commands sparingly because their use causes nonportable code to be generated with the compiler. So far in this text, we have used two significantly different compilers. Each compiler manufacturer claims that their compiler complies to ANSI. In the case of the C6805 compiler from Byte Craft, it probably conforms as closely as can be expected for such a primitive machine. Both the MC68HC11 and the MC68HC16 compilers comply more closely to the standard than the C6805. Both of these machines are so much more computer that one should expect very close compliance. Any extension to the basic language should be used with care. The above two commands are desirable and provide useful functions for the embedded control field.

The Cosmic compilers also have an extension that has not been used in this text. The way in which they define the internal registers to the machine is not standard, and it does not permit very efficient use of bit manipulation by the compiler. The approach used here is shown in the various header files written for the parts. The compiler writer assumes that the use of a construct like

```
#define ABLE (*(Register *) 0x2000)
```

is too complicated for most programmers to understand. The approach that they use is nonportable, and the latter approach is completely portable among ANSI compliant compilers. It is recommended that, even though the header files contain some code that might be difficult to explain, you should use the approach presented here to create code that is as portable as possible.

Some of the sections that follow will show how portable code can be used. Examples from both the MC68HC11 and the MC68HC05 will be used on the MC68HC16, and you will see that much of the code will be transferred with little change. Where would you expect changes? Recall the recommendation that each program be broken into three sections: the initialization section, the applications section, and the asynchronous service section. Each section will be subject to some change when moving from one machine to another, but the applications section will probably suffer little change and the other two sections will see the most changes when the code is moved. For example, you will see that the initialization of the MC68HC16 is somewhat different from that of the MC68HC11, but the interrupt service routines will be nearly the same. In fact, in some cases the isr for the two parts is identical. On the other hand, the way that things are handled on the MC68HC05 is so different that the initialization and isr will probably have to be completely rewritten when moving code to one of the larger machines. However, here the machine-independent portions of the applications routine can be moved with little change.

System Integration Module (SIM)

A brief examination of the names of most of the modules will reveal their use. There is one notable exception—what is a system integration module (SIM)? The SIM is sort of the interface between the IMB and the outside world. It is very useful, and contains much of the circuitry that a hardware designer would have to incorporate to make a computer out of a microprocessor. The object of the chip designer with the introduction of the SIM was to make it possible to use the MC68HC16 in a system with a minimum of external circuitry. This section is not to provide you with a complete description of the SIM. The SIM Reference Manual is a 200-page document, and I'm not intending to duplicate that manual here. The following paragraphs each contain a brief description of the several blocks found within the SIM. The set up and control of these blocks are all controlled by the registers described in the SIM book on the CD-ROM.

System Configuration and Protection

This module monitors many of the things that can go wrong with the operation of the MC68HC16. Internal and external signals can be generated that signal an error has occurred. Reset signals can be originated from several sources. The reset status monitor keeps track of the source of the latest reset to help with debug operations. The halt monitor responds to a HALT signal on the internal bus. This monitor, if properly enabled, can request a reset. The bus monitor and the spurious interrupt monitor can each request a bus error. The bus monitor responds primarily to an unanswered asynchronous bus transfer request. Such a sequence is usually the result of a program access to unimplemented memory. If properly enabled, the spurious interrupt monitor can initiate a bus error.

There are two time-based functions in the system configuration and protection section of the SIM. The first is a software watchdog timer. This timer requires that the program access a memory location with a code sequence. This access resets a timer. With a proper program, the special location in memory is accessed routinely in the normal execution of the program and the timer should never overflow. If it does, there is probably a system error that is corrected by a system reset. The periodic interrupt timer is used whenever a simple clocking sequence is needed. We will see use of this timer in a later section.

System Clock

The system clock provides timing signals for the IMB and all of the internal modules of the microcontroller. The time base for the microcontroller can be a 32768-kHz reference crystal , a 4.194-MHz crystal, or an external clock signal. If either of the crystal oscillators are used there is an internal phase-locked loop frequency multiplier that will multiply the operating frequency to the final system clock frequency.

External Bus Interface

The external bus interface transfers information between the IMB and external devices. This interface supports a 16-bit data bus, up to a

24-bit address bus, a three-line function control signal bus, control signals for dynamic bus sizing, and handshaking for external bus arbitration.

Interrupts

The CPU16 contains a three-wire, seven-level interrupt system. These interrupts are interfaced to the SIM through the IMB. The outside world is interfaced into the SIM as seven individual interrupt lines which are multiplexed onto the three-wire line within the SIM. There are also two sources of interrupt from within the SIM itself. These interrupts are from the periodic interrupt timer and the software watchdog timer.

Chip Selects

There are many signal lines on the microcontroller that might not be needed with a typical system. For example, address lines 20 through 23 all follow the condition of address line 19. Perhaps not all of the external interrupt lines are needed. Often the function control lines that can be used to decode the nature of a bus cycle—i.e., either data or program access—are not used. Altogether there are 12 signal lines that can be implemented as chip selects. This line will assert when there is an access within a memory range specified.

Reset and System Initialization

The microcontroller has several sources of reset. The reset and system initialization section directs the several resets to the proper operation, and records the source of the resets. Also, when the system is reset initially, the state of several pins on the system bus is analyzed to determine the mode of the part when it exits the reset sequence.

General Purpose I/O

There are 16 SIM pins that can be configured as general-purpose input/output signals. These ports are ports E and F, and the pins are all multiply assigned. Port E pins, for example, are bus control pins, and Port F pins have a second use as the external interrupt inputs to the system.

A Pulse Width Modulation Program

While the MC68HC16 has a built-in pulse width modulation (PWM) system, it is sometimes desired to achieve more flexible resolution than can be obtained with the built-in system. Therefore, it is not unreasonable that one would want to use the general-purpose timer to create a PWM. Prior to writing the code for a PWM in this manner, we should look at the basic time period of the processor. Unless otherwise directed, all of the timers in the GPT are driven by the system clock. This signal is passed through a prescaler controlled by the bits CPR2 through CPR0 in the register TMSK2. The default prescaler value will cause a clock rate to the GPT of the system clock divided by four.

The question now is the clock rate. Recall that the clock control is a portion of the SIM. Within the SIM there is a register named SYNCR. The clock frequency is controlled by the three bit fields named W, X, and Y in this register. When operating at a low crystal frequency—between 25 and 50 kHz—the formula for the system frequency is given by

$$F_s = F_r \left[4(y+1)\left(2^{2w+x}\right) \right]$$

where y has a value between 0 and 63, and both w and x can have values of 0 or 1. With a frequency f of 32767 and the W, X, and Y default values of 0, 0, and 63, respectively, the device will come out of reset with a system frequency of 8.388 MHz. To be able to get the finest resolution for our timing functions, let us plan to operate the system clock frequency at its maximum value by changing the value of X from its default value of 0 to 1. This change will cause the system frequency to be 16.776704 MHz. The frequency of the input into the GPT is one-fourth this value or 4.194176 MHz. The time period for this frequency is 238 nanoseconds.

The code for a PWM on the MC68HC11 was shown in Chapter 5. A program that implements a PWM on the MC68HC16 is shown below.

```
/* This program provides a pwm output to OC3. The period
   will be the integer value found in pwm_period, and the
   on time will be the integer value found in pwm_count.
   Keep pwm_count less than pwm_period. */
```

```c
#include "hc16.h"
#include "gpt.h"
#include "sim.h"

#define PERIOD 0x1000
#define ON_TIME   0x0800
#define GPT_IARB  5
#define GPT_IRL   6
#define GPT_VBA   4

/* function prototypes */
@port void OC3_Isr( void); /* the PWM isr */

WORD pwm_period=PERIOD, pwm_count=ON_TIME;

main()
{
    /* The initialization portion of the program */

    SYNCR.X=ON; /* set the clock freq to 16.78 MHz */
    SYPCR.SWE=OFF; /* disable the watchdog */
    GPT_MCR.IARB=GPT_IARB;    /* pick an IARB for the Timers */
    ICR.IRL=GPT_IRL;       /* interrupt level 6 */
    ICR.VBA=GPT_VBA;       /* vectors start at 0x40 */
    OCONM.OCONM3=ON;       /* sent OC1 out to pin */
    CONM.OCONM5=ON;        /* couple OC1 to OC3 */
    TMSKON.OC3I=ON;        /* enable the OC3 interrupt */
    OCOND.OCOND5=ON;       /* turn on OC3 when OC1 occurs */
    TCTLON.OL3=ON; /* toggle OC3 when OC3 occurs */
    TOC1=TCNT+pwm_period;/* set OC1 to the period */
    TOC3=TOC1+pwm_count; /* set OC3 time on */
    cli();   /* enable the system interrupts */

    /* the applications portion of the program */

    FOREVER
    {
    }
}

/* The asynchronous service portion of the program */

@port void OC3_Isr( void) /* the PWM isr */
```

```
{
    TFLG1.OC1F=OFF;          /* reset OC1 interrupt flag */
    if(OC1D.OC1D3==ON)  /* compliment OC1D3 */
        OC1D.OC1D3=OFF;
    else
        OC1D.OC1D3=ON;
    TFLG1.OC3F=OFF;          /* reset OC3 interrupt flag */
    TOC1+=pwm_period;
    TOC3=TOC1+pwm_count;
}
```

Listing 6-2: Elementary PWM Program For The MC68HC16

In keeping with the new usage of header files for writing code for the MC68HC16, the header files hc16.h, gpt.h, and sim.h are included in the above program. These files contain definitions of all registers needed for the implementation of this program. You should include hc16.h with every program or function that you write for this part. In this case, most of the program involves registers within the general purpose timer, so gpt.h is included. There is one register accessed from the system integration module. Therefore, the header sim.h is also included.

The MC68HC16 contains a software watchdog. The part comes out of reset with the watchdog enabled. Therefore, unless a program periodically accesses the watchdog, the part will execute a watchdog reset. This periodic reset will make debugging of timing operations difficult. The watchdog is disabled in the first instruction of the initialization of the program. The next three instructions are MC68HC16-specific instructions.

The MC68HC16 has a seven-level interrupt system similar to the MC68000 family of parts. The seventh level is the highest priority and is the only nonmaskable interrupt for the part. The remaining levels are progressively lower priority until the level 0 is found. At level zero, no interrupt is being processed, and this level is where the processor usually operates. When an interrupt occurs, the hardware level of the interrupt is placed in the interrupt priority field of the code register. Further interrupts of the designated level or lower will remain pending until a RTI instruction restores the IP field to a lower level. This approach provides for priority selection among several internal or external interrupt sources.

Another potential problem occurs when two internal modules are assigned the same priority level. In this case, a second level of arbitration is set up to choose among these several modules, in the event that more than one module requests an interrupt at the same time. The module control register of each module has a field called IARB. This 4-bit field is assigned by the programmer and it is an arbitration level that will be applied when the processor must select between two equal priority interrupts that occur simultaneously. The interrupting module with the largest IARB value will be given control of the processor. The IARB field can contain values from 0 to 15. When the module is being used, its IARB field must be assigned a non-zero value, and no two modules can contain the same IARB value. The code line

```
GPT_MCR.IARB=GPT_IARB;/* pick an IARB for the Timers */
```

places a value 5 into the IARB of the general purpose timer.

The interrupt level 6 is assigned to the GPT by the code line

```
ICR.IRL=GPT_IRL;      /* interrupt level 6 */
```

and the vector base address is assigned a value of 40 by the code line

```
ICR.VBA=GPT_VBA;      /* vectors start at 0x40 */
```

The means by which the vector assignment is accomplished in the GPT is different from that for the remaining modules in the MC68HC16. For the GPT, a 4-bit vector base address field is found in the interrupt configuration register. A vector is an 8-bit value.

The vector assignment is accomplished when the value placed in the VBA field of the ICR is used as the high nibble of an 8-bit number. The lower four bits are specified by the contents of Table 6-2. There you will note that the vector address of OC3 is at 0xV6. When the contents of the VBA field is 4, then the vector for OC3 is 0x46. The vector address is twice the value of the vector or 0x8c in this case. That is the reason that the address OC3_Isr is placed in the address 0x8c in the vector initialization routine.

You will note that each interrupt in the timer will be assigned a vector with the most significant nibble of the value placed in the VBA field of the ICR. There is a prearranged priority among these several interrupts. One interrupt can be moved to the highest priority among the several timer interrupts if desired. The priority adjust bits

Table 6-2: GPT Interrupt Priorities And Vector Addresses

Name	Function	Priority Level	Vector Address
	Adjusted Channel	0 (highest)	0xV0
IC1	Input Capture 1	1	0xV1
IC2	Input Capture 1	2	0xV2
IC3	Input Capture 3	3	0xV3
OC1	Output Capture 1	4	0xV4
OC2	Output Capture 2	5	0xV5
OC3	Output Capture 3	6	0xV6
IC4	Output Capture 4	7	0xV7
IC4/OC5	Input Capture 4/Output Capture 5	8	0xV8
TCF	Timer Overflow	9	0xV9
PAOVF	Pulse Accumulator Overflow	10	0xVA
PAIF	Pulse Accumulator Input	11 (lowest)	0xVB

PAB field in the ICR allows this shift. For example, if the number 6 were placed in the PAB field, then the priority of OC3 would be shifted from 6 to 0, where 0 is the highest priority of the 11 levels with the GPT. In this case, the vector for OC3 would be located at 0x40, and the vector address would be 0x80. None of the other interrupt vectors or priorities would be changed by this operation.

The remaining code of the initialization section of the above program is almost the same as that found in Listing 5-5. The register and bit naming conventions used with the MC68HC16 are such that the code written for the MC68HC11 can be used directly on the MC68HC16. There is one change. In the MC68HC11, it was necessary to set the DDRA7 bit to allow the output from OC1 to show up on the pin PA7. The GPT has a different output pin arrangement on the MC68HC16 and it does not require the use of the DDRA register at all.

One additional modification: A cli() instruction was used in Chapter 5 to enable the system interrupts. There is no single bit in the MC68HC16 that can be used to enable the interrupts. The 3-bit field in the condition code register named IP sets the level of interrupt that can be acknowledged. Since there is no equivalent instruction, a macro definition of an instruction

```
#define cli() ("andp $ff1f \n")
```

is included in the header file hc16.h. There is also a macro

```
#define sei() ("orp $00e0\n")
```

These two macros accomplish the equivalent of the same instructions for the MC68HC16. The `cli()` instruction clears the bits of the IP to zero, so that any interrupt will be acknowledged by the processor. The `sei()` instruction set the IP bits so that only a nonmaskable interrupt will be acknowledged. Additional macros can be written that enable the programmer to set the interrupt level anywhere between 1 and 7 if needed.

Recall that with the MC68HC11 the two registers `TFLG1` and `TFLG2` were different from the usual registers in the part. To reset bits in these registers, it is necessary to write ones to the designated bits rather than zeros. On the MC68HC16, this anomaly has been corrected. On the MC68HC16, to reset bits in the `TFLG1` and `TFLG2` registers the program must write zeros to the appropriate bits. That change shows up in two locations in the `OC3_Isr` routine. The much more logical

```
TFLG1.OC1F=OFF;
    .
    .
TFLG1.OC3F=OFF;
```

instructions are used here. Otherwise, the remainder of the interrupt service routine shown in Listing 6-2 is the same as that found in Listing 5-5.

This portability is what you should expect when changing between the MC68HC11 and the MC68HC16 family of parts. Care has been used in the design to assure that register names and bitfield names are common between the families. Therefore, code written for the MC68HC11 should move to the MC68HC16 with little change. The need for change at all is caused by the fact that architectures of the basic machines are different.

EXERCISES

1. Modify the program shown in Figure 6-2 to allow the PWM range to vary from 1 to 0XFFF. Compile this program and test the code.

2. Write a macro that will permit the program to put an arbitrary value between 0 and 7 into the IP field of the condition code register.

Cosmic MC68HC16 Compiler

The Cosmic compiler for the MC68HC16 is quite similar to the MC68HC11 compiler. Its operation is the same. Several command files are used in the course of executing a compilation, and these files will be shown here. The first file is the program that completely compiles and links the program. This file is shown below:

```
c -dlistcs +s +o %1.c
lnkh16 < %1.lnk
hexh16 -s -o %1.hex %1.h16
pause
```

This command file requires as an input a file with a .c extension. For example, if you were to compile the program newpwm.c, you would enter

```
c:\>comp newpwm
```

The first line will invoke the compiler and create a listing file, a source assembly listing, and an object file named newpwm.o . The second line invokes the linker named lnkh16, and requires an input file named newpwm.lnk . The .lnk file is similar to that one discussed in Chapter 5, and a listing of the one for this program is shown below.

```
# Link command file for NEWPWM.c
+h                       # multi-segment output
-max 0xfffff             # maximum size
-ps16 -pc.               # set up banking options
-o newpwm.h16            # output file name
+text -b 0               # reset vectors start
                         address
vector.o                 # vectors
+text -b 0x400           # program start address
+data -b 0x700           # data start address
crts.o                   # startup routine
newpwm.o                 # application program
c:/cc16/lib/libi.h16     # C library (if needed)
c:/cc16/lib/libm.h16     # machine library
+def __memory=__bss__    # symbol used by library
```

Here the entries are fairly well explained by the comments. Note that two additional input files are required by this link file. The vector table `vector.o` is a compiled version of the vector listing given in Listing 6-1. The `crts.o` object module is an assembled version of the start-up routine for this machine. Both of these routines must be written and compiled or assembled for the specific program. Otherwise, the link command file is as discussed in Chapter 5. A version of `crts.s` is shown below:

```
; C STARTUP FOR 68HC16
; Copyright (c) 1991 by COSMIC (France)
;
  .external _main, __memory
  .external ._main, .__bss__
  .public _exit, __stext
  .psect _bss
sbss:
  .psect _text
__stext:
  ldk #.__bss__
  tbek
  tbxk
  tbzk
  ldab #0fh          ; start of the i/o memory space
  tbyk               ; put it in y
  ldx #sbss          ; start of bss
  clrd               ; to be zeroed
  bra mtest          ; start loop
bcl:
  std 0,x            ; clear memory
  aix #2             ; next word
mtest:
  cpx #__memory      ; end of memory ?
  blo bcl            ; no, continue
  aix #1000h         ; 4K stack
  txs                ; for instance
  jsr _main,#._main  ; call application
_exit:
  bra _exit          ; loop here if return
```

```
;
.end
```

In the above code, the EK, XK, and ZK registers are initialized to the value found in `.__bss__` . The initial value of YK is set to `0xf`. The Y register will be used by the compiler to contain an offset to all of the data contained in the control registers found in the header files. Therefore, the YK register must be set to the top memory block in the computer memory space.

Using the SCI Portion of the Queued Serial Module

The queued serial module (QSM) contains two parts. The first is a convenient serial communications interface (SCI) which provides asynchronous serial communications that is used between computers and other devices. The remainder of the QSM is a queued serial peripheral interface that is often used for high-speed communications between computers and peripheral devices. This interface is strictly synchronous.

Let's examine an interface between the program and a terminal much like that found in Chapter 5. Here, the interface will simply read in a number from the screen and put that number into the pwm_count value for the PWM program. With this operation in place, the operator can type in a value and change the PWM on time at will. Listing 6-3 contains a program that will accomplish this end.

```
#include "hc16.h"
#include "gpt.h"
#include "sim.h"
#include "qsm.h"
#include "defines.h"

#define PERIOD 0x1000
#define ON_TIME   0x0800
#define SIM_IARB  4
#define GPT_IARB  10
#define GPT_IRL   6
#define GPT_VBA   5
/* set the baud rate=fclock/32*baud_rate */
#define BAUD_SET     (32768*512)/(32*38400)

/* function prototypes */
@port void OC3_Isr(void);
```

```
/* External variables */
WORD pwm_period=0x1000, pwm_count=0x0800,new_input=0;
BYTE new_character;

main()
{
/* The initialization portion of the program */

  /* initialize the SIM registers */
  SYNCR.X=ON;            /* set the system freq to 16.78 mHz */
  SYPCR.SWE=OFF;         /* disable the watchdog */
  /* initialize the GPT */
  GPT_MCR.IARB=GPT_IARB;/* pick an IARB for the timers */
  ICR.IRL=GPT_IRL;       /* interrupt level 6 */
  ICR.VBA=GPT_VBA;       /* vectors start at 0x40 */
  OC1M.OC1M3=ON;         /* sent OC1 out to pin */
  OC1M.OC1M5=ON;         /* couple OC1 to OC3 */
  TMSK1.OC3I=ON;         /* enable the OC3 interrupt */
  OC1D.OC1D5=ON;         /* turn on OC3 when OC1 occurs */
  TCTL1.OL3=ON;          /* toggle OC3 when OC3 occurs */
  TOC1=TCNT+pwm_period;      /* set OC1 to the period */
  TOC3=TOC1+pwm_count;  /* set OC3 time on */

  /* initialize the SCI       */
  SCCR0.SCBR=BAUD_SET;        /* set baud rate to 9600 */
  SCCR2.TE=ON;                /* enable the transmit and */
  SCCR2.RE=ON;                /* receiver of the SCI */

  cli();   /* enable the system interrupts */

  /* the applications portion of the program */

FOREVER
{
    if (SCSR.RDRF==ON) /* read in data if it is there */
    {
      new_character=SCDR;   /* get new byte, reset RDRF */
      while(SCSR.TDRE==OFF)       /* wait until transmit
buffer empty */
      SCDR=new_character; /* send out byte and reset TDRE */
      /* got an input, process it */
      if(new_character>='0'&&new_character<='9')
         new_input=10*new_input+new_character-'0';
      else if(new_character=='\r')
      {
      /* reject any number out of range */
```

```
                /* and start over again */
                if(new_input>=1 && new_input<=4048)
                   pwm_count=new_input;
                new_input=0;
                }
                else
                new_input=0;     /*reject everything else*/
            }
        }
}


/* The asynchronous service portion of the program */

@port void OC3_Isr( void) /* the PWM isr */
{
    TFLG1.OC1F=OFF;                 /* reset OC1 interrupt flag */
    if(OC1D.OC1D3==ON)              /* toggle OC1D3 */
        OC1D.OC1D3=OFF;
    else
        OC1D.OC1D3=ON;
    TFLG1.OC3F=OFF;                 /* reset OC3 interrupt flag */
    TOC1+=pwm_period;
    TOC3=TOC1+pwm_count;
}
```

Listing 6-3: PWM System With Keyboard Input

The qsm.h header file is included to add all of the register and defines needed for the operation of the SCI. We are going to place the operation of the SCI into the applications portion of the program. Implementation of the SCI requires that a baud rate be selected and the transmit enable along with the receive enable bits be set in the serial communications control register number 2. The baud rate is set to 38400 by placing a properly calculated value into the SCBR field of SCCR0. With these lines of code, the serial communications interface is set up and ready to work.

The code inside the application portion of the program is essentially identical to that found in Chapter 6. The only difference is that the variable names are changed to be more compliant with this application. Here is another interesting case where the code written for the MC68HC11 will move directly to the MC68HC16 with minimal change.

This program is a minimum SCI system. The SCI portion of this chip has comprehensive capabilities that are not exploited in the program above. These capabilities and error-checking devices are all available on the chip, and their access and use is explained in detail in the QSM Reference Manual. Access of the various capabilities is exactly the same as was shown above in this program.

Periodic Interrupt

To make the problem a little more interesting, let's add another load to the machine by making use of the periodic interrupt capability provided in the SIM to create an interrupt. With this interrupt, we will build a time-of-day clock similar to that developed in Chapter 4. In this case you will see that the applications code developed for the MC68HC05 will work fine for the MC68HC16. Of course, the initialization code and the interrupt service routine will be completely different for the MC68HC16. This additional code will be put right into the program listed in Listing 6-3. It is asserted that there will be no adverse interactions between the various sections of the code.

```c
#include "hc16z1.h"
#include "gpt.h"
#include "sim.h"
#include "qsm.h"
#include "defines.h"

#define PERIOD 0x1000
#define ON_TIME    0x0800
#define SIM_IARB   4
#define PIC_PIRQL 6
#define PIC_PIV    0X38
#define PIT_PITM   16
#define GPT_IARB      10
#define GPT_IRL    6
#define GPT_VBA    5
/* set the baud rate=fclock/32*baud_rate */
#define BAUD_SET     (32768*512)/(32*38400)
#define TIME_COUNT        1000
#define MAX_MIN    59
#define MAX_HOURS 12
#define MIN_HOURS 1
#define NO_WAIT    0
#define OC2_OFFSET        5
#define MAX_SEC    59
```

```c
#define MAX_MIN    MAX_SEC
#define MAX_HOURS 12
#define MIN_HOURS 1

/* function like macros */
#define get_hi(x) ((x)/10+'0')
#define get_lo(x) ((x)%10+'0')

/* function prototypes */
@port void OC3_Isr(void);
@port void PIT_Isr(void);
void output_time(void);
void putch(int);
void send_out(WORD);

/* External variables */
WORD pwm_period=PERIOD, pwm_count=ON_TIME,new_input=0;
WORD hrs,mts,sec,been_here=0;
char new_character;

main()
{
    /* The initialization portion of the program */

  /* initialize the SIM registers */
  SYNCR.X=ON;              /* set the system freq to 16.88 mHz */
  SYPCR.SWE=OFF;           /* disable the watchdog */
  SIM_MCR.IARB=SIM_IARB;   /* IARBs for each module        is
different */
  PICR.PIRQL=PIC_PIRQL;    /* put all timers at level 6 */
  PICR.PIV=PIC_PIV;        /* vector is 0x38, address is 0x70 */
  PITR.PTP=ON;             /* 512 prescaler */
  PITR.PITM=PIT_PITM;      /* divide by 16*4, 1 tic per
second */

  /* initialize the GPT */
  GPT_MCR.IARB=GPT_IARB;   /*pick an IARB for the timers */
  ICR.IRL=GPT_IRL;         /* interrupt level 6 */
  ICR.VBA=GPT_VBA;         /* vectors start at 0x40 */
  OC1M.OC1M3=ON;           /* sent OC1 out to pin */
  OC1M.OC1M5=ON;           /* couple OC1 to OC3 */
  TMSK1.OC3I=ON;           /* enable the OC3 interrupt */
  OC1D.OC1D5=ON;           /* turn on OC3 when OC1 occurs */
  TCTL1.OL3=ON;            /* toggle OC3 when OC3 occurs */
  TOC1=TCNT+pwm_period;    /* set OC1 to the period */
  TOC3=TOC1+pwm_count;     /* set OC3 time on */
```

```
/* initialize the SCI */
SCCR0.SCBR=BAUD_SET; /* set baud rate to 9600 */
SCCR2.TE=ON;          /* enable the transmit and */
SCCR2.RE=ON;          /* receiver of the SCI */
cli();                /* enable the system interrupts */

  /* the applications portion of the program */
  FOREVER
  {
     if (SCSR.RDRF==ON) /* read in data if it is there */
     {
        new_character=SCDR;  /* get new byte,
                                    reset RDRF */
        while(SCSR.TDRE==OFF)
           ;                     /* wait until transmit
                                    buffer empty */
        SCDR=new_character;  /* send out byte and
                                    Reset TDRE */
        /* got an input, process it */
        if(new_character>='0'&&new_character<='9')
           new_input=10*new_input+new_character-'0';
        else if(new_character=='\r')
        {
           /* reject any number out of range */
           /* and start over again */
           if(new_input>=1 && new_input<=4048)
              pwm_count=new_input;
           new_input=0;
        }
        else
           new_input=0; /*reject everything else*/
     }

     if(sec>MAX_SEC)
     {
        sec=0;
        if(++mts>MAX_MIN)
        {
           mts=0;
           if(++hrs>MAX_HOURS)
              hrs=MIN_HOURS;
        }
     }

     if(been_here && !new_input)
     {
```

```
               been_here=OFF;
               output_time();
            }
        }
}

void output_time(void)
{
  int i;
  putch('\r');          /* send out a carriage return */
  putch('\t');
  putch('\t');          /* tab over the pwm_count */
  putch('\t');          /* on the screen */
  send_out(hrs);
  putch(':');
  send_out(mts);
  putch(':');
  send_out(sec);
  putch('\r');
}

void putch(int x)
{
    while(SCSR.TDRE==OFF)
            ;                       /* wait until data
                                    register is empty*/
    SCDR = (char) x;
}

void send_out(WORD data)
{
    putch(get_hi(data));
    putch(get_lo(data));
}

/* The asynchronous service portion of the program */

@port void OC3_Isr( void) /* the PWM isr */
{
    TFLG1.OC1F=OFF;         /* reset OC1 interrupt flag */
    if(OC1D.OC1D3==ON)
        OC1D.OC1D3=OFF;
    else
        OC1D.OC1D3=ON;
    TFLG1.OC3F=OFF;         /* reset OC3 interrupt flag */
    TOC1+=pwm_period;
```

```
        TOC3=TOC1+pwm_count;
}


@port void PIT_Isr( void) /* the PIT isr */
{
    been_here++;
    sec++;
}
```

Listing 6-4: Clock Routine Added to PWM

If you compare this listing with that shown in Listing 6-3, you will find that there are few structural changes to the program. The code used to initialize the SIM is changed by the addition of the initialization of the periodic timer interrupt. This code is shown below.

```
SIM_MCR.IARB=SIM_IARB;    /* IARBs for each module
                             is different */
PICR.PIRQL=PIC_PIRQL;     /* put all timers at
                             level 6 */
PICR.PIV=PIC_PIV;          /* vector is 0x38,
                             address is 0x70 */
PITR.PTP=ON;                /* 512 prescaler */
PITR.PITM=PIT_PITM;       /* divide by 16*4, 1
                             tic per second */
```

The interrupt arbitration level field in the SIM module control register is set to 4. Recall that the value here can be anywhere between 1 and 15, with 15 the highest priority. All active internal modules that are to use an interrupt must have a unique IARB value. The IARB value for the GPT was set to 5. Note that the interrupt level for both the GPT and the PIT is set to the level 6. Therefore, both sources of timing have the same interrupt priority; however, since the IARB of the GPT is higher than that of the PIT, in the event of a simultaneous occurrence of the two interrupts, the GPT service routine will be executed before the PIT.

The interrupt vector for the PIT is placed at 0x38. Because the address of the vector is twice the value of the vector, the interrupt vector address is 0x70. A pointer to the PIT interrupt service routine will be placed at this address in the vector.c routine. The periodic timer itself is set up by the next two lines of code. This clock is driven by the EXTAL signal. In our case, the frequency of the EXTAL signal is 32768 Hz, not some number around 16 MHz. The formula to calculate the periodic interrupt time is

$$T_{pit} = 4\,PITM\,(511\,PTP + 1)/F_{extal}$$

Here `PITM` is the 8-bit field with the same name found in the `PITR`. `PTP` is a single-bit field in the `PITR` that can have a value of either 0 or 1. As such, if `PTP` is 1, the prescaler value of 512 is used. Otherwise, when there is no prescaler, the value in the parentheses reduces to 1. With the values placed in these fields in the above code, i.e., `PTP` of 1, and `PITM` of 64, and with a 32768-Hz external crystal, the periodic interrupt time should be one second.

Two additional blocks of code are added to the applications section of the code. This code is shown below:

```
if(sec>MAX_SEC)
{
   sec=0;
   if(++mts>MAX_MIN)
   {
      mts=0;
      if(++hrs>MAX_HOURS)
         hrs=MIN_HOURS;
   }
}
if(been_here && !new_input)
{
   been_here=OFF;
   output_time();
}
```

The first eight lines of code here are taken directly from similar clocking code found in Chapter 4. This code merely counts the time in seconds, minutes, and hours. The second block of code determines if a `PIT` has been serviced, and if it has, it resets the `been_here` flag that indicated that the `PIT` service routine has been entered and then sends the time out the serial port when `output_time( )` is executed.

Here is a case where several subroutines are used in the applications portion of the program. `output_time( )` calls functions `putch( )` and `send_out( )`. In turn `send_out( )` calls convert_bcd( ). putch( ) is straightforward. This routine waits until the transmit data register is empty and then stores the character to be transmitted into the serial communications data register. The routine send_out ( ) takes the character parameter passed to it and causes it to be converted from integer format to two binary-coded decimal BCD characters. These two characters are then converted to ASCII characters by the addition of the character '0' to each and then sent to the output. (Recall that convert_bcd( ) was shown in Chapter 4 in Listing 4-8.) Another version of this function is shown in Listing 4-7. This alternate version has been tried in the program above and it works as well as the function used above.

With the functions putch( ) and send_out ( ) available, it is a simple matter to write the code that will output the time to the center of the top line of the screen. It is assumed that the cursor is on the top line when the program begins to run. The last remaining modification is the interrupt service routine PIT_Isr( ). Within this function, the been_here flag is set, and the value in sec is incremented. Since the applications portion of the program will process sec and reset it whenever it reaches a value of 60, there is no need for other code in this isr. For interest, listed below is the output from the compiler for both OC3_Isr( ) and PIT_Isr( ).

```
; 150 @port void OC3_Isr( void) /* the PWM isr */
; 151 {
 .even
_OC3_Isr:
 pshm k,z,y,x,d,e
 tskb
 tbek
 tbxk
 tbyk
 tbzk
; 152   TFLG1.OC1F=OFF; /* reset OC1 interrupt flag */
 ldy   #0
 bclr -1758,y,#8
; 153   if(OC1D.OC1D3==ON)
 brclr    -1783,y,#8,L102
; 154   OC1D.OC1D3=OFF;
 bclr -1783,y,#8
```

```
;155   else
 bra   L112
 L102:     ; line 155, offset 33
;156   OC1D.OC1D3=ON;
 ldy   #0
 bset  -1783,y,#8
 L112:     ; line 156, offset 41
;157   TFLG1.OC3F=OFF; /* reset OC3 interrupt flag */
 ldy   #0
 bclr  -1758,y,#32
;158   TOC1+=pwm_period;
 ldd   _pwm_period
 addd  -1772,y
 std   -1772,y
;159   TOC3=TOC1+pwm_count;
 addd  _pwm_count
 std   -1768,y
;160   }
 pulm  k,z,y,x,d,e
 rti
;161
;162 @port void PIT_Isr( void) /* the PIT isr */
;163 {
 .even
_PIT_Isr:
 pshm  k,z,y,x,d,e
 tskb
 tbek
 tbxk
 tbyk
 tbzk
;164   been_here++;
 incw  _been_here
;165   sec++;
 incw  _sec
;166 }
 pulm  k,z,y,x,d,e
 rti
```

Listing 6-5: Interrupt Service Routines

Lines 150 through 160 above are the interrupt service routine OC3_Isr( ) and lines 161 through 166 comprise the PIT_Isr( ). The main item that is observed here is the quality of the optimizer for the compiler. In general, interrupt service routines, ISR, must save the complete status of the machine prior to executing any code. The CCR along with the PC are both saved by the interrupt sequence. The remainder of the registers must also be saved if the ISR can use any of the additional register resources of the computer. Usually this case will be found. Note, for example, in OC3_Isr( ) the first instruction is

```
pshm k,z,y,x,d,e
```

This instruction causes the contents of all significant registers to be saved on the stack, so that the status of the machine at the time the interrupt occurred can be restored before control is returned to the portion of the program that was interrupted. This restoration is completed by the two instructions

```
pulm k,z,y,x,d,e
rti
```

which refills all of the registers with the values they contained when the ISR was entered, and the rti instruction restores the condition code register to the value it had and the program counter is then restored. Thus, the status of the machine is restored and the program control is returned to the interrupted instruction. Note that the routine PIT_Isr( ) compiles to a much simpler routine. This routine has simply two increment instructions and requires no register resources. The optimizer recognizes this simple operation and does not save the status beyond that saved when the interrupt occurred.

This routine has probably been pushed further than it should be, for demonstration purposes. This routine demonstrates multiple interrupts working simultaneously, keyboard input and output, and generates a simple pulse-width modulation signal whose on period is determined by a number entered from the keyboard. All of these operations are quite similar to those shown in Chapter 5; however, some major modifications in approach are required to meet the several requirements of the peripheral components on the MC68HC16. Let us now look at some other applications often used with microcontrollers.

Table Look-Up

Often in the implementation of a practical problem it is necessary to implement a conversion of data in a completely heuristic manner. Observations are made and a curve of sorts is fit to the data. It is then desired to put a few samples of this curve into a data table and then calculate values between these samples with an accuracy that usually requires interpolation between the points contained in the data table. Such tables are usually sparse in that the number of measured points across the range covered by the table are few. The curve in Figure 6-1 shows an example of such a conversion table, and the table itself is shown as Table 6-3.

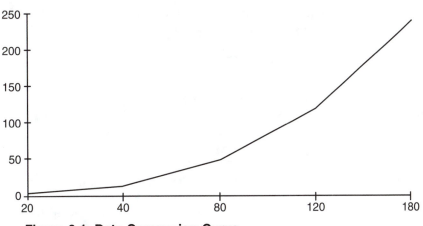

Figure 6-1: Data Conversion Curve

X	Y
20	5
40	15
80	50
120	120
180	240

Table 6-3: Look-Up Table

The object of the program is to deliver to the program a number like 67 and get the proper result back from the table look-up routine. The x value of 67 lies between the 40 and the 80 entry in the table. Therefore, the interpolation calculation needed in this case is

$$Result = 15 + (50 - 15)\frac{(67 - 40)}{80 - 40} = 38.625$$

Notice that there is a product and a division that is the same for all input values between 40 and 80 in this case. This calculation is the slope of the line between the two points being interpolated. Usually table look-up operations are used where time constraints on the program are great so that these operations, multiply and divide (especially divide), are unwelcome in these programs. There is a way to avoid the divide operation. Note that between each set of points the slope is a constant. Therefore, if the slope of the line is built into the table, the calculation would involve one subtraction, one multiply, and one addition. Let us modify the table above to contain the slopes of the lines between each point.

20	5	0.5
40	15	0.875
80	50	1.75
120	120	2
180	240	

Table 6-4: Look-Up Table With Slopes

Note that the slope of the line runs from the lower point to the next point. Therefore, there is no slope for the last point. When this table is built in memory, it consists of a header followed by a four-byte entry for each entry in the table. The header consists of a single byte that indicates the length of the table. In this case, since the table starts with an independent variable value of 20, there are expected to be no input values less than 20. If there are, however, the value of 5 will be used for these output values. Also, if the value of the independent variable is greater than the maximum of 180, a value of 240 will be returned from the routine. With data from the above table, the calculation above would reduce to

$Result = 15 + 0.875(67 - 40) = 38.625$

which requires a multiply, a subtract and one add. Of course, the routine will truncate the fractional part of the result so that the value returned will be 38.

Each table entry contains four bytes. The first byte is the independent variable value, often designated as x. The second byte contains the dependent variable value, usually called y, and the final two bytes contain a floating point format of the slope. The slope must be floating point. The third byte contains the integer position of the slope, and the fourth byte contains the fractional value of the slope. With the binary point placed between the third and fourth bytes of the table entry, we can accomplish some interesting multiply operations. Prior to looking at the coding of this problem, let us complete the table and convert the slopes into hexadecimal format.

5			
20	5	0x00	0x80
40	15	0x00	0xe0
80	50	0x01	0xc1
120	120	0x02	0x00
180	240		

Table 6-5: Look-Up Table with Hexadecimal Slopes

How does one build a table of this nature in C? Of course, it is an array of structures, or it could be a structure that contains an array of structures. The latter approach would be

```
struct entry
{
char x;
char y;
int slope;
};

struct lut
{
char entries;
struct entry data[5];
} _lut = {
5,
20, 5, 0x0080,
40, 15, 0x00e0,
```

```
80, 50, 0x01c0,
120, 120, 0x0200,
80, 240
};
```

Note that in the table above, the slopes are entered in the table as two-byte integers. There is no binary point in the slope. The conversion of these integers to floating-point numbers is accomplished easily. We must merely recognize that these numbers are a factor of 128 too large, so after the slope is used as a multiplier, the result must be divided by 128 to get the correct answer. Of course, division by 128 can be accomplished by a shift right by 8, or more practically merely choosing the left byte of the product.

A function that will make use of this table is

```
char table_look_up( char x_in)
{
    int i;
    for(i=0;x_in>_lut.data[i].x && i<=_lut.entries; i++);
    if(i>=_lut.entries)
        return _lut.data[_lut.entries-1].y;
    else if (i==0)
        return _lut.data[0].y;
    else
        return _lut.data[i-1].y+(((x_in-_lut.data[i-1].x)*
            _lut.data[i-1].slope)>>8);
}
```

Listing 6-6: Table Look-Up Routine, Version 1

The compiler optimizer will recognize the shift right by 8 in the above function and will merely select the upper byte of the result rather than executing the shift operation indicated.

This function was checked on an evaluation system for the single input value of 67 and the result was the expected value of 38. Of course, that did not check the function over its full range of operation. The check over the full range was accomplished on a host machine. It is a simple matter to include the above function and table in the following program.

```
#include "tlu.h"
void main(void)
```

```
{
    int i;
    for(i=0;i< 256; i++)
      printf(" i = %d r = %d\n",i,table_look_up(i));
}
```

where `tlu.h` contains the function and the table above. This code compiles and runs on a MS-DOC PC, and the result is as expected. Here, the program scans the entire input range and prints out the resultant value at each point. A check of the outputs will show that the function hits the break value at each break, and creates a linear interpolation between all points. For values above or below the range, the proper output is observed.

In a normal control system, there will often be need for several table look-up operations. The above code is not too good in this case because the code to do the look-up must be repeated with each table. This extra code can be eliminated if the function `table_look_up( )` is passed two parameters: the first parameter is the interpolation value, and the second value is a pointer to the look-up table as is shown below:

```
char table_look_up( char x_in, struct lut* table)
{
    int i;
    for(i=0;x_in>table->data[i].x && i<=table->entries; i++) ;
if(i>=table->entries)
        return table->data[table->entries-1].y;
    else if (i==0)
        return table->data[0].y;
    else
        return table->data[i-1].y+
            (((x_in-table->data[i-1].x)* table->data[i- 1].slope)>>8);
}
```

Listing 6-7: Table Look-Up, Version 2

This form of the table look-up should probably be used in all cases. Here is another example of where it is wise to examine the code generated by a compiler. It will not be listed here, but the assembly code required for Listing 6-6 is 182 bytes long, and that for listing 6-7 is but 142 bytes. This greatly improved utility automatically gives a substantial savings in code.

It was pointed out that the compiler optimizer will sometimes change the basic code dictated by the programmer. Perhaps, it should be stated that the optimizer changes the code suggested by the programmer. Often when you examine the code generated by the compiler, the operations that take place will not even resemble those that the programmer had in mind. This alteration of the code will show up with shifts, and divides or products of powers of two. A proper optimizer should determine if a multiply or a shift is better and provide the best code. Often you will find that the compiler optimizer will provide either fastest execution or minimum code. Usually the two will be different. Faster code will almost always take more memory.

It may seem that the lost time is small, but you will always find that if you want fast code, you will not use any looping constructs. The code that increments a counter and tests the counter must be executed each loop. Without the looping construct, this code is completely eliminated, and its execution each loop will be completely eliminated. Therefore, if you want fast code, you should eliminate looping constructs and repeat the code within the loop the desired number of times. The cost of this move is more code. We have also seen that recursive code can require an inordinant amount of time. Again, recursive code merely creates hidden looping constructs that require frequent generation of stack frames prior to new function calls. The construction of these stack frames and return from functions require time which is often masked by the elegant appearance of the code. Usually you will produce faster code if you figure a way to accomplish the same end operation without calls to the executing function.

Today, we are at the very beginning of a completely new set of microcontrollers. These machines are based on RISC (reduced instruction set computer) techniques. Do not be misled. RISC does not really mean reduced instruction set. The instruction sets are complete and extensive. However, the basic architecture of a RISC machine is quite different from the older machines like those we have been working with. One of the main differences is that the design of the machine is to provide for the execution of one or more instructions with each clock cycle. A standard RISC will allow almost one instruction per clock cycle, and a super scaler architecture will allow two or more instructions per clock cycle. This operation is enabled by the use of instruction pipelines and multiple arithmetic logic units (ALUs). Sometimes, an instruction must use more than one clock to complete an instruction. For example, an integer multiply instruction might require five clock cycles. In that case, an instruction pipe five instructions long would be implemented within the integer ALU. Therefore, a multiply instruction would be launched into the instruction pipe for execution. In the meantime, other instructions could execute while the multiply is progressing through the pipe.

Compiler optimization for these machines must take into account all of the pipes, the separate ALUs, and other unique operations when creating the required assembly code. Clearly, a straightforward creation of code in the order that might seem to be natural to a programmer will not necessarily create the quickest code for a RISC machine. Here, speed optimization consists of arranging the code so that, as nearly as possible, the maximum number of instructions are launched each clock cycle. This rearrangement of your code will make it extremely difficult to examine the assembly code version of the program and even make sense of it. So long as the results are not altered, the optimizer for a RISC machine will move instructions around in the code stream to accomplish this end. Therefore, from a practical sense, any debugging on a RISC machine will probably be done with a source level debugger, and not an assembly language version of the program.

The chip used in Chapter 8 is of the MCORE family of RISC microcontrollers. We will see more of the above comments in that chapter.

EXERCISE

1. Sometimes two input values are needed to specify a parameter. For example, if you recall from Chapter 5, the change in pulse on time to properly control the motor speed was given by

$$\Delta p_c = -3809500 \frac{\Delta p}{p^2}$$

create a sparse two-dimensional look-up table whose inputs are p and Δp and which provides Δp_c as the result of three interpolations. Is the look-up table better than the calculation in any way—less code, quicker, etc.? Why would you use a look-up table in a problem like this one?

Digital Signal Processor Operations

Most microcontrollers, regardless of their basic speed, are not really able to process signals in real time. The basic speed of the processors is fast enough to accomplish most signal processing; however, the set of things needed to do digital signal processing is not usually available in the regular microcontroller. The basic digital signal processor (DSP) function that is required is summarized by three actions: 1) the processor must multiply two values, 2) the processor must add the product into a value, and 3) it must prepare for the next multiply. This set of operations must execute quickly enough that the computer can keep ahead of the real-time input of data being processed. Almost all signal processing operations are based on the multiply-accumulate sequence. Filtering, correlation operations, scalar products of vectors, Fourier and other transformations, and convolutions are but a few of the operations that are built around the DSP multiply and accumulate sequence above. The MC68HC16 family has an extension to its core that can execute basic DSP operations fast enough to support real-time filtering, correlation, and so forth.

Unfortunately, C compilers do not know of these added functions, so the C programmer would seem to be unable to include DSP operations in programs. We will see here that, while it is rather inconvenient, it is possible to write assembly functions that will permit the programmer access to the complete DSP capabilities found in this family of devices.

One of the good features of C is that it is not necessary for the programmer to have detailed knowledge of the nature of the basic computer being programmed. So far, we have had little to say about accumulators or index registers or the like. No more! We must now get inside of the computer to create functions that will accomplish our DSP needs. When these functions are complete, we should be able to treat the DSP operations in much the same manner that we would any other function call. The DSP contains four registers and controls three bits in the condition code register (CCR). The first two registers are called the MAC multiplier input registers H and I, respectively. These registers must be loaded with the multiplier and the multiplicand to be executed. Data stored in these registers are signed fractional binary numbers with the radix point between bits 15 and 14. The product will be accumulated into the MAC accumulator M register. This register is 36 bits long. For convenience, the register is broken into two portions, bits 35 through 16 and bits 15 through 0. We will see later that different portions of this register are moved by different instructions, so that breaking this register into the two parts is logical.

The last register in the DSP register model is the MAC XY mask register. Automatic selection of the addresses for the next multiply needs modulo arithmetic. We will see more modulo arithmetic later. The mask register contains the modulo base for both the x and y index registers which will allow fixed coefficient tables, that can be manipulated as either single-or two-dimensional arrays, to be traversed automatically during successive multiply and accumulate instructions. The DSP register model is shown in Figure 6-2 below.

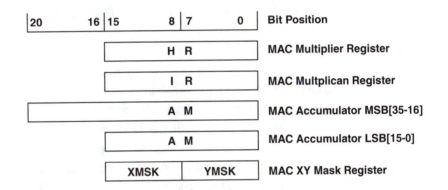

Figure 6-2: MC68HC16 DSP Register Model

The CCR contains three bits that are associated with the DSP. Bits 14 and 12 are DSP overflow flags. These flags will be discussed in detail later. Bit 4 is called the SM bit and is the DSP saturation mode control bit.

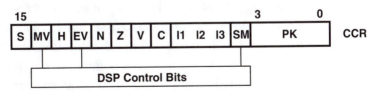

Figure 6-3: DSP Control Bits In The CCR

The accumulator is 36 bits with the radix point between bits 31 and 30. When accumulating into the MAC accumulator AM, there are two types of overflow that can occur. The first is when an addition operation, which should always add fractions, causes an overflow from bit 30 to 31. This type of overflow is reversible because the arithmetic that caused the overflow cannot cause more than a 1-bit overflow error, and bits AM[34-31] are there to absorb this type of overflow. The contents of this register are signed, so that bit 35 of AM is the sign bit. The maximum value that can be placed in the 36-bit AM register is 0x7ffffffff. This value has a decimal value of 15.999969482. The minimum value is 0x800000000 corresponding to -16. Whenever arithmetic involves the bits AM[34-31] the EV bit will be set, and the program will know that the bits in AM[35-31] are the signed integer part of the number. If successive operations cause the overflow to disappear, the EV bit will be reset.

Another situation is less tractable. Suppose an overflow occurs as a result of an arithmetic operation into the AM that causes bit 34 to overflow into bit 35. This type of error is not reversible as is the case above. When this overflow occurs, the MV bit is set to notify the program of the result. An internal latch known as the sign latch SL will contain the value of A[35] after the overflow has occurred. Therefore, SL is the complement of the sign-bit when the overflow occurred.

To communicate with the DSP portion of the MC68HC16, we have to use assembly language. In keeping with the basic rule to use C whenever possible, the approach to be taken will be to create C-callable functions that access these important capabilities. What features should we include in these functions? Obviously, all possible functions cannot be conceived. A set of functions that will embody the most important features of the DSP capabilities will be written.

Let us examine how this compiler transfers parameters to a function. When more than one parameter is passed, the parameters are pushed on the stack starting with the right-most parameter in the function argument list. The leftmost parameter is put into the D register. If a single parameter is passed, it is placed in the D register prior to the function call. Within the function, the D and X register is saved on the stack, and the stack pointer is decremented by an amount needed to provide space for all of the function variables. At that point, the stack pointer is transferred into the X register. An example of this code is shown below. This little function receives three parameters and merely puts these values into three local variable locations.

```
void dot_product(char length, int* xdata, int* ydata)
{
        int i,*xp, *yp;
        xp=xdata;
        yp=ydata;
        i=length;
}
```

Listing 6-8: Sample Parameter Handling Function

The compiled version of the above function is shown below. The first important instruction is the .even assembly directive that causes the code to follow to start on an even boundary. Code must be started at an even address. Note in the program above that three parameters are passed to the function. After the contents of the X and the D registers are saved, the stack pointer is decremented by 6 to provide space for the variables i, *xp, and *yp. The two pointer parameters require 16 bits each, and the character parameter needs only 8 bits. The program, in favor of greater speed, will store the variable i in a 16-bit location even though i is only 8 bits wide. At this time, the value contained in the stack pointer is transferred into the X register. During this transfer, the value will be automatically incremented by two so that the contents of the X register will be pointed to the last value stored on the stack rather than to the next empty location on the stack as is found in the stack pointer.

```
1  ; Compilateur C pour MC68HC16 (COSMIC-France)
2  .include "macro.h16"
3  .list +
4  .psect _text
5  ; 1 void dot_product(char length, int* xdata, int* ydata)
6  ; 2 {
7  .even
8  _dot_product:
9  pshm x,d
10 ais #-6
11 tsx
12 .set OFST=6
13 ; 3 int i,*xp, *yp;
14 ; 4
15 ; 5 xp=xdata;
16 ldd OFST+8,x
17 std OFST-4,x
18 ; 6 yp=ydata;
19 ldd OFST+10,x
```

```
20 std OFST-6,x
21 ; 7 i=length;
22 ldab OFST+3,x
23 clra
24 std OFST-2,x
25 ; 8
26 ; 9 }
27 ldx 6,x
28 ais #10
29 rts
30 .public _dot_product
31 .even
32 .psect _data
33 .even
34 .psect _bss
35 .even
36 .end
```

Listing 6-9: Compiled Version Of Parameter-Handling Function

The diagram shown in Figure 6-4 will help you visualize what is happening here. On the right side of this diagram you will find the location at which the stack pointer is pointed at various times during the function call and its execution. On the left side of the diagram, you will see the locations pointed to by the X register. An offset named OFST is established by the program. This offset will have a value equal to the space emptied on the stack with the ais instruction: in other words, in this case OFST will be 6. In every case from this point forward in the program, variable and parameter accesses will be indexed relative to the X register, and the total offset from the X register will be a value OFST+k where k is a positive or negative value that corresponds to the address of the parameter being accessed. For example, the instruction

```
ldd OFST+8,x
```

will load the value at x+OFST+8 which you can see in Figure 6.4 is the value *xdata. This value is stored at the location x+OFST-4 by the instruction

```
std OFST-4,x
```

which is the location where xp is stored on the stack. Remember that each word on the stack is two bytes, so that all of the offsets and address will be even numbers.

The code in lines 16 through 20 above saves the values of xdata and ydata in xp and yp respectively. The operations shown in lines 21 through 24 save the 8-bit value found in the B register to the 16-bit location at x+OFST-2. To be certain that the value is not corrupted by some garbage value in the A register, the clra instruction is inserted prior to the time that the value is saved.

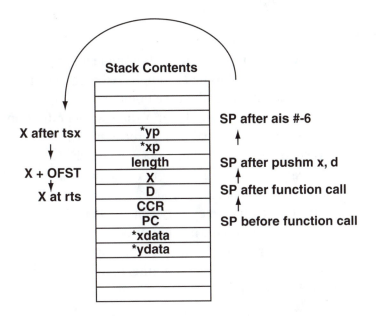

Figure 6-4: Stack Contents During Function Operation

After the closing brace of the function in the above listing, two important operations take place. First, the contents contained in the X register on entry to the function are restored by the instruction

```
ldx 6,x
```

and the stack pointer content is restored to the value it contained when the function was entered. At this point in the program, an rts instruction will return the program control to the instruction following the jsr or bsr instruction used to enter the function code originally.

When the function return is executed, the registers D, E, Y, Z, and CCR are all undefined. A 16-bit return from the function will be returned in the D register, and a 32-bit return will be contained in the E and the D register. The least significant portion of the return is in the D register.

What is a `dot_product( )`? In vector algebra, a dot product, or a scalar product, operates on all of the members of two vectors and returns a single scalar result. This value is the magnitude of the projection of one vector on the other. The familiar arithmetic form for a dot product is

$$c = \sum_{0}^{n-1} a_k b_k$$

Note that all corresponding members of the two vectors are multiplied and summed. The result is a single number. Another important calculation needed to be accomplished by a DSP is called convolution. A convolution is the time domain operation of a filter. Most of the time, a designer thinks of a filter as operating on the different frequencies of the signal being processed. In the frequency domain, at every frequency the filter has a gain which is complex. "Complex" in this case means the gain has two dimensions that can be thought of as magnitude and phase. The signal also has a similar two-dimensional description in frequency. At each frequency, the magnitude of the filter gain multiplies the magnitude portion of the signal, and thephase of the filter gain adds to the corresponding phase of the signal. There are easy ways to treat this operation in the frequency domain. In fact, the design of most filters takes place in the frequency domain.

However, the frequency domain is an artifact that we can never really get our hands on. In reality, the signals we must deal with are varying voltages or currents. These varying signals can be continuous, or when converted to a tractable form for operation in a computer, they are a series of samples. Let us call them x_k. Here x is the value of the signal at sample points k. Now k might be thought of as related to time, and in fact different values of k do correspond to samples taken at different times. Usually, k corresponds to samples taken periodically at carefully spaced, equal intervals.

A filter in the time domain has what is called a weighting function. The weighting function is indeed the Fourier transform of the complex frequency response of the filter. In the continuous domain, there is a mathematical trick that allows the weighting function to be shown. A function called a Dirac Delta function is defined as a function that is 0 everywhere except at one point. The integral across this point is one. Such a function really does not exist. However, if such a function were delivered into the input of a filter, the output of the filter would be the filter weighting function.

As we move to the digital realm, the Dirac Delta function is replaced by the Kroniker *Delta* function. This simple function, δ_k, is 0 for all values of k except for k = 0 where its value is 1. If this function is sent through a digital filter, the output observed is the weighting function of the filter.

In both cases, analog and digital, the frequency response of the filter is the Fourier transform of the filter weighting function. This duality between the frequency response and the time domain response makes it possible to design filters to accomplish what is really desired. Usually, the filter specification is best established in the frequency domain. The designer knows what frequencies are to be passed or rejected by the filter. There has been a long history in the field of passive network synthesis devoted to the "approximation problem." How does one specify a filter to meet accurately a desired frequency response? This problem has led to many sophisticated approaches to the specification by mathematics of a frequency response to meet the system need. More important, these frequency responses have a nature that can be realized by a finite collection of passive electrical components—resistors, capacitors, and inductors. In other words, these frequency responses are realizable.

A series of mathematical transformations exist that can be used to transform frequency responses directly to filter weighting functions. We will not go into these transformations here, but will refer you to Elliott for practical means to specify the weighting functions for digital filters.[7] A more general text on this subject is by Antoniou.[8] The time domain response calculation is called a convolution. If there is a signal x_k applied to a filter with a weighting function h_k there will be an output from the filter at each time sample, and this output will be called y_k. The relations between these parameters are given by the equation

$$y_k = \sum_{i=0}^{n-1} x_{k-i} h_i$$

[7] Elliott, Douglas F., Handbook of Digital Signal Processing and Applications: Academic Press Inc. 1987

[8] Antoniou, Andreas, Digital Filter Analysis and Design: McGraw Hill, 1979

The convolution is little more than a series of dot products, one dot product for each output sample. Therefore, if you have a function that will calculate a dot product, it can also calculate a convolution.

An item of consequence is the use of modular arithmetic in calculation of the data addresses. Often times, it is desirable to traverse an array and return to its beginning automatically when the end is reached. Modular arithmetic allows this type of operation. When working in combination with an unusual step value, modular arithmetic will permit the collection of coefficients from a rectangular array placed in linear memory space. Here the step value refers to one of the numbers associated with the mac or the rmac instruction. These two values are called xo and yo. This value is used to increment the address of the corresponding register whenever data is loaded into either the H or the I register. The location of the array in memory should be placed at special location in memory. This location is discussed below. The effective address for the next value to be placed in the X after the value is incremented by the xo register is given by

IX = (IX)&~XMASK | ((IX)+xo)&XMASK

Note that XMASK here is an 8-bit mask that defines the length of the circular array. The array length must be a power of two less than or equal to 256. The value placed in XMASK is one less than the array length. When the contents of IX, or (IX), is anded with the complement of the sign extended value of XMASK, the value that is left is the starting address of the array. The second term above causes the contents in IX to be incremented. As the value in IX is replaced by ((IX)+ xo)&XMASK after each multiply and accumulate operation. Since XMASK must contain a number that is one less than a power of two raised to the n: its least significant n bits are 1. The value placed in IX is the address with the n least significant bits masked off. When these two values are ORed together, an address is created that will range from the beginning to the end of the array and then back to the beginning in steps of xo. The requirements to make this scheme work are:

1. The array length must be a power of two less than or equal to 256.

2. The array must begin on a specific address. This address is any value where the least n bits are zero. Here n is determined by

$n = \log 2$ (array length)

or

array length $= 2^n$

3. The value in `xo` must be equal to the step between adjacent data samples in the array.

This scheme can be used to move, in a modular manner, around multiple-dimensional arrays as well as one-dimensional arrays. Let us now modify the earlier code to show how a circular buffer can be used to advantage.

The above compilations use the default memory model for the Cosmic C compiler for the MC68HC16. This computer has a memory addressing space of 20 bits. In the design of the part, the basic address registers were all made 16 bits wide, and the additional bits required to address the total space were placed in extension registers. Therefore, for each of the addressing registers, the stack pointer, the program counter, and the index registers X, Y and Z, there is a 4-bit extension register. These registers are called SK, PK, XK, YK, and ZK respectively. As mentioned earlier, several of these extension registers are initialized upon reset, and the remainder must be initialized prior to the execution of the main program. Usually, when a calculation alters a value in an extension register, this change takes place seamlessly. It is typically not necessary to worry about the contents of the extension registers. Recall in the code for the initialization routine listed above, `crts.s`, that the values for XK and ZK were set to 0 and the value placed in YK was `0xf`. The reason for this choice is that the compiler automatically uses the Y register as an offset when calculating the addresses listed in the various header files. Since all of these registers are in the highest memory page, the YK value of `0xf` is appropriate.

The default memory model is called the compact model, and the code is compiled in the compact model so that all code, data, and stack memory space is contained within one 65 kilobyte (K) memory bank. Therefore, the initialized values of the extension registers need never change. On the other hand, it is sometimes necessary to change the values in these registers, and it is desirable to have the compiler take care of this bookkeeping when needed. All of the additional memory models will provide this tracking of the extension registers. These models are: *small*, one 65 K bank for code and one 65 K bank for data and stack; *program*, multiple 65 K banks for code and one 65 K bank for data and stack; *data*, one 65 K bank for code and multiple 65K banks for data and stack; and finally *far*, multiple 65 K banks for code, data, and stack. With each of these memory models, it is necessary to keep track and often change the contents of the extension registers. As a result, with these models, it is necessary to alter slightly the stacking sequence. Such a sequence is shown in Figure 6-5. The function call that will cause the stack to be arranged as is shown in the figure below has four arguments. From left to right these arguments are `xlen`, `*xdata`, `ylen`, and `*ydata`. The lengths are simple character values, and the pointers in this case must be 20-bit values. If a function call to this routine is compiled with the small memory model, or any other model with the exception of the compact model, the stacking will include the extension registers as shown below.

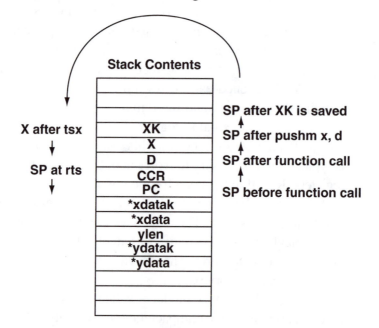

Figure 6-5: Stacking Sequence That Passes Extension Registers

The B register, which is the righthand side of the D register, will contain the left-most parameter when the function is called. Therefore, the parameter `xlen` will be found in the least significant bits of the location labeled D in the above stack outline. The routine shown below will make use of both the X and the Y registers. The compiler allows for all registers with the exception of the X register to be altered by any function call. Therefore, it is necessary to save the contents of the X register prior to altering this register.

```
; int circular_conv(char xlen, int* xdata, char ylen, int* ydata)
    .even
    .set xlen=5
    .set xk=11
    .set xdata=12
    .set ylen=15
    .set yk=17
    .set ydata=18


_circular_conv:
    pshm x,d
    txK  ; save old xk
    pshm d
    tsx
    clrm ; clear the m register
    clra ; put ylen into e register
    ldab ylen,x ; ylen
    addb #-1  ; decrement count to get correct
    tde  ; number of iterations
    ldaa xlen,x ; xlen is the circular part
    asla ; of the convolution. Create
    coma ; a mask that is the compliment
    clrb ; of twice the length.
    tdmsk   ; put xlen in XMASK
    ldab yk,x
    tbyk
    ldy ydata,x  ; ydata with yk extension
    pshm x ; save x
    ldab xk,x
    tbxk
    ldx xdata,x ; xdata with the xk extension
    ldhi ; load H and I registers for mpy
    rmac 2,2  ; do rmac ylen times
    tmer ; send result to e and then
    pulm x ; restore x
    ldx 2,x ; restore it to its original value
```

```
pulm d ; restore old xk
tbxk
ais #4 ; fix the stack
ted   ; data to be returned
rts   ; return
.public _circular_conv
.end
```

Listing 6-10: Circular Convolution Routine In Assembly Language

Variable names get lost from the program when executing an assembly language program called from C. Therefore, when a function is entered, the names of the variables are replaced by offsets from a table saved by the compiler. Programming functions without the aid labels and variable names requires careful attention to the details of stacking and unstacking these data. It is recommended that when a program is to be prepared, an abbreviated version like the function dot_product( ) above—which will guide the programmer in setting up the variable locations in memory—be written. In the above routine, the offset OFST was not used, and the various offsets were assigned names that correspond to the variables. This approach makes it easier to understand what the program is doing and easier to write the code correctly.

It is important that the contents of the XK register be restored to its initial value when a function call is returned to the calling program. The content of the X and D registers is saved on the stack and the value or XK is then saved. The contents of the stack are placed in the X register, and this register is incremented by two so that it will point to the last data pushed on the stack. The M register is cleared, and the parameter ylen is moved into the E register. This value is used with the rmac instruction to tell how many times the mac instruction is to be repeated. The value of xlen is the length of the buffer that contains input data. This buffer is fed circularly so that when the calculation reaches the last entry in the buffer the pointer into the buffer will be returned to the top of the buffer to get its next entry. The length xlen is the number of entries in the buffer, but the length for the calculation must be the number of bytes in the buffer. Therefore, the value of xlen must be doubled prior to the creation of a mask to be used in XMASK. In this case, XMASK value is calculated, and YMASK which is stored as the B register content of the D register is made zero. Therefore, there will be a circular buffer calculation on X, but there will be no circular calculation on the Y register.

After the MASKs are set up, the two 20-bit pointer registers, X and Y, are assigned the values passed as parameters, and then the product data are put into place. The rmac instruction is then executed the number of times indicated by the contents of the E register. The register contents are then restored and the result of the single-point convolution is moved into the D register before control is returned to the calling program.

A simple program was written to test the operation of this function. This function does not attempt to make a digital filter, but it rather sets up to show how the circular convolution works. In this case, it was intended that the program run on an EVB16 board that has a normal serial interface to an RS232 port. Therefore, the data being tested can be sent out of the serial port to a terminal. The purpose of the program is really quite simple. It is to execute a convolution between a short set of coefficients and a long set of data. The coefficient set is 32 integers long, and the data set is 64 integers long. The coefficient data is a descending array of numbers that are in the upper byte of the number. These numbers start at 31 and reduce successively to zero. The data that will be used here are merely the numbers 0 through 63 in the upper byte of the number. The system is set up, the system frequency is set to 16.78 MHz, and the watchdog is disabled. The serial port is set to 9600 baud and the SCI transmitter is enabled.

```
#include "hc16.h"
#include "sim.h"
#include "qsm.h"

int putchar(char new_character);
void dprint (int c);

void main(void)
{
   int circular_conv(char, int*, char, int*);
   int data[64], coef[32],point[64],i,j,*ip;

   /* initialize the SIM */
   SYNCR.X=1; /* set the system freq to 16.78 MHz */
   SYPCR.SWE=0; /* disable the watchdog */

   /* initialize the SCI */
```

```
    SCCR0.SCBR=55; /* set baud rate to 9600 */
    SCCR2.TE=1; /* enable the transmit of the SCI */

    for(i=0;i<64;i++)
        data[i]=i*0x100;
    for(i=0;i<32;i++)
        coef[i]=(31-i)*0x100;

    for(i=0;i<64;i++)
    {
        ip=data+i;
        point[i]=circular_conv(64,ip,32,coef);
    }

    for(i=0;i<8;i++)
    {
        for(j=0;j<8;j++)
        {
            dprint(point[j+8*i]);
            putchar(' ');
        }
            putchar(0xd);
            putchar(0xa);
    }

}

int putchar(char new_character)
{
    while(SCSR.TDRE==0); /* wait until transmit buffer empty */
    SCDR=new_character;/* send out byte and reset TDRE */
}

void dprint (int c)
{
    if(c<0)
    {
        putchar('-');
        c=-c;
    }
    if ( c/10)
    dprint(c/10);
        putchar(c%10+'0');
}
```

Listing 6-11: A Test Program for Circular Convolution

Next, the circular convolution program is executed 64 times. These results are stored in the array `point[]`. This array is then sent out the terminal eight numbers at a time. A few lines of code are extracted from Figure 6-2 to make the `putchar()` function. Finally, the function `dprint()` sends the data out of the serial port to the terminal. The output from this program is as follows :

64 64 64 64 64 64 64 64
64 64 64 64 64 64 64 64
64 64 64 64 64 64 64 64 .
64 64 64 64 64 64 64 64
64 64 64 64 64 64 64 64
64 64 64 64 64 64 64 64
64 64 64 64 64 64 64 64
64 64 64 64 64 64 64 64

This program is merely a test program to show that the circular convolution does indeed work. The fact that the output data is always the same value, decimal 64, shows that the addresses are handled correctly inside of the circular convolution. Let us examine first to determine why the answer should be 64. The coefficients and data are each `0x100`. Two of these values are multiplied and summed 32 or `0x20` times. Therefore, one would expect that the result would be `0x20000`. However, you must remember that each of the above products is in fact the product of two binary fractional numbers. The binary point in each case is between bit numbers 14 and 15. The product of these numbers will yield `0x1000`, but in that case, with the binary point between bits 29 and 30. Actually, the binary point dictated by the microcontroller is between bits 30 and 31. The answer is corrected to this binary point location in the M register, and the result of the product is then `0x20000`. This number is summed 32 times and the final result is `0x400000`. When the M register is moved into the E register and then into the D register, the value that is saved is `0x40`, or 64, as the test program showed.

Remember Equation 6-1 for the convolution:

$$y_k = \sum_{i=0}^{n-1} x_{k-i} h_i$$

In this expression, it seems possible that the x subscript can have a negative value. Actually, such a case is not possible because a negative subscript implies that data is used before it is available. Positive subscripts correspond to time that has already passed. Values of x will be zero for negative subscripts. If the kth sample corresponds to "now," increasing values of i will get older samples of x. This operation can lead to a little problem in creating the code for the convolution. The looping construct within the assembly program above selects the different values of i in the above equation. The coefficients h_i are placed in memory in successive order so that an increase in the value of i will select the correct next coefficient. However, if the data values x were placed in memory as one would naturally expect, the newest value of data would be at the current array index, and old values of the data would be at lesser index values. This arrangement will not work correctly. The data must be placed in the array backwards in order to get the convolution to work. Older data must be at higher indices than the current data sample. Also, when filling the array initially, the program should start at the top of the array rather than the bottom. The routine listed below will store the data properly in the array.

```
int data[64];

int handle_data(char new_data)
{
    static int i=63;
    if(i<0)
    i=63;
    data[i-]=new_data;
    return i+1;
}
```

Listing 6-12: Convenient Data Storage For DSP Use

This routine is integrated into the code shown in Listing 6-11 and used to test the circular convolution. This resultant program is shown in Listing 6-13. In this case, the variables data and coef are moved outside of the function main() to make them global.

```c
#include "hc16.h"
#include "sim.h"
#include "qsm.h"

int putchar(char new_character);
void dprint (int c);
int handle_data(int new_data);

int data[64] @0x2000;
int coef[32] @0x2100;

void main(void)
{
    int circular_conv(char, int*, char, int*);
    int point[128],i,j,*ip;

    /* initialize the SIM */
    SYNCR.X=1;  /* set the system freq to 16.78 MHz */
    SYPCR.SWE=0;    /* disable the watchdog */

/* initialize the SCI */
SCCR0.SCBR=55; /* set baud rate to 9600 */
SCCR2.TE=1; /* enable the transmit of the SCI */

for(i=0;i<64;i++)
    data[i]=0x00;
for(i=0;i<32;i++)
    coef[i]=0x100;

for(i=0;i<64;i++)
{
/* get new data—use 0x100 for this test */

    ip=data+handle_data(0x100);
    point[i]=circular_conv(64,ip,32,coef);
}

for(i=0;i<8;i++)
{
        for(j=0;j<8;j++)
        {
           dprint(point[j+8*i]);
           putchar(' ');
        }
```

```
        putchar(0xd);
        putchar(0xa);
    }


}
int handle_data(int new_data)
{
    static int i=63;

    data[i--]=new_data;
    i &=63;
    return (i==63)?0:i+1;
}
```

Listing 6-13: Test Program For The Circular Convolution

Then the code to send the data to the `circular_conv()` is modified slightly to that shown below.

```
for(i=0;i<64;i++)
{
/* get new data — use 0x100 for this test */

    ip=data+handle_data(0x100);
    point[i]=circular_conv(64,ip,32,coef);
}
```

The value `0x100` is sent into the function `handle_data`, and the return is the index into the array where the data was stored in the array data. When this integer is added to data, which is a pointer to a type `int`, the pointer `ip` will point to the location into the array where the latest value was stored in memory. Under normal circumstances, this piece of code would be entered under control of a clock and the data sent into the routine `handle_data()` would be new input from an analog to digital converter. Also, the `for` statement is not expected in a practical application. The output from this program is:

```
 2  4  6  8 10 12 14 16
18 20 22 24 26 28 30 32
34 36 38 40 42 44 46 48
50 52 54 56 58 60 62 64
64 64 64 64 64 64 64 64
64 64 64 64 64 64 64 64
64 64 64 64 64 64 64 64
64 64 64 64 64 64 64 64
```

The data array starts out empty, and the coefficients are each 0x100. The output is as you would expect—it starts and increases linearly from 0 to 64 in the first 32 samples. The output then remains at 64 until the end of the test. This linear ramp is exactly what one would expect when sending a step function into the constant weighting function.

There is one important point to be found in this program. It was stated earlier that the programmer should attempt to keep the code in C whenever possible. In the above case, we introduced a rather simple function in assembly language. The fact that the program was a function, and it was necessary to use a large memory model, the code required to handle the data as it was passed into the function is much longer than you would expect. Also, the implied C code in preparing the machine for the function call and the handling of the extension registers in the main program makes the overall program larger than expected. Unfortunately, it is not possible to access the DSP with C, so if you wish to do DSP operations, assembly language access is all that you can use.

Other MC68HC16 Considerations

The discussion in this chapter has been dedicated to the MC68HC16 microcontroller. Of course, there are components of this microcontroller that are not covered in this chapter. No attempt was made to outline the access to the analog-to-digital converter, several features of the general purpose timer, the serial peripheral interface, or the static RAM. However, the programs shown here do cover enough of the part to demonstrate that the on-board peripherals can be accessed from the C language. In the approach used, the header files contain all of the bit field definitions needed to access any bit or bit field in any control register in the part.

Chapter 7

Advanced Topics in Programming Embedded Systems (M68HC12)

During the past few years, we have seen an unbelievable proliferation of embedded systems products. Devices that could only be imagined ten years ago are commonplace today and their very existence demonstrates the importance of C programming for small microcontrollers. Let's take a look at one such application and see how easily you can develop rather complicated applications on microcontrollers.

Throughout this text, much emphasis has been placed on the construction of small functions and then integrating these functions into a working package. This approach is about the only way that you can really hope to create a complicated piece of firmware in a sensible time. My early projects would always start with careful design of the whole project, partitioning of the project into sensible modules, design of each module, writing the code for each module, integration of the whole project and then, after cleaning up syntax errors, I'd begin to test the whole program. What a disaster! These programs would never work and there would be no hope of ever getting the package to run as a unit.

As I gained experience, I found that the top-down approach I used was probably satisfactory if I tempered the integration of the system. Today, I start with a careful design for the whole system. This design is partitioned into constituent components. At that point, another look at the design is in order to see if the existing components can be further reduced into sensible components. At that point, the lowest-level components are coded. Often these functions are so simple that they work when first tested. Whenever there is a required interface for a component, I take the time when it is being created to write stub functions to provide any necessary interface to test the function. Every small function is tested as it is written.

The hierarchy of these functions is built upward. Those functions that make use of the lower-level functions are coded and tested, and so forth until the whole project is completed. Every function is written and tested as it is created so that, as the program builds upward, it is always based on known working modules. And this approach is used to complete the whole program.

At every phase in the construction of the project the key words are test, test, test. Every function is tested at its creation and, therefore, the whole project is built upon relatively simple, small functions that have been tested and work. Does this approach guarantee that no bugs will be built into the program? Does it guarantee that the final integrated version of the program will work as desired? In both cases, the answer is a resounding NO. However, you have at least *some* chance of creating a program that can be debugged and will meet the desired specification. Also, you will find that writing and testing a series of small functions always requires less time than writing and testing the aggregate, more complicated function.

In summary, make your functions small, test them until there is no possibility of hidden errors, revise them and retest them always with the intent of reducing the function size. Then, when you have completed the function, parade it in front of your peers and see if they can suggest ways to improve the functions. Build your project with such blocks and you will have a reasonable chance of meeting your deadlines.

In this chapter on the M68HC12, we will discuss programming into the chip a part of the features of a telephone, or perhaps of an electronic phone book. Note that the problem discussed here is but a small part of the control of a telephone. A telephone book function is one complete module that is a part of the telephone control.

As it stands, the HC12 has a small amount of EEPROM into which a phone book can be stored. The chip also has a large amount of FLASH ROM in which the program is stored. The FLASH memory has some disadvantages that discourage its use to store the telephone book data. On these chips, the FLASH memory is broken into two parts. The smaller portion of the FLASH is called BOOT FLASH and the larger portion is specified for general program storage. If you need to erase any memory, the whole block, either the BOOT or the general program storage, must be completely erased. Therefore, if you should want to change the contents of the phone book, the whole block would have to be erased and rewritten. Another approach that could be used is just to assume that you will have enough FLASH to write over and discard the memory used and move the entry to new memory whenever it is changed. This approach, however, is wasteful of the memory.

The EEPROM does not suffer this problem. Any individual byte in the EEPROM memory block can be written, read, or erased without a problem. Also, the EEPROM can be erased and rewritten at least 10,000 times without deterioration. FLASH, on the other hand, is specified to be able to withstand 100 erase/write cycles during its lifetime. All in all, FLASH makes a good program storage memory and EEPROM makes a better changeable nonvolatile memory.

The C compiler used in this chapter is that provided by Cosmic. This compiler is essentially the same as that seen in Chapters 5 and 6 except, of course, it creates code for the M68HC12 family of parts. One very nice extension of the compiler is the ability to identify EEPROM in the code. The approach followed here uses the following #pragma:

```
#pragma data[768] @eeprom
```

It identifies data as an array 768 bytes long that is stored in EEPROM. At link time a command like

```
+seg .eeprom -b 0xd00 -m768
```

indicates that the EEPROM will be found at the address 0xd00 and it will be 768 bytes long. The nice feature derived from this extension to the C language is that any assignment to the data array will first erase the byte and then store the data into the specified location automatically.

The fact that the HC12 component that is to be used for this project contains a large amount of FLASH and little RAM leads to some difficulty in writing code for use on this part. There are no development environments that contain sufficient RAM to test a significant program. Therefore, prior to executing the first byte of code on the target chip, you must be more certain than normal that the code will work as intended. I solve this problem by developing most of the code for the final configuration to execute in a DOS environment and only after I have a complete working program is there any attempt to move it into the target system. Code developed in this manner was discussed in the previous paragraphs. The main difference is that the development flow will be for a DOS-based system and the microcontroller-based code will be very carefully designed and tested and integrated into the program as it is developed. The final tests after the code has been transferred to the target system will be limited to those items that are specific to the target system only.

Let us look now at what we would like our phone book code to do.

The purpose of the program is to allow storage of names and telephone numbers in the EEPROM section of an M68HC912B32. This chip has 768 bytes of EEPROM and 32K bytes of FLASH EEPROM. It has an on-board UART through which all of the communications with the chip are conducted. One of four single-letter commands can be entered into the system:

Command	Response
n	Receive a NAME terminated by an <enter> followed by a phone number also terminated by an <enter>.
s	Display the entire directory contents.
a	Display the next directory entry.
r	Delete the entire contents of the directory.

Nonvolatile storage is at a premium. Therefore, all data stored in EEPROM will be encoded to compress the data as much as practical. All numbers will be stored in BCD form. This approach requires 4 bits per stored number when, in fact, 3.32 bits per digit is required if it is assumed that the use of each number is equally likely. Confusion can result when an empty number is stored, so the value stored for the number 0 will be `0xa` rather than `0x0`.

Alpha, or letter, data will be compressed using a Huffman code as was shown in Chapter 5. This code will be written specifically to compress data from the names found in a telephone directory. Frequency of letter usage here is different from that found with general English text. The decode scheme to be used here will follow the general approach given in Chapter 5.

As a first estimate, the following functions will be required in putting this program together:

Monitor	This function executes all of the time and receives data from the keyboard. It interprets the entries and passes control to the appropriate function to execute.
Encode	Encodes the alpha data read from the keyboard with a Huffman code.
Decode	Decodes the Huffman encoded data stored in FLASH when needed.
Numdup	Converts numeric data passed in an array to the modified BCD format and saves these data in the FLASH array, also a passed parameter.
Putbcd	Converts the encoded numeric data contained in the passed array to ASCII form. Places the converted data in an array that is passed to the function.
Getchar	Reads in a character from the serial port. This function and `putchar()` below work with the standard library input/output functions that will be used by the program.
Putchar	Sends a character to the serial port.
Get	Reads in a character string, either numeric or alpha from the serial input.
Printout	Prints the contents of the phone book stored in EEPROM.
Printafter	Prints the next entry in the phone book.
Saveit	Saves the phone book entry in the proper EEPROM location.
Reset	Erases the contents of the EEPROM.

Most of these functions have nothing to do with the underlying computer. Therefore, we will write code that is completely independent of the computer. If there is ever a potential modification in this code when changing to the embedded microcontroller, standard compiler control commands will be used.

Numeric Encoding

Data will be entered from a keyboard that is assumed to provide an ASCII character stream. In operation, the first entry will be the telephone number. As this number is received the coding will be converted from ASCII to a modified BCD format. The modification is needed to eliminate trouble with the occurrence of zeros in the number. The conventional BCD encoding for a zero is a 0x0 that is 4 bits wide. If you should have a double zero, the encoded version would be an 8-bit zero or '\0', which is interpreted in C as an end of a character string. That confuses the issue enough that it was decided to encode the digit zero as 0xa. The literal interpretation of this number is the value ten. But, with our BCD encoding, the number ten will never occur, so it is safe to use this value for the value zero.

```
/* The ascii data in the constant array s[] con
tains a number. These data are converted to a
modified BCD form and stored in the array
array[]. The number zero is stored as 0xa. The
series is terminated with a null character
followed by an enter character. */

#include <ctype>

int numbdup(char * const s, unsigned *array, int len)
{
    char *pq,*sp;
    int i;

    sp=s;    /* use local pointer */
    pq=(char *)array; /* convert pointer to character */
    for(i=0;i<len;i++)  /* empty the array */
      pq[i]=0;
    i=0;
    while(*sp!='\0'&&*sp!='\n') /* read until termination char*/
    {
      *pq=0;
      if(isdigit(*sp)) /* convert the inputs two at a time*/
```

```
{
  if(*sp=='0') /* handle a zero input */
       *sp='0'+0xa;
  *pq|=(*sp-'0')<<4 ;
}
else
  pq|=0xf0; /* non digit, mark it */
if(isdigit(*(sp+1))) /* the next input */
{
  if(*sp=='0')  /* treat a zero */
       *sp='0'+0xa;
  *pq|=(*(sp+1)-'0');
}
else
  pq|=0xf; /* another non digit */
sp+=2; /* Increment the input data pointer, */
pq++;  /* the output data pointer, */
i++;     /* and the data count */
}
return i; /* length of the array */
}
```

Listing 7-1: Numeric Encoding Routine

All of the storage arrays in the EEPROM are of the type `unsigned int`. Here the storage of both numeric data and alpha data requires that every bit of every storage location be used so that unsigned is the norm. In the coding routines, the data are passed in as characters and the destination arrays are unsigned. Therefore, to aid the local bookkeeping, the destination array pointer is immediately assigned to a type `char *` and this pointer will be used to store the encoded data.

The following program provides a simple test for the numeric coding. This program has a serious problem though. All computers configure memory in either a big endian or a little endian order. In the big endian order, the most significant byte, 8 bits, of a 16-bit address is given the smaller address and the least significant byte goes to the larger address. The little endian order is just the reverse.

Here the least significant byte of data is assigned to the smaller address and the most significant byte goes to the larger address. Almost all Motorola chips use big endian, and almost all Intel chips use little endian. There can be some confusion when developing code to run on one style of data storage on a machine with the opposite. This problem is seen in the following program.

```c
#include <stdio.h>

main()
{
    unsigned array[25];
    int i;

    numbdup("123456789098765",array,25);
    for(i=0;i<8;i++)
      printf("%x",array[i]);
    putchar('\n');
}
```

Listing 7-2: Numeric Encode Test

If this program is compiled with a PC (Intel-based) compiler, the result will not appear to be correct. However, if the program is compiled on an HC12, or 68HC16, or 683XX, or 68HC11, or 68HC05 compiler, it will seem to work correctly. In fact, both results are correct, only the numeric representation in memory is different.

Numeric Decoding

Once the numeric data are encoded and stored, they must be decoded to be used by other parts of the program. The decode routine is called putbcd(). This function is shown below.

```c
void putbcd(char *s,char *number)
{
    int c,i=0;
    char *sa;
    sa=s;
    while(*sa!='\0')
    {
```

```
      if(isdigit(c=(*sa>>4)+'0'))
        number[i++]=c;
      else if(c=='0'+0xa)
        number[i++]='0';
      if(isdigit(c=(*sa&0xf)+'0'))
        number[i++]=c;
      else if(c=='0'+0xa)
        number[i++]='0';
      sa++;
    }
    number[i++]='\n';
    number[i++]=0;
}
```

Listing 7-3: Numeric Decoding Routine

In this function the output data is called `number[]` and the input is `s[]`. The encoded data in `s[]` is converted one BCD 4-bit field at a time to ASCII characters. In the event that a character received has a value '0'+0xa, it is then a character zero or '0'. Each byte is converted from two 4-bit BCD values to two ASCII characters that represent the proper digits.

The test program for this routine is a combination of the encode test routine with one to decode the encoded data. As one would expect, when both the encode and the decode routine are used together, the result is correct. This observation is true on either the Intel or the Motorola style chip. The endian-ness of the chip is immaterial when the entire encode/decode operation is completed.

```
#include <stdio.h>
#define ARRAY_SIZE 100
int decode(unsigned M[],char *s);
int encode(char *a,unsigned *array,int length);

main()
{
    char a[ARRAY_SIZE] ;
    int c,i=0;
    unsigned array[ARRAY_SIZE];
    char s[ARRAY_SIZE];
```

```
while((c=getchar())!='\n')
  a[i++]=c;
a[i]='\n';
encode(a,array,ARRAY_SIZE);
decode(array,s);
printf("%s",s);
}
```

Listing 7-4: Numeric Decode Test

Coding the alpha data

For encoding and decoding the name data to be stored in our phone book, we will use a Huffman code. We saw the decoding of a Huffman code in Chapter 5 and the decoding approach used here will be almost the same as was used there. In the discussion in Chapter 5 there was no encoding and that feature must be added here. To do justice to the encoding technique, it is necessary to try to build the code to encode the type of text that a phone book represents. There is no reason to suspect that a collection of English names will contain the same character frequency as standard English text. It is necessary to understand the frequency of occurrence of each letter in the text to be encoded. With this understanding, you can write a code that assigns few bits to frequently occurring letters and more bits to letters that occur less frequently.

The program below reads in data, counts the occurrences of letters, both upper and lower case, in a document. The occurrence of letters is sorted in order of decreasing occurrence. These data are printed out. The program calculates the average theoretical entropy, bits per character, of each character in the document and displays this number. Also included in the calculations are the space, ' ', character and the new line, '\n', character. These characters cause the output to be distorted, so the character '>' is used to indicate a space character and a '<' indicates a new line. These data will then be used to create a Huffman code used to compress the data prior to storage in the internal EEPROM.

This particular program, along with many variations, has been an exercise used for years in classes. It does demonstrate some important considerations. The shell sort was used in Chapters 2 and 5 to sort data, and here we will use it again. In this case, the data to be sorted is contained in an array of structures. The structure is `typedefed` and called a type `Entry`. Each instance of an `Entry` contains an integer value count and a character named `letter`. The `letter` is the actual letter being recorded and count is the number of occurrences of the letter in a document.

Our alphabet consists of the normal 26 letters plus the space and new line characters. Therefore the constant LETTERS is given a value of 28.

In the main program, the necessary variables are defined. Note that the array of `Entrys` named `letters[]` contains LETTERS entries. The variables used to count the input data are initialized. The variable `characters` is initialized to zero. The character member of the array `letters` is initialized to the actual character values in order and the `count` value is initialized to zero. The character values in the last two entries in the array are initialized to '>' and '<' respectively to correspond to the space character and the new line character.

When reading the data in, each character is operated on by the `tolower()` function. This operation converts any upper-case letter to a lower-case letter, but it does not alter any other characters. If the character returned is a letter, a space or a new line character, it will be processed by the following block. Otherwise, the character is discarded and a new character is read in by the argument of the `while()` loop. As the characters to be processed are detected, the corresponding `letters.count` is incremented in the array. After all of the data are entered, an EOF is detected, the data are sorted and then printed out.

The modifications to the earlier shell sort are minimal. First of all, the array passed to the routine is identified as a type `Entry` rather than an `int`. Also, the `temp` variable is a type `Entry`. Then the comparison in the test argument of the innermost `for()` loop is converted to compare the two `v[].count` entries. The swap operation that follows needs no modification.

```
#include <stdio.h>
#include <math.h>
#include <ctype.h>

#define LETTERS 28

typedef struct{
```

```
      int count;
      char letter;
}Entry;

int main()
{
    int c,characters,i;
    Entry letters[LETTERS];
    double a,sum;

    characters=0;
    for(i=0;i<LETTERS;i++)
    {
     letters[i].count=0;
     letters[i].letter=i+'a';
    }
    letters['z'-'a'+1].letter='>';
    letters['z'-'a'+2].letter='<';

    while((c=getchar())!=EOF)
    {
     c=tolower(c);
     if(isalpha(c)||c==' '||c=='\n')
     {
      characters++;
      if(c>='a'&&c<='z')  /* count the letters */
          letters[c-'a'].count++;
      else if(c==' ')
          letters['z'-'a'+1].count++; /* count the spaces */
      else if(c=='\n')
          letters['z'-'a'+2].count++; /* count the new lines */
     }
    }
    /* got all of the data in and processes, print it out */
    shellsort(letters,LETTERS);
    printf("\n\n");
    printf("Char    Frequency    Char
```

```
Frequency\n\n");
  for(i=0;i<LETTERS/2;i++)
    printf(" %c      %7.4f     %c      %7.4f\n",
       letters[i].letter,
       100.*letters[i].count/characters,
       letters[i+LETTERS/2].letter,
       100.*letters[i+LETTERS/2].count/characters);
  printf("There are %d characters\n",characters);
  sum=0.;
  for(i=0;i<LETTERS;i++)
  {
    a=1.*letters[i].count/characters;
    a=(a!=0)?a:0.00001;
    sum += -a*log(a);
  }
  sum /=log(2);
  printf("The theoretical average bits per char-
acter is %f\n",sum);
}


/* shellsort: sort v[0] ... v[n-1] into increasing
order */
void shellsort(Entry v[], int n)
{
  int gap,i,j;
  Entry temp;

  for(gap=n/2;gap>0;gap /= 2)
    for(i=gap;i<n;i++)
      for(j=i-gap;j>=0 && v[j].count<v[j+gap].count;j-=gap)
      {
          temp=v[j];
          v[j]=v[j+gap];
          v[j+gap]=temp;
      }
}
```

Listing 7-5: Letter Analysis Program

The purpose of this code is to calculate frequency of occurrence of letters in a document and provide some guidance as to how well the compression approach developed works. This program was run with a ten-page instruction manual and then with a telephone book with 200 entries. The results of these two executions are shown below.

Char	Frequency	Char	Frequency
>	30.5977	l	2.1504
e	9.1109	p	2.0276
t	6.5040	f	1.8257
o	5.2664	u	1.2727
r	5.1523	y	1.2288
i	4.6959	g	1.1762
a	4.6169	b	0.8514
s	4.5730	w	0.5793
n	4.2746	k	0.5617
h	3.0194	v	0.2984
c	2.8263	x	0.2721
d	2.6946	q	0.0351
<	2.2031	j	0.0088
m	2.1768	z	0.0000

```
There are 11393 characters
The theoretical average bits per character is
3.797302
```

Output 7-1: Calculation of entropy for the document manual.doc

The outputs shown above follow very closely the expected occurrence of letters found in the typical technical text. The bits per character should be about 4.5, but this value is distorted because the space character is included in the count, and its very frequent occurrences distort the overall averages and hence the entropy per character found in the document.

Shown below in Output 2 is a repeat of the same calculation on the contents of a phone book. Note here that occurrences of the letters and other characters are quite different from those found above. Even though the phone book used to create the table below contained only about 200 entries, these data will be used to create a Huffman code to compress the data when storing names into the microcomputer EEPROM.

Char	Frequency	Char	Frequency
e	9.2042	d	2.7331
<	7.7572	k	2.3312
>	7.5563	b	2.2106
a	7.5161	y	1.9695
r	7.3151	p	1.8891
n	6.3505	u	1.7685
o	5.6270	g	1.7283
i	5.3859	w	1.2862
l	5.0241	f	1.2460
s	4.5418	j	1.2058
t	4.0595	v	0.8039
c	3.8585	x	0.1206
h	3.2958	z	0.1206
m	3.0547	q	0.0402

There are 2488 characters
The theoretical average bits per character is
4.392597

Output 7-2: Calculation of Entropy for Phone Book

Next, a Huffman code will be created to encode the data from the phone book. A Huffman code is built into a complete binary tree. Such a tree always has two descendents from every node unless the node is a leaf node. As such, whenever a Huffman tree is created to encode n characters, there will be $2n-1$ nodes in the tree. Figure 7.1 shows an instance of such a tree. This tree encodes the data shown in Output 2 above. As with most trees, analysis, or encoding, starts at the root node at the top of the page. Whenever you traverse to the left, a code value of zero is recorded. When traversing to the right, a code value of 1 is recorded. For example, the character R will be encoded as 0010, and the character M will be 111100. This table is constructed and filled to keep the most frequently occurring letters at the top of the tree and the least frequently occurring letters at the bottom. Therefore, the number of bits for each character is inversely proportional to its frequency of occurrence. This choice for the letter codes requires less than the number of bits one would expect when using the standard 8 bits per character.

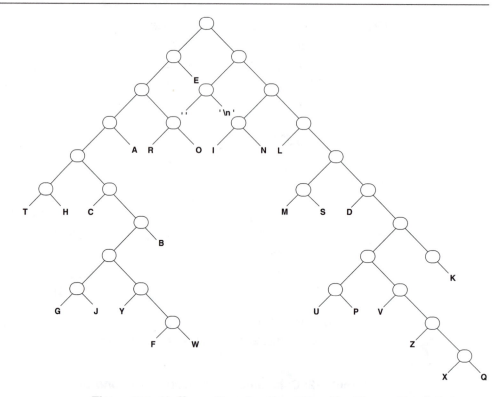

Figure 7-1: Huffman Tree for Encoding the Phone Book Data

We have seen above that the minimum number of bits per character for the telephone book is 4.39. We cannot expect to reach this level, but we should expect to be significantly fewer than 8 bits per character. The code corresponding to the tree in Figure 7-1 is shown in the following table:

Character	*Code*
' '	100
'\n'	101
a	0001
b	0000111
c	000010
d	111110
e	01
f	0000110110
g	000011000

h	000001
i	1100
j	000011001
k	1111111
l	1110
m	111100
n	1101
o	0011
p	111111001
q	11111101111
r	0010
s	111101
t	000000
u	111111000
v	111111010
w	0000110111
x	11111101110
y	000011010
z	1111110110

Table 7-1: Huffman Code for Compressing Telephone Book Names

The encoding routine is shown in Listing 7-6. Contained in the listing of the encoding routine is a look-up table that contains all of the codes. In this table, the first two entries correspond to a space character and a new line character. The following entries correspond to the letters in the alphabet. In other words, the third entry corresponds to the letter A and the seventh entry corresponds to the letter E. Notice that this table is defined as external, but it is labeled static so that there is no linkage to the table outside of the file encode.c.

In operation, this function receives three parameters. The first is a pointer to an array that contains the data to be encoded. This array contains a zero terminated string. The second array of unsigned integers is named array. Its length is the third passed parameter length. Encoded data are all loaded into this array. All of the local variables used by encode are straightforward. The variable bitbase is an unsigned int with its most significant bit set to one and the remainder of its bits zero. When the function is executed, the array[] is first filled with zeros. The variable i is initialized to zero and `bit` is given the value `bitbase`. Then the code for each character is retrieved successively. If the character read is a space, the first entry in the table is used; if it is a new line character, the second entry in the table is used; and if it is any other letter, the letter is converted to an index into the alphabet and that particular code, offset by two, is used as the code string for the input character.

```
static char *code[]={
"100","101","0001","0000111","000010","111110",
"01","0000110110","000011000","000001","1100",
"000011001","1111111","1110","111100","1101",
"0011","111111001","11111101111","0010","111101",
"000000","111111000","111111010","0000110111",
"11111101110","000011010","1111110110"
};

#include <ctype.h>
int encode(char *a,unsigned *array,int length)
{
   unsigned i,bit,bitbase=~(~0u>>1);
   int c=1;
   char *ptr,*pa;

   pa=a;
   for(i=0;i<length;i++)
    array[i]=0;      /* initialize the array */
   i=0;
   bit=bitbase;
   while(c!='\n')
   {
    c=*pa++; /* assumes file is not empty */
    if(isalpha(c=toupper(c))||c==' '||c=='\n')
    {
      if(c==' ')
           ptr=code[0];
      else if(c=='\n')
           ptr=code[1];
      else
           ptr=code[c-'A'+2];
```

```
        while(*ptr!='\0')
        {
                if(*ptr++=='1')
                        array[i]|=bit;
                bit>>=1;
                if(bit==0)
                {
                        bit=bitbase;
                        i++;
                }
        }
    }
  }
}
    return ++i;    /* the length of the coded array */
}
```

Listing 7-6: Encode Function

The program then enters a while loop that examines the contents of the code received. If the leftmost entry is a character '1' a value of bit is ORed into the location array[i]. In either case, *ptr=='1' or *ptr=='0', the value of bit is replaced by bit shifted right by 1. Whenever bit has been shifted until its value becomes zero, it indicates that the unsigned int value pointed to by ptr has been filled and bit is reinitialized to bitbase. Also at this time i, the index into array[], is incremented to get the next character to decode.

Decoding the alpha data

The above function encodes the alpha data entered in the array s[] into a Huffman code of the same data and returns the encoded data in the array array[]. Perhaps the easiest way to test the encode routine is to execute it in conjunction with its corresponding decode routine. The decode operation essentially recreates the tree shown in Figure 7-1. Rather than a two-dimensional rendition, it must be a single-dimension list. The list will have built-in mechanisms for traversing the tree from its root node to the encoded character based on the 1 and 0 patterns in the encoded data.

Recall in Chapter 5 the decode scheme involved intermixing jump distances in with the decoded characters in a table. There, the decode operation started at the zero entry in the table. The code being decoded was examined a bit at a time. If the code bit was zero, the table index was incremented by one. If the code bit was one, the value in that table location would be added to the table index. Whenever the table index fell to a location that contained a character, that character would be output and the index would be returned to the value zero.

That approach is fine for relatively small alphabets, as used in Chapter 5. Here, we are using the full alphabet, which makes the creation of the table above extremely complicated. Another approach was used this time. We still have a table that contains jump instructions intermixed with characters to be output. The characters to be output are each ORed with the hex value 0x80. The printable characters here are all identified with the least significant 7 bits of the character. Therefore, a test for a character is to determine if the value found in the table has a value when ANDed with 0x80.

Each numeric entry in the node table is broken into two nibbles. The left 4 bits correspond to the jump when a code 0 is found and the right 4 bits correspond to the jump when a code 1 is found. In other words, the decode operation starts at the beginning of the node table. If the first bit of the encoded data is a 0, the value found in the most significant 4 bits of the data is added to the node table index and the decoding is continued from that point in the node table. If the encoded data is a 1 the contents of the least significant 4 bits is added to the node table index. Whenever the node table index is changed, the value of that location is tested to see if the most significant bit is turned on. If so, that bit is turned off and the result is saved in output array. Otherwise, the process is repeated from that location until an output character is found. At that time, the node table index is reset to zero and the process repeated until a new line character is detected. Then the null character is put on the end of the output data and control is returned to the calling program.

One little problem with this approach: The number that contains the jump data can never have its most significant bit turned on. Therefore, the maximum jump when the encoded bit is 0 is seven. This restriction did not cause any difficulty when writing this tree. In fact, most of the time the jump caused by a 0 bit was 1 or 2. This restriction did cause a few longer jumps corresponding to 1. Overall, the table was quite easy to construct.

```c
static const char node[]={
 0x1d,0x21,'E'|0x80,0x41,0x21,'O'|0x80,'R'|0x80,0x21,
 'A'|0x80,0x1e,0x21,'H'|0x80,'T'|0x80,0x14,0x21,'\n'|0x80,
 ''|0x80,0x14,0x21,'N'|0x80,'I'|0x80,0x1f,'L'|0x80,0x12,
 'C'|0x80,0x21,'B'|0x80,0x14,0x12,'G'|0x80,'J'|0x80,0x12,
 'Y'|0x80,0x12,'F'|0x80,'W'|0x80,0x14,0x12,'M'|0x80,
 'S'|0x80,0x12,'D'|0x80,0x15,0x15,0x12,'U'|0x80,'P'|0x80,
 'K'|0x80,0x12,'V'|0x80,0x12,'Z'|0x80,0x12,'X'|0x80,
 'Q' |0x80
};


int decode(unsigned M[],char *s)
{
    unsigned mask,maskdo = ~(~0u>>1);
    char i=0,k=0,l=0;       /* l is the node pointer,
                            i is the byte pointer,
                            M is the message pointer */
    mask=maskdo;
    while(k !='\n')
    {
       if((mask & M[i])==0)
            l+=node[l]>>4;
       else
            l+=node[l]&0xf;
       if(node[l]&0x80)
       {    /* if a printable, send it out */
         *s++=(k=node[l]&0x7f);
         l=0;       /* also go to the root node */
       }
       if((mask>>=1)==0) /* if the mask is 0, turn
                          on MSB */
       {
       mask = maskdo;
       i++; /* and get the next byte from message */
       }
    }
```

```
     *s=0;
     return i;
}
```

Listing 7-7: Huffman Decoding Data

The above function is tested in conjunction with the encode routine with the following relatively simple program. In this code, provision is made to enter a line of text from the computer keyboard. This text is terminated when a new line character is detected. These data are then sent to the encode routine. The encode routine returns the encoded data in the array `array[]`. This array is passed to the decode routine. The return information from decode is contained in the array `s[]`. This string is then printed out to the screen.

```c
#include <stdio.h>
#define ARRAY_SIZE 100
int decode(unsigned M[],char *s);
int encode(char *a,unsigned *array,int length);

main()
{
 ' char a[ARRAY_SIZE] ;
   int c,i=0;
   unsigned array[ARRAY_SIZE];
   char s[ARRAY_SIZE];

   while((c=getchar())!='\n')
      a[i++]=c;
   a[i]='\n';
   encode(a,array,ARRAY_SIZE);
   decode(array,s);
   printf("%s",s);
}
```

Listing 7-8: Encode / Decode Test Routine

The above program echoes the input string to the computer screen. All lower-case letters are converted to upper case in the process.

Read data from the keyboard

A function get() is used to read data from a keyboard into a data buffer. This function is used in the monitor. A problem with many such functions is that they do not provide proper protection from a buffer overflow as the data are read in. The standard library function fgets() almost meets the needs of this function and more. The "more" in this case is the reason that we should not use the fgets() in this case. This function is part of the standard library and as such, it requires the definition of an input. The most often used input file here is the one named stdin. When we construct this system, we do not want to include all of the side effects of adding the standard library to our system. Therefore, in this case, it is probably best to write the function get() from scratch.

The function get() is shown below. This function takes two parameters. The first is a pointer to a character string where the input data are to be stored and the second is the length of this array. In the event that the input data size exceeds the array size, the data array is filled with zeros. Otherwise, the new line character is placed on the end of the string and the string is terminated with a null character.

```c
void get(char* a,int n)
{
    /* read in field and terminate the read with an '\n' */
    int i=0,c;

    while((c=getchar())!='\n' && i<(n-1))
       a[i++]=c;
    if(i<n-1)
    {
       a[i++]='\n';
       a[i]='\0';
    }
    else   /* input did not terminate soon enough */
       for(i=0;i<n;i++)
            a[i]=0;
}
```

Listing 7-9: get() Input Data Routine

This function is tested with the following program:

```
#include <stdio.h>

void get(char *, int);

#define LENGTH 15

main()
{
    char data[LENGTH];

    get(data,LENGTH);
    if(data[0]=='\0')
      printf("Buffer overflow\n");
    else
      puts(data);
}
```

Listing 7-10: get() Test Routine

This program reads in a line of data and echoes the string to the computer screen. If the length of the input data is longer than the specified length, the Buffer overflow message is printed to the screen.

The Monitor Program

The next program to be written is the monitor routine. This function executes all of the time and receives data from the keyboard. It will interpret the entries and pass control to the appropriate function to execute. In building all of the functions, `monitor()`, `printafter()`, `printout()`, and `reset()`, there are a large number of constants and function prototypes that must be included in each function. All of these items will be collected together into a single header file to be included in each function. This header file is shown below as Listing 7-11. This file starts with the usual multiple inclusion protection. The code for this program will be tested completely on a DOS-based system before it is compiled for use for the final microcontroller. There are a couple of items needed for the DOS-based system that are not needed for the microcontroller. Therefore, the parameter DOS is defined at the beginning of the header file and certain lines of the file will be included or excluded depending on the definition of this parameter.

It is intended to store the data in a linked list in memory. The linked list will have a node called an Entry. An Entry contains two characters that will be indices into the data array. The first member is the index to the data in the data array. The second member is an index to the next Entry for the next data entry. An array of 35 Entrys will be stored in EEPROM along with an array of 698 (=768–35*2) chars to store the nonvolatile data. There are 768 bytes of EEPROM on the particular M68HC912B32 chip that we are using here.

A structure type named Epro is created to hold the collections of Entrys and the remaining data for the data storage area. This structure will be forced to the address 0xd00 at link time. It will also be identified as EEPROM so that assignments to this memory area will compile to storage to EEPROM rather than writes to normal data memory. The first three entries in the data[] array are devoted to special uses. data[0] contains the next open index into the data[] array, data[1] contains an index to the beginning of the list and data[2] contains the number of entries in the list. To simplify both the code writing and remembering of the uses of these memory locations, I used macros to define useful names for these locations. Also included here is an old favorite, the FOREVER loop.

```
#define DOS
#ifndef PHONE_H
#define PHONE_H

#ifdef DOS
#include <stdio.h>
typedef unsigned int WORD;
enum Bool{FALSE,TRUE};
#define FOREVER while(TRUE)
#endif

#include <stdlib.h>
#include <string.h>

typedef struct {
    unsigned dataindex;
```

```
    unsigned next;
}Entry;

#define ALEN        30
#define NLEN        16
#define DLEN        35
#define EEPROMLEN   768
#define DATAPROM    EEPROMLEN-DLEN*sizeof( Entry)
#define END         0xff

typedef struct {
    Entry header[DLEN];
    unsigned data[DATAPROM];
}Epro;

#define NEXT_OPEN epro->data[0]
#define START_OF_LIST epro->data[1]
#define LIST_ENTRIES epro->data[2]

void saveit(char *,char *,Epro *);
void printout(Epro *);
void printafter(Epro *);
void reset(Epro *);
int encode(char *,unsigned *,int);
int decode(unsigned *,char *);
void get(char *,int);
int numbdup(char * const, unsigned *, int);
void putbcd(char *,char *);
#ifndef DOS
void inituart(void);
void putchar(int);
int getchar(void);
void puts(char *);
#endif

#endif
```

Listing 7-11: Phone Book Header File

The final portion of this header file is a collection of all of the function prototypes needed for this program. When this program is moved from the DOS-based system to the microcontroller-based system, it is necessary to remove the first line of the above header file. There are three functions found in the standard input/output library that will be rewritten for this program. These functions are `inituart()`, `putchar()`, `getchar()` and `puts()`. The function prototypes for these functions are included and will be discarded when the parameter DOS is not defined.

The monitor program is shown below in Listing 7-12. In the header file above, a structure was `typedefed` as an `Epro`. This structure is the size of the EEPROM on board the chip. An external instance of an `Epro`, named `able`, is created and it will be used as a destination for all of the nonvolatile stored data in the program.

Inside the `main()` program, two arrays are created: one array to store the name entered from the keyboard and the other to store the phone number entered from the keyboard. Also, the external structure will be passed around from function to function via a pointer. This pointer is created and initialized to the structure `able`. Also, if the parameter DOS is not defined, the function `inituart()` is executed to enable the use of the UART on the microcontroller when it is needed.

After this initialization is completed, control is passed into a FOREVER loop where it will remain so long as the computer continues to run. Within this loop, an input is read from the keyboard. It is assumed that the keyboard input will read in the data and return ASCII characters. If the system is a part of a telephone or a PDA, the input routine will have to read in the keyboard data and convert it to the correct ASCII value prior to its use in the following program. Therefore, `getchar()` in the following program can be a function that reads data from a serial port or some other program that will input the data from whatever keyboard is used with the system. Inside of the FOREVER loop, a character is read in and then it is tested in a `switch()`/`case` sequence. The current test values are 'n', 'a', 's', and 'r'. These inputs are commands that determine what the program will do:

Command	Action
n	Read in a number/name sequence, encode these data and save them in the nonvolatile array.

r	Reset the system and erase the EEPROM array which destroys all stored information.
s	Display all stored data in sequence.
a	Display the next stored number and name.

The functions that perform these various operations are discussed in the following sections. All of the case choices merely call the required function with the exception of the 'n' command. The 'n' command calls the function get() twice to read in the number followed by the name. These data are stored in the designated arrays and the two arrays containing the data are passed to the function saveit() along with a pointer to the nonvolatile structure able. Otherwise, the program remains in the FOREVER loop, exiting the loop only to execute commands entered from the keyboard.

```
/* monitor.c is the initial program that is being
   developed to use on the HC12.The end product
   can be used as a part of a PDA or a telephone
   that requires that you enter names and phone
   numbers.These entries are to be saved in
   nonvolatile RAM such as flash or EEPROM,
   perhaps both.  In order to save memory space,
   encoding of the stored data will be used. All
   numbers will be stored as BCD digits and
   letters will be stored as a Huffman code.It
   is assumed that letter fields and number
   fields will not contain mixed data.
       The code in this program is programmed
   for the DOS based system. No printf is used,
   but other i/o functions are used as needed. It
   is assumed now that any input will be
   received as serial data from a serial port
   on the chip.
       This particular program reads data in
   from the keyboard. The numeric field is
   written first and the alpha field is written
   second. If it is a numeric field, the data
   are converted to BCD and stored in an
   allocated memory field. If it is alpha data,
   it is written to an allocated memory field.
```

Any encoding or decoding is done in the storage routines.

T. Van Sickle August 1, 2000 */

```c
#include "phone.h"

Epro able;

main()
{
   char name[ALEN];
   char number[NLEN];
   Epro *epro;
   int c;

   #ifndef DOS
   inituart();  /* initialize the uart to 9600 b/s */
   #endif
   epro=&able;
   reset(&epro);
   FOREVER
   {
     c=getchar();
     fflush(stdin); /* needed for the PowerC compiler */
     switch(c)
     {
     case 'n':   /* new entry */
                 puts("Enter new number and name\n");
                 get(number,NLEN);
                 get(name,ALEN);
                 saveit(name,number,&epro);
                 break;
        case 'a': /* print the next entry */
                 printafter(&epro);
                 break;
        case 's': /* show all */
                 printout(&epro);
```

```
                    break;
    case 'r':  /* reset the system */
                    reset(&epro);
                    break;
    }
  }
}
```

Listing 7-12: Monitor Program

The SAVEIT() Routine

The `saveit()` function receives the data entered from the keyboard in `monitor()`, encodes these data and saves the result in EEPROM for later use. This program is set up to save the data in a linked list. The linked list is an array `DLEN`, found in `phone.h`, long and each member of the list is of the type `Entry`. These data are stored in the main data array and access to these data is made through the values stored in the corresponding `Entry` for each set of data.

In the header file `phone.h`, several macros are defined that make it somewhat easier to code this function. The first three entries in the data array are used for special purposes. Therefore, these values are renamed as macros to make the code more understandable. These macros are:

```
#define NEXT_OPEN epro->data[0]
#define START_OF_LIST epro->data[1]
#define LIST_ENTRIES epro->data[2]
```

Useful names can now be used rather than the cryptic actual names of these various memory locations.

On entry to this function, both the name and number parameters are encoded and the result is saved in the array `name[]` and `number[]` respectively. The next block of code determines if there is enough room in the EEPROM array to store all of the data. In the event that there is not enough room, the message "`*** buffer full***`" is sent to the output and control is returned to the monitor program when return is executed.

```
/*********************************************
```
This program works in conjunction with moni-
tor(). It receives the input found in monitor

which then sends the information to this function, saveit(). The data that arrives is in the form of two binary strings. The first string, the name, is Huffman encoded and the second string, the phone number, is bcd encoded. These data are to be stored in the EEPROM. A linked list that uses indicies rather than pointers is used. The EEPROM is broken into two fields. The first field is an array of the type Entry. This array is DLEN long. The next field is an array that contains the remainder of the EEPROM. This array is an array of type unsigned. It will mostly contain data that has been encoded. The first three entries in the array are special. The first field contains the index to the next unused entry in the array. The second field is the index to the start of the list and the final field contains the number of entries in the list.
```
************************************************************************/
#include "phone.h"

void saveit(char *s,char *n,Epro *epro)
{
    int i,j,sl,nl;
    unsigned name[ALEN];
    unsigned number[NLEN];

    sl=encode(s,name,ALEN); /* encode both the name and */
    nl=numbdup(n,number,NLEN); /* the number */
    /* store encoded data on the end of the array */
    if(NEXT_OPEN+sl+nl>DATAPROM)/* do not overwrite the array */
    {
      puts("***buffer full***\n");
      return;
    }
    for(i=NEXT_OPEN,j=0;j<sl;i++,j++)
      epro->data[i]=name[j]; /* save the name then the number */
```

```
    for(j=0;j<nl;i++,j++) /* at NEXT_OPEN  */
      epro->data[i]=number[j];
    epro-
>header[LIST_ENTRIES++].dataindex=NEXT_OPEN;
    NEXT_OPEN+=sl+nl;

    if(LIST_ENTRIES>=1) /* no entries in list never 0 */
      epro->header[LIST_ENTRIES-1].next=LIST_ENTRIES;
}
```

Listing 7-13: The `saveit()` Function

If there is enough room in the EEPROM array to store the new data, the encoded name and the encoded number are both written into the array at the appropriate location. Recall that the encode routines return the lengths of the encoded data. NEXT_OPEN is the next unused entry in the array. After the data are written to the array, the new value for NEXT_OPEN is calculated by adding the length of the two encoded arrays to the old value.

Finally, if there is more than one entry in the array the index for the new entry, LIST_ENTRIES, will be put in the next location of the previous entry, epro->header[LIST_ENTRIES-1].next.

The printout() and the printafter() Functions

These two functions are almost the same. Therefore, they will be discussed together. In both cases, control is returned to the calling program if there are no data to be printed out. The `printout()` routine starts at the beginning of the list and prints the entire contents of the EEPROM. Prior to the printout, the contents are decoded. Both decode routines return a new line character at the end of the data stream along with a null character to indicate the end of the line. Notice in the `printout()` function, the data starts at the start of the list and cycles through all of the contents of the list.

```
#include "phone.h"

void printout(Epro *epro)
```

```
{
  char na[ALEN],nu[NLEN];
  int i,j,k,count=0;

  if(LIST_ENTRIES==0)
    return;          /* no entries */
  k=START_OF_LIST;
  do
  {
    i=epro->header[k].dataindex;
    j=decode(&epro->data[i],na);
    puts(na);
    putbcd(&epro->data[i+j+1],nu);
    puts(nu);
    k=epro->header[k].next;
    count++;
  }
  while(count<LIST_ENTRIES && count<DLEN);
}
```

Listing 7-14: The `printout()` Function

Listing 7-15 contains the `printafter()` routine. Here, the routine cycles through the data, decodes it and prints it out one `Entry` at a time. In this case, the parameter k is static and it starts with the value 0. This is the value of the index to the first `Entry` in the array. After each output, k is incremented and so is `count`. Whenever `count` attains the value LIST_ENTRIES, all of the data in the memory has been printed out. Then `count` is restored to 0 along with the value of k being set to 0. This action causes the data in the array to be printed out one field at a time and when all of the data are sent out, it is restored to the beginning of the array and recycled.

```
#include "phone.h"

void printafter(Epro *epro)
{
  char na[ALEN],nu[NLEN];
  int j,i;
  static k=0,count=0;
```

```
 if(LIST_ENTRIES==0)
   return;                /* no entries to decode */
 i=epro->header[k].dataindex;
 j=decode(&epro->data[i],na);
 puts(na);
 putbcd(&epro->data[i+j+1],nu);
 puts(nu);
 k++;
 if(++count==LIST_ENTRIES)
 {
   count=0;
    k=0;
 }
}


#ifdef DOS
void puts(char *s)
{
   char *sp;
   sp=s;

   while(*sp!='\n' && *sp!='\0')
    putchar(*sp++);
   putchar('\n');
}
#endif
```

Listing 7-15: The `printafter()` Function

When developing and testing this program with the DOS system, I found it necessary to include the new puts() function. When the program is moved to operate on the HC12, it will be necessary to include separate i/o functions to be discussed below. The i/o functions are not included in the DOS version of the program. Therefore, the puts() function that is added to the end of the printafter() function above is included. This little function will be discarded whenever the parameter DOS is not defined. In that case, the i/o functions should be included.

Reset

The reset function is dependent on the system being used. When programming for the DOS-based system, the array that represents the EEPROM is put into a state that simulates erased EEPROM. Then the first three entries in the array data[] are initialized to the proper values. Recall that data[0] will always contain the index to the next open entry in the data[] array. This index is defined by a macro in the phone.h file to be NEXT_OPEN. The next two members data[0] and data[1] contain the index to the starting index into the header array, START_OF_LIST and the number of entries in the list LIST_ENTRIES respectively. The first three members of this array are used as described above. The first available member for storage of data is at the index 3. Therefore, NEXT_OPEN is assigned a value of 3 and both START_OF_LIST and LIST_ENTRIES are initialized to 0. Remember, that this data array is the second member of a structure of the type Epro. An external instance of this structure named able is defined in the file monitor.h, and a pointer to this structure epro is also defined there.

The reset() function is the first in this program that requires code for the DOS implementation that differs from the HC12. When the EEPROM is initialized, all bits in the memory are turned on: the memory is filled with 0XFFFF. Therefore when simulating this memory, the initialization will fill the memory similarly. There is a library function in the Cosmic library, eepera(), that erases the memory. Therefore, when coding for the HC12, this function will be used. The code for the reset() function is shown in Listing 7-16.

```
#include "phone.h"

void reset(Epro *epro)
{
    int i;
    #ifdef DOS
    memset(epro,0xff,EEPROMLEN);
            /*make arrays look like EEPROM */
    #else
    eepera();/* erase the EEPROM with library function */
```

```
    #endif
    /* Start the linked list */
    NEXT_OPEN = 3;/* next open entry in the list */
    START_OF_LIST =0;/* start of the list */
    LIST_ENTRIES=0;/* number of entries in the list */
    epro->header[0].dataindex=NEXT_OPEN;/* start
the list */
    for(i=0;i<DLEN;i++)
      epro->header[i].next=END;


}
```

Listing 7-16: The Reset Function

Input/Output Functions

There are three input/output functions that are usually found in the standard I/O library that we will replace with microcontroller specific code here. These functions are putchar(), getchar(), and puts(). In the first two instances, rather than sending and receiving data from devices like stdout and stdin, all output will go to the serial port on the HC12 and input will come likewise from the serial port. Therefore, these routines will have to be written from scratch. In addition to the direct input/output functions, an initialization function that enables the serial port will be needed. This function must set the bit rate for the serial port and enable both the UART transmitter and receiver. This function is as follows:

```
#include "hc12.h"

#define BAUD9600    52

#define BAUDREG *(BYTE *)&SC0BDL

void inituart(void)
{
    BAUDREG=BAUD9600;   /* Set the bit rate */
    SC0CR2.TE=ON;       /* turn on transmitter */
    SC0CR2.RE=ON;       /* and the receiver */
}
```

The choice for bit rate for this system is 38.4 kbits per second. To achieve this rate, the baud rate divisor value is given by

$$BR = E_{clk}/16\ BaudRate$$

The E clock for the system is 8 MHz. Therefore, the divisor, BR, is 52 when rounded to the nearest integer value.

The register SC0BDL is defined in the header file hc12.h as a type Register, or a collection of eight individual bits. In this case, it is more understandable to put the data into this register as a char rather than as a set of bits. The type of this location can be changed to a type BYTE easily, using the line of code

```
#define BAUDREG *(BYTE *)&SC0BDL
```

Read this line of code from right to left. It says to cast a pointer to the memory location SC0BDL onto a pointer to a type BYTE and then dereference it. Therefore, whenever the defined name BAUDREG is used, it accesses the BYTE contents found at the address SC0BDL. This address is defined in the header file hc12.h.

The next two lines of code turn the bits TE and RE in SC0CR2 on. When these bits are ON, both the UART transmitter and receiver will work.

The next function is putchar(). This routine sends the designated BYTE to the serial port. It is necessary to wait until any data in the transmit data register has been completely processed before sending new data to this register. The first line of code does not allow the value of SC0DRL to be altered until the transmit data ready bit, TDRE, is set. Then the value x is stored in the location SC0DRL, which causes the data to be sent to the serial port.

```
void putchar(BYTE x)
{
   while(!SC0SR1.TDRE)
    ; /* wait until register is ready */
   SC0DRL=x; /* send the data out */
}
```

The last of the I/O functions is getchar(). The getchar() function is a little longer than the putchar() function. This difference is caused by the fact that when a character is read in from the serial port it should be immediately echoed back. Therefore, the entered data are stored in the memory location a and sent out with the instruction putchar(a) before it is returned to the calling program.

```
BYTE getchar(void)
{
   BYTE a;
   while(!SC0SR1.RDRF)
    ; /* Wait for data ready */
   a=SC0DRL;
   putchar(a);  /* echo the data and */
   return a;    /* then return it*/
}
```

Finally, the puts() function from the standard library does not terminate transmission when it detects a new line character. For proper operation in this program, it should, so the following function was written. Notice that this function terminates whenever either a new line character or a zero character is detected in the input string. Like the standard library puts() it outputs a new line character regardless of the termination of the data.

```
void puts(char *s)
{
   char *sp;
   sp=s;

   while(*sp!='\n' && *sp!='\0')
    putchar(*sp++);
   putchar('\n');
}
```

An interesting observation: If you look in Chapters 4, 5, 6, and 8 you will find similar routines for other chips. In every case, the functions, even though they are for a broad range of different chips, are the same—a testimonial to the use of a high-level language like C to program our microcontrollers. A listing of these three functions is collected together in Listing 7-17 shown below.

```
#include "HC12.H"

#define BAUD9600  52
```

```c
#define BAUDREG *(BYTE *)&SC0BDL

void inituart(void)
{
   BAUDREG=BAUD9600;/* Set the bit rate */
   SC0CR2.TE=ON;      /* turn on transmitter */
   SC0CR2.RE=ON;      /* and the receiver */
}

void putchar(BYTE x)
{
   while(!SC0SR1.TDRE)
     ; /* wait until register is ready */
   SC0DRL=x;   /* send the data out */
}

BYTE getchar(void)
{
   BYTE a;
   while(!SC0SR1.RDRF)
     ; /* Wait for data ready */
   a=SC0DRL;
   putchar(a);        /* echo the data and */
   return a;          /* then return it*/
}

void puts(char *s)
{
   char *sp;
   sp=s;

   while(*sp!='\n' && *sp!='\0')
    putchar(*sp++);
   putchar('\n');
}
```

Listing 7-17: The Input/Output Routines Used with the M68HC912B32 Chip

You will notice that the file shown in Listing 7-17 is the only file in which the header file `hc12.h` is included. None of the other code depends in any way on the bit field structures that control all of the peripherals on the chip. There are a few places throughout the code where there are some specific differences between the two programs. The first line of code in the header `phone.h` is

```
#define DOS
```

This line is left intact whenever the program is compiled to run under DOS. When the program is compiled to run on the HC12, this line is removed. Then sequences like the following found in `monitor.c`

```
#ifndef DOS
    inituart(); /* initialize the uart to 9600 b/s */
#endif
```

will cause the execution of `inituart()` to be ignored when compiled for DOS but it will be included when the program is compiled for the HC12. Such inclusions will be found throughout the various functions that are linked to form the program.

Putting It All Together

The above programs were all compiled, linked and tested with the DOS-based compiler provided by MIX Software[1]. This reliable compiler created satisfactory test code to run on any PC-style computer. When everything was working as desired, the code was moved to the HC12 system. The compiler used in this case was the COSMIC compiler[2]. This compiler was provided by the same company that provided the compilers used in Chapters 5 and 6. Cosmic provides two software packages that are very useful. The first is called IDEA12. This package is a so-called Integrated Development Environment, IDE. Within the IDE, you can specify things like the default directory, and provide a list of files for a make utility. The make utility will compile all source files that are newer than the corresponding object files. Therefore, as you debug various files in

---

[1] Mix Software, 1132 Commerce Drive, Richardson, TX 75081, (972) 783-6001.
[2] Cosmic Software, 400 W. Cummings Park, Ste. 6000, Woburn, MA 01801-6512, (781) 932-2556 x15.

your program, you can automatically compile only those files modified since they were last compiled.

There are a couple of additional files needed to complete the program for the phone book. The first is the interrupt vector table. The vector table is stored in nonvolatile memory on this system, and as such, it is best to write a C program that creates this table. This little program is compiled and linked just like any other module in the program. It will be linked as a constant section at the address 0xFFCE. The interrupt vector table program is shown in Listing 7-18. There is only one external module to be linked to this particular table. It is _stext(). The function prototype for this function is included at the beginning of the program, and its name is placed in the proper vector location. Recall from our earlier discussion of complicated declarations that the line of code

```
void (* const _vectab[]) ()
```

tells us that _vectab is an array of constant pointers to functions that return the type void. The addresses of the various entries in the array of pointers to functions begin at the memory location 0xFFCE. The remainder of the vectors that follow each have a specific use, and if there were an interrupt service routine needed for the program, its address would be placed in the corresponding location.

```
/* INTERRUPT VECTORS TABLE 68HC912B32
*  Copyright (c) 1997 by COSMIC Software
*/
void _stext();        /* startup routine */

void (* const _vectab[]) () = {   /* 0xFFCE */
    0,        /* Reserved         */
    0,        /* BDLC          */
    0,        /* ATD              */
    0,        /* SCI 1            */
    0,        /* SCI 0            */
    0,        /* SPI              */
    0,        /* Pulse acc input  */
    0,        /* Pulse acc overf  */
    0,        /* Timer overf      */
    0,        /* Timer channel 7  */
```

```
0,        /* Timer channel 6 */
0,        /* Timer channel 5 */
0,        /* Timer channel 4 */
0,        /* Timer channel 3 */
0,        /* Timer channel 2 */
0,        /* Timer channel 1 */
0,        /* Timer channel 0 */
0,        /* Real time       */
0,        /* IRQ             */
0,        /* XIRQ            */
0,        /* SWI             */
0,        /* illegal         */
0,        /* cop fail        */
0,        /* cop clock fail  */
_stext    /* RESET           */
};
```

Listing 7-18: M68HC912B32 Vector Table

Listing 7-19 shows the next function that must be included for operation on the M68HC12. This start-up program is named `crts.s`. The `.s` extension indicates that the function is an assembly language function. This is the sum total of all assembly code needed for the program discussed in this chapter. Rather than give a line-by-line description of the code, we will see that the initial portion of the program initializes the section named `bss` to all zeros. This section is where all static and external memory are stored. Then the stack pointer is initialized to the value designated as `__stack` and then control is passed to the function `main()`. If `main()` should return, which it should not, the instruction following `main()` forms an infinite loop that branches to itself and does nothing.

```
;  C STARTUP FOR MC68HC12
;  Copyright (c) 1996 by COSMIC Software
;
   xdef    _exit, __stext
   xref    _main, __sbss, __memory, __stack
;
__stext:
```

```
    clra                    ; reset the bss
    clrb
    ldx     #__sbss         ; start of bss
    bra     loop            ; start loop
zbcl:
    std     2,x+            ; clear word
loop:
    cpx     #__memory       ; up to the end
    blo     zbcl            ; and loop
    lds     #__stack        ; initialize stack pointer
    jsr     _main           ; execute main
_exit:
    bra     _exit           ; stay here
;
    end
```

Listing 7-19: Crts.s C Program Start-Up Function

Any linker for an embedded system requires some type of linker command file. The command file for this application is as follows.

```
#  link command file for test program
#  Copyright (c) 1996 by COSMIC Software
#
+seg .text -b 0x8000 -n .text# program start address
+seg .const -a .text # constants follow program
+seg .data -b 0x800  # data start address
+seg .eeprom -b 0xd00 -m768 #identify EEPROM block
+def __sbss=@.bss    # start address of bss
crts.o               # startup routine
monitor.o            # applications programs
saveit.o
encode.o
decode.o
reset.o
numbdup.o
putbcd.o
priout.o
priafter.o
serial.o
```

```
get.o
"C:\COSMIC\CX12\lib\libi.h12"
"C:\COSMIC\CX12\lib\libm.h12"
+seg .const -b 0xffce # vectors start address
vector.o                # interrupt vectors
+def __memory=@.bss     # symbol used by library
+def __stack=0xc00      # stack pointer initial value
```

Listing 7-20: Linker Command File

This file can be broken into three sections. The first section specifies all of the important memory locations needed for the program. The first instruction identifies the text section as beginning at the hex address 0x8000. That is the beginning of the FLASH EEPROM on this chip. The designation text identifies the executing portion of the program. The next line says that all program constants shall be placed in the text section. Since the text section is placed in internal nonvolatile FLASH, nothing in this section can be changed by the program.

The internal RAM on this part is the 1024 bytes beginning at the hex address 0x800. The ending address of RAM is 0xBFF. The next instruction places data, variables and stack at the starting address 0x800. Jump to the end of this command file. Note that the parameter __stack is given a value of 0xC00. This value is used because the M68HC12 stack pointer points at the top of the stack rather than the next open location on the stack as was seen with the M68HC11 family of parts.

The final segment designation is that of the EEPROM location. This segment has a beginning address of 0xD00 and it contains 768 bytes.

The next section of the file contains the files to be linked. There are twelve applications files to be linked. With the exception of the crts.o file, these files are those created and tested earlier in this chapter. After the application program files, two standard compiler libraries are included. These libraries contain all functions needed to complete the program.

The last section links the vector table discussed above to the correct address in the program, establishes the value for the initial stack pointer, and remembers the address of the end of the bss section for use in the program.

The program was compiled, linked and an srecord of the code created. This code was loaded into an M68EVB912B32 evaluation board. Loading the code into this board requires the use of another development board, an HC12 FLASH PROGRAMMER. This board contains an M68EVB912B32 board also. The code is downloaded into the programmer board and from there it is transferred to the FLASH memory on the second evaluation board. This combination worked to program the FLASH, and it must be noted that the final program worked as designed after one error was corrected.

Summary

In writing this chapter, I have attempted to show that modular development of a program can yield very satisfying results. The program here was reduced to several relatively small functions that could each be developed and tested separately. Then these functions were integrated one-by-one to build the whole program. This is not to say that this approach is less work. Several of the modules listed above are very complicated and require careful design to make certain that they work as desired. Some of these modules require almost invention. For example, the means used to express the Huffman table in the decode routine needed several different starts before a satisfactory one was found. Recall that the complete binary tree needed to express a Huffman code requires $2n-1$ nodes where n is the number of items being encoded. I felt that it was desirable to express this tree by an array with no more than $2n-1$ members. The array shown in Listing 7-7 contains exactly $2n-1$ bytes. It took six different tries to arrive at this particular representation, and hence the code to decode the data.

The reward was that the final code worked in the embedded product with almost no error.

Chapter 8

MCORE, a RISC Machine

Reduced Instruction Set Computer (RISC) machines are the new architecture rage. First, the R in RISC is an absolute misnomer. When first learning the M68000 machine, a CISC (Complicated Instruction Set Computer), I found a controller with about 70 instructions. The MMC2001, a RISC, has about 110 instructions, and it has no instructions for the complex addressing modes that make the CISC machine so easy to program. The RISC has its advantages though. Most of the RISC instructions require only one clock cycle per instruction while the CISC typically requires six clocks per instruction with a range of two clocks to twenty-four clocks per instruction. Therefore, a RISC chip with a 33-MHz clock will execute more than thirty-one million instructions per second, obviously much faster than a similar speed CISC, which would probably execute less than six million instructions per second. However, the RISC machine will require more instructions to execute the same program. Overall, the RISC machine is almost always faster than the corresponding CISC, even though the RISC requires more memory to implement the same code.

The RISC/CISC dichotomy will be the source of nearly religious debate until the next big architecture change is introduced. This text is not aimed at comparison of different architectures. The goal is to help you write code for the chips in the C language. Now the problem becomes one of finding an appropriate compiler for the chip.

The compiler/debugger combination used in preparing the code found in this chapter is provided by DIAB and SDS.[1][2] These software

---

[1] Diab Data, Inc., (650) 571-1700, fax (650) 571-9068, email info@ddi.com, www.ddi.com
[2] Software Development Systems, Inc., (630) 368-0400, fax (630) 990-4641, email sales@sdsi.com, www.sdsi.com

systems are available as demonstration systems from the respective manufacturers. Contact them directly for information on demonstration systems.

A photo of the development system is shown in Figure 8.1. The host computer in this case is a laptop computer. The Extended Background Debug Interface, EBDI, is connected to the computer serial port. A flexible cable connects the EBDI to the AXIOM demonstration board that contains the MCORE chip, the MMC2001. Additional peripheral devices such as a keypad or an LCD display panel can be connected directly to the AXIOM board. There are two serial ports on the Axiom board. These serial ports can be used by any program that needs serial access.

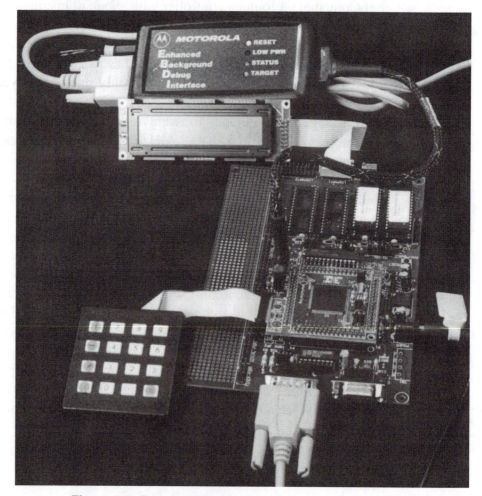

Figure 8-1: Development System

Delay Routine

In the course of this chapter, we are going to see several small routines and programs that can be used to accomplish some useful tasks. One routine that is often useful is a delay routine. There are several ways to implement a delay. A most important feature of any delay routine is that it must be accurate and it must not depend on counting a number of instruction cycles to measure the delay. Almost all microcontrollers have timer subsystems that can be used to control these delays.

A delay program has to cause an executing program to suspend execution for a specified time. What is the program to do during this time? The simplest approach is to have the delay program merely execute a loop until the time has past, and this is the first approach that we will use. If your program consumes essentially all of the computer resources, such an approach is very wasteful. The computer will merely execute a tight loop during the entire delay and exit the loop when the delay is completed. The computer is unable to do anything else during the delay.

There is a second approach that returns control of the computer to the calling program and allows other operations to execute during the delay time. This approach uses a semaphore to control execution of the calling program. We will see this approach shortly.

Listing 8.1 is the code for the function void delay(int time). This function makes use of the programmable interval timer, PIT, portion of the on-board timer found on the MMC2001. The timer is driven by a 32768-Hz crystal oscillator. The frequency generated by this oscillator is divided by four. Therefore, the time period of clocks entering the PIT is 1/8192 seconds, or 122.070312 microseconds. This is the resolution of the PIT on this chip. There are two registers, the PIT data register, ITDR, and the PIT alternate data register, ITADR. The data written to the ITDR is retained and is transferred to the ITADR the next time that the ITADR underflows, indicating completion of the specified time sequence. At that time, the contents of the ITDR are written to the ITADR, and the PIT interrupt flag ITIF is set. If the interrupt sequence is enabled, a core processor interrupt is requested. Otherwise, the system can be polled to determine when the time has expired. This approach is used in the following program.

```
#include "mmc2001.h"
#include "timer.h"

/******************************************************
    Delay by t milliseconds. The data here are passed
    as an int because, this routing ties up the
    computer the whole delay time. An alternate means
    should be used if a long delay is needed. This
    routine uses the PIT which counts at about 122 us
    per ticks. It is not very accurate because, the
    count is at 1/8192 seconds, more nearly 122.070312
    us ticks. Not good enough to make a clock, but
    good enough to control a few milliseconds.

    This modification assumes that the pit is enabled
    and shall remain enabled.  TVS 6/2000
*******************************************************/
void delay(WORD t)
{
    UWORD now,next;
    UWORD count;

    count=(long)t*1000; /* make the delay in us */
    count/=122;    /* 122 us per tick */
    ITCSR.EN=ON;   /* enable the pit */
    while(count>0) /* there are probably faster ways to do */
    {              /* this, but who cares, you are killing time */

        now = ITADR; /* and don't care for speed */
        do
        {
            next = ITADR;
        } while(now==next); /* wait until next tick */
        count--;
    }
    ITCSR.EN=OFF;   /* delay is done don't need the pit now */
}
```

Listing 8-1: Delay routine

The function parameter time is the delay time in milliseconds. The clocking rate for the counter is each 122.0... microseconds. Therefore the time is converted to counts when it is divided by 122. The PIT is then enabled with the instruction

```
ITCSR.EN = ON;        /* enable the pit */
```

This rather clumsy loop reads the value contained in `ITADR` into the variable `now`. This register is then read into the location `next` until the value in `ITADR` is changed by the clock. That should occur 122 microseconds later. At this time, `count` is decremented. When `count` becomes 0 the time has expired.

The variables `now` and `next` are both declared to be volatile. This declaration is necessary to avoid loss of the whole loop when the code is optimized. The variable `ITADR` is declared to be volatile in the header file. That declaration does not guarantee to the compiler that the variables will ever change, so the optimizer could well remove the code

```
now = ITADR; /* and don't care how fast your test is */
do
{
    next = ITADR;
} while(now == next); /* wait until next tick */
```

during the optimization phase. That would make the whole function rather pointless.

Semaphore

This delay function is used in later code to implement debounce routines. A more practical delay routine can be implemented with the use of a semaphore. Here, we are going to develop a semaphore that follows that developed in *Reusable Software Components* .[3] We will not create a semaphore object as done in that text, but the program will follow the items shown there.

A semaphore is merely a flag that is attached to a process like a program or a function. The semaphore is set and the calling program is not able to proceed until the called program resets the semaphore. This reset is usually done in an interrupt service routine implemented in the called program. Let's look at the elements of a semaphore first.

When a semaphore is set, it usually marks a resource as busy. It is interesting, but you will find that many semaphores are used around interrupt service routines and while interrupts are being used. Often a strange race condition can occur that will cause a semaphore to be improperly set. Suppose that I want to set a flag, semaphore, to indicate that a resource is busy. The process of setting the semaphore is to first

3 *Reusable Software Components*, Ted Van Sickle, Prentice Hall, Upper Saddle River, NJ, 1997

read the contents of the semaphore and then, if it is in the proper condition, it will be set. Interrupts can come asynchronously at any time. Suppose that in the process of attaching a semaphore, the status of the interrupt is read to be tested, and, prior to marking the semaphore as used, an interrupt occurs. Within this interrupt routine, it also could require an interrupt. In this case, the interrupt that was being processed by the earlier routine will seem to be available, but it is not. In fact when control is returned to the initial portion of the program, the semaphore will be again marked as busy, and now two completely different processes will both assume control over the same semaphore.

```
/* A call to attach_semaphore('a') will attempt to attach
   safely a semaphore. If a semaphore can be attached, a
   semaphore number will be returned. Otherwise, a -1 is
   returned and no semaphore can now be attached. When
   attach_semaphore(n), where 0<=n<10, is executed, an
   attempt is made to attach the semaphore specified by the
   number. If it is not available, a -1 is returned. After
   the semaphore is attached, use the semaphore number as a
   parameter on all other semaphore function calls. Function
   release_semaphore() returns the semaphore to the avail-
   able semaphore pool. The function semaphore_status()
   returns TRUE when the semaphore is NOT available, and the
   function wait_for_semaphore() waits in a tight loop until
   the semaphore becomes available */

#include "mmc2001.h"

#define Number_of_semaphores 10
#define Minus_one -1

/* function prototypes */
int attach_semaphore(void);
void release_semaphore(int);
int semaphore_status(int);
void wait_for_semaphore(int);

static volatile int semaphore[Number_of_semaphores];
int save;

int attach_semaphore(int r)
{
    int i;
```

```c
    /* save interrupt status */
    asm(" mfcr R1,PSR\n lrw R2,save\n stw R1,(R2,0)\n");
    /* disable all interrupts */
    Disable_Interrupts();
    Disable_Fast_Interrupts();

    /* request specific semaphore */
    if(0<=r && r<Number_of_semaphores&&semaphore_status[r]&&r!='a')
        return Minus_one;    /* it is not available */
    else
    {
        semaphore[r]=TRUE;
        /* reenable interrupts */
        asm(" lrw r2,save\n ldw R1,(r2,0)\n mtcr R1,PSR\n");
        return r;            /* return semaphore number */
    }

    for(i=0;(i<Number_of_semaphores) && (semaphore[i]!=0);i++)
        ; /* find an unused semaphore */
    if(i>=Number_of_semaphores)
    {
        /* reenable interrupts */
        asm(" lrw r2,save\n ldw R1,(r2,0)\n mtcr R1,PSR\n");
        return Minus_one;  /* no semaphore available */
    }
    else
    {
        semaphore[i]=TRUE;  /* mark semaphore as used */
        /* reenable interrupts */
        asm(" lrw r2,save\n ldw R1,(r2,0)\n mtcr R1,PSR\n");
        return i;            /* return semaphore number */
    }
}


void release_semaphore(int i)
{
    semaphore[i]=FALSE;
}

int semaphore_status(int i)
{
    return semaphore[i];
}

void wait_for_semaphore(int i)
{
```

```
while(semaphore[i]==TRUE)
    ;

}
```

Listing 8-2: Semaphore functions

This problem is avoided by disabling all interrupts prior to reading the semaphore state during the attachment process. Interrupts are disabled and enabled by setting and clearing bits in the PSR register on the core processor. If you examine the `mmc2001.h` header file, there are four macros, `Enable_Interrupts()`, `Enable_Fast_Interrupts()`, `Disable_Interrupts()` and `Disable_Fast _Interrupts()` defined. These macros are all written in assembly language because the C language cannot access core control registers directly. In the `attach_semaphore()` routine above, there is one line of assembly code as follows

```
asm(" mfcr R1,PSR\n lrw R2,save\n stw R1,(R2,0)\n");
```

These three assembly instructions save the contents of the PSR register in the memory location named `save`. The code

```
asm(" lrw r2,save\n ldw R1,(r2,0)\n mtcr R1,PSR\n");
```

moves the contents of the memory location `save` back into the PSR. The bits that enable both the fast interrupts and the conventional interrupts are contained in this register. Therefore, the code as shown in the `attach_semaphore()` routine saves the currently enabled interrupts, disables all interrupts, and after the status of the semaphore is established and set properly, restores the PSR to its original value.

When the semaphore is attached, it is marked true in the array of available semaphores. The function `release_semaphore()` marks the semaphore as FALSE. The program can query the semaphore to determine if it is being used. This function returns the value saved in the specified location in the array of semaphores that is TRUE when the semaphore is being used. The final function is to `wait_for_semaphore()`. When this function is entered, control will remain in the function until the specified semaphore is released. It is important that you do not attempt to wait for an unattached semaphore. In that case, control will never be returned to the calling program.

Delays Revisited

With semaphores available, we can improve on the delay function by removing the wait loop from within the function to the calling function, where its use can be dictated by the program. In this case, the calling function will create a semaphore and send the semaphore to the delay function. The proposed delay function is

```
include "mmc2001.h"
#include "timer.h"
#include "intctl.h"


/*******************************************************
   Delay by t milliseconds. This delay controls a semaphore
   that will be examined by the calling program. When the
   semaphore is released, the delay is completed. It is
   assumed that fast interrupts are enabled when this func-
   tion is called and that the interrupt handler has been
   appropriately set up.
   *****************************************************/

static int sem; /* this variable needs to be file global */

void delay(WORD t, int semaphore)
{
   long count;

   sem=semaphore;    /* save semaphore so isr can see it */
   ITCSR.EN = ON;    /* enable the pit */
   ITCSR.OVW=ON;     /* enable write through */
   /* an interrupt will occur in t milliseconds */
   count =(long)t*1000; /* make the delay in us */
   ITDR =count/122;     /* 122 us per tick */
   ITCSR.ITIE = ON;     /* enable pit interrupt */
   FIER.EF8=ON;         /* enable the fast interrupts */
} /* return to the calling program, we are done here */

void pit_isr(void) /* interrupt occurs when delay time expires */
{
   ITCSR.ITIF = ON;   /* turn interrupt flag off */
   ITCSR.ITIE = OFF;  /* disable the PIT interrupt */
   ITCSR.EN = OFF;    /*  disable the PIT */
   FIER.EF8 = OFF;    /* disable the pit fast interrupt */
   release_semaphore(sem); /* release semaphore to calling program */
}
```

Listing 8-3: Alternate Delay Routine

The program that calls the delay() function above has some responsibilities in making this function work. This delay is using the PIT interrupt. Most of the code necessary to implement the interrupt is contained in the delay function. There are, however, a couple of items that have to be taken care of outside the delay routine. The first is to place the address of the interrupt handler into the fast interrupt vector. Also, the system fast interrupts must be enabled. You will see these matters are taken care of in the following test function.

In the previous routine, the semaphore number is saved externally so that it can be accessed by the interrupt service routine. Next the PIT is enabled and the ITDR to ITADR write through is enabled so that the value written to the ITDR is the value to be counted down in the timer. The count value is next calculated. Remember, the counter is driven at 8192 Hz. This value is obtained by counting the output from a crystal-controlled oscillator running at 32768 Hz by 4. The time per count is approximately 122 microseconds. The delay time passed to the function is in milliseconds, so to calculate the count value to be placed into the ITDR/ITADR, the program first converts the delay time to microseconds by multiplying it by 1000. Then, the count value is calculated by dividing the microseconds by 122. This value is written to the ITDR and it is automatically written through to the ITADR where it is counted down by the hardware.

The remaining code in the delay() routine enables the PIT interrupt and also enables the fast interrupt, bit 8, that is connected to the output from the PIT. Control is then passed back to the calling program.

```
#include "mmc2001.h"
#include "serial.h"

#define FAST_AUTOVECTOR 0x3000002c

/* function prototypes */
void handler(void);
int attach_semaphore(void);
void wait_for_semaphore(int);

main()
{
        UWORD count=0;
        int semaphore;
```

```
inituart(38400);
Enable_Fast_Interrupts();
vector(handler,FAST_AUTOVECTOR);
while(count++<300)
{
        if((semaphore=attach_semaphore())==-1)
        {
                puts("semaphore attachment error\n");
                exit(0);
        }

        delay(1000,semaphore);
        wait_for_semaphore(semaphore);
        printd(count);
        putchar('\r');
}
}
```

Listing 8-4: Delay Test Routine

The delay test routine first initializes the on-board UART so that signals can be sent to a terminal and then enables the fast interrupts and places the handler address in the FAST_AUTOVECTOR location as needed. The main test attaches a semaphore, executes a delay of 1000 milliseconds, waits for the delay to expire and prints a value to the screen. This sequence is placed in a loop that is executed 300 times. Note that it is important that `attach_semaphore()` be checked for a −1 return. If not, you could attempt to attach too many semaphores, and your code would not catch the error. In the above case, the program was exited with the `exit()` function. It is not usually necessary to abort the program when there is no semaphore when one is needed. You can merely wait for a semaphore with any value between 0 and 9. When that semaphore is released, you can proceed with a request to attach a semaphore and it should then succeed.

The attach semaphore is in the calling program in this case. The semaphore number is sent to the delay program where it is in turn passed to the pit interrupt service routine. When the interrupt service routine, after the proper delay, is executed, the specified semaphore is released. Therefore, in the calling program, the wait for semaphore routine is executed until the semaphore is released. If needed, the semaphore status can be polled synchronously by the program to determine when the delay time has expired. In other words, the computer does not have to sit in a wait loop until the delay time is over to implement the delay.

The two delay functions, those shown in Listings 8-1 and 8-3, are examples of code that are very close to the microcontroller. The application program shown in Listing 8-4 is the type of program that could almost be viewed as divorced from the microcontroller. In 8-4 the two operations

```
Enable_Fast_Interrupts();
vector(handler,FAST_AUTOVECTOR);
```

are clearly processor dependent, but these processor dependent functions are written as functions, or function-like macros. Therefore, if it is necessary to move to another processor, it is necessary only to adjust the contents of these functions to make the program usable. The two delay functions, on the other hand, are a mass collection of processor-specific commands that would have no meaning if the processor were to be changed. It is a good idea when writing code for embedded microcontrollers to collect together all of the processor specific code into functions by themselves and allow the general code to be as free from processor specifics as possible. We will see this idea in the following section where several general functions interface with several processor-specific functions.

Serial Input/Output

Serial input/output is an important capability of almost all microcontrollers that allows the programmer or user to communicate with the processor. Most of these functions are rather simple to implement. Some of the functions are quite processor-specific and others are more generally applicable to many microcontrollers. As mentioned above, it is important that you split these functions apart so that as much of the code that you generate as possible is portable. Let us look at the functions contained in `serial1.c`.

```
/**********************************************************
    These routines implement a serial port on the MMC2001.
The UART1 is used and the default is set to 9600 b/s, 8 bit,
no parity, and 1 stop bit. The baud rate is passed as an
integer to the inituart() function. There are several i/o
functions included. These are:
```

```
inituart()    Initializes the uart to the passed baud
              rate and sets 8 bit, no parity, and one
              stop bit.
getchar()     tests for receive ready and echo
getch()       no echo
getce()       no ready test or echo
kbhit()       returns TRUE if a key has been hit,
              FALSE otherwise
putchar()     Tests for transmit ready
puts()        sends out the indicated string
gets()        reads in a string into the buffer
getse()       reads in a string and echos
printd()      convert an unsigned long integer to an
              ascii string representation of a decimal
              integer and send it out the serial port
printx()      convert an unsigned long integer to an
              ascii string representation of a
              hexidecimal integer and send it out the
              serial port

*********************************************************************/

#include "mmc2001.h"
#include "uart.h"
#include "serial.h"

#define U1SRint (*(volatile unsigned short *)(0x1000a086))
enum {TRDYint=8192};
#define U1RXint (*(volatile unsigned short *)(0x1000a000))
enum {CHARRDYint=32768};

#define CLOCK_FREQ 32000000L

/* the functions */
/* Initialize the UART1 */
void inituart(int baud)
{
   /* Parity is disabled at reset by default */
   /* stop bits is set to 1 by default */
   /* assumes a 32 mHz system clock */
   U1CR1.UARTEN=ON;   /* Turn the uart on */
   U1CR1.TXEN=ON;     /* enable transmitter and receiver */
   U1CR1.RXEN=ON;
   U1CR2.IRTS=ON;     /* ignore request to send */
   U1CR2.WS=ON;       /* 8 bit word */
   /* Baud rate=CLOCK_FREQ/16/baud= */
   U1BRGR.CD=(UHWORD)(CLOCK_FREQ/16/baud);
```

```
    U1PCR.PC3=ON;    /* connect all i/o pins to the uart */
    U1PCR.PC2=ON;
    U1PCR.PC1=ON;
    U1PCR.PC0=ON;                           •
    U1DDR.PDC1=ON;  /* make output pin for the uart */
}                    /* probably not needed */

/* send a character out the serial port when it is ready */
static void put(BYTE x)
{
    while((U1SRint & TRDYint)==0)
        ;   /* wait until character is ready */
    U1TX.DATA =x; /* send the data out */
}

/* read in and echo a character through the serial port */
BYTE getchar(void)
{
    BYTE a;

    while((U1RXint & CHARRDYint)==0)
        ;   /* wait till character is ready */
    a=U1RX.DATA;
    putchar(a);
    return a;
}

/* read in a character with no echo */
BYTE getch(void)
{
    BYTE a;

    while((U1RXint & CHARRDYint)==0)
        ;   /* wait till character is ready */
    a=U1RX.DATA;
    return a;
}

/* read in a character from the serial port. Do not
check for the character ready. This routine should be used
with kbhit(). */
BYTE getce(void)
{
    return U1RX.DATA;
}
```

```
/* return TRUE when a key has been hit and FALSE otherwise */
int kbhit(void)
{
    return U1RX.CHARRDY;
}


/* convert a long to a series of ascii characters
   and send it out the serial port */
void printd(unsigned long x)
{
    if(x/10)
        printd(x/10);
    putchar(x%10+'0');
}

/* same as above but output hexidecimal */
void printx(unsigned long x)
{
    if(x/16)
        printx(x/16);
    putchar(x%16+((x%16>9)?('a'):('0')));
}
```

Listing 8-5: Serial Input/Output Functions

At the beginning of the program, the usual headers are included. In this case, `mmc2001.h` is needed to identify the address of the UART, the header `uart.h` contains all of the definitions needed for the UART registers, and the `serial.h` header contains function prototypes for all of the serial input/output functions. These functions are for demonstration purposes only. They all work. I have used these functions at 115200 bit rates and had no problems. None of them have any built-in tests for errors on reception like parity tests, framing errors, or overrun errors. These tests need to be included and the errors handled properly if you want to have a production quality input/output library.

In the text that follows, each of the several small functions contained in the above listing will be described individually. The first function is the `inituart()` function.

```
#include "mmc2001.h"
#include "uart.h"
#include "serial.h"
```

```
#define CLOCK_FREQ 32000000L

/* Initialize the UART1 */
void inituart(int baud)
{
    /* Parity is disabled at reset by default */
    /* stop bits is set to 1 by default */
    /* assumes a 32 mHz system clock */

    U1CR1.UARTEN=ON;    /* Turn the uart on */
    U1CR1.TXEN=ON;      /* enable transmitter and receiver */
    U1CR1.RXEN=ON;
    U1CR2.IRTS=ON;      /* ignore request to send */
    U1CR2.WS=ON;        /* 8 bit word */
    /* Baud rate=CLOCK_FREQ/16/baud= */
    U1BRGR.CD=(UHWORD)(CLOCK_FREQ/16/baud);
    U1PCR.PC3=ON;       /* connect all i/o pins to the uarts */
    U1PCR.PC2=ON;
    U1PCR.PC1=ON;
    U1PCR.PC0=ON;
    U1DDR.PDC1=ON;      /* make output pin for the uart */
}
```

Listing 8-6: The inituart() Function

Each of these functions will have the same three header functions shown in Listing 8-6. In this case, the function requires the clock speed of the chip. This speed can be anything, and in our particular case, it is 32 MHz. The macro definition CLOCK_FREQ makes this value 32000000. If you ever change the clock speed of the computer, this value should be changed. The several instructions in the function are all commented with their operations. This function should be executed whenever the serial I/O system is to be used by a program. Pass the desired baud rate to the function as an integer. The integer size on the DIAB compiler used for this system is 32 bits; therefore, there will be no overflow problem for any of the usual choices for baud rates.

Following the inituart() function above, there is the series of five functions shown in Listing 8-7. These functions are all placed together because they each access on-board features of the MCORE chip. Almost always, it is best to collect operations that access on-board registers that control peripheral devices into individual functions that can be called from the applications portion of the program. This way, the detailed features of the chip are tightly contained in functions that can be easily changed if the chip itself is changed. The `defines` and `enums` set at the beginning of Listing 8-7 are from the top of the `serial1.c` program. The values established by these devices are used in the functions of Listing 8-7. These `#defines` and `enums` would not be necessary if the compiler worked as expected. Generally, if you want to wait until it is safe to send a character out the serial port, the code that you might use is

```
while(U1SR.TDRY ==0)
    ;    /*   wait here until TDRY is 1 */
```

This code will cause the computer to sit in a loop while the TDRY bit found in U1SR is 0. Unfortunately, the code created by the compiler failed to reload the value of the bit inside the loop, so the computer went into an infinite loop when it executed this code.

There are always ways around such problems and with C it is not usually necessary to jump to assembly language when such a problem is found. Notice the code below. Rather than the argument shown above, the argument was converted to a simple bit-wise AND, which did compile correctly. Remember, the addition of a `#define` or an `enum` in your code does not add or subtract from the executable code in your program.

The first two functions shown below are `put()` and `getchar()`. The function `put()` sends a character to the serial port number 1 on the board. It works with the function `putchar()`, which works exactly the same as the `putchar()` function that you are used to using in your programs. Here `putchar()` sends a character to the serial port rather than to the device `stdout`. The function `getchar()` receives a character from the serial port 1. In C, when a `'\n'` character is sent to the output the program executes a carriage return and a line feed. Therefore, to make `getchar()` here the same, you will see in the next group of functions that the data received by `getchar()` is tested. If it is a `'\n'` character, two characters, `'\n'` and `'\r'`, are sent to the serial port. Otherwise, the character passed to the function is sent to the serial port. This operation mimics the operation of the normal `getchar()`, but it also requires a special function to output the character to the serial port. The function `put(BYTE x)` is that function. Since the function `put()` is never to be used outside of this file, it is designated as `static`.

The function `getchar()` reads a single character from the serial port. This function also mimics the `getchar()` that you are used to using, because it echoes the character received to the serial port output. There might be an occasion in which you want to read in a character without echoing it to the serial output. In that case, you can use `getch()` shown below. This function works exactly the same as `getchar()` but it does not echo the data received.

```c
#define U1SRint (*(volatile unsigned short *)(0x1000a086))
enum {TRDYint=8192};
#define U1RXint (*(volatile unsigned short *)(0x1000a000))
enum {CHARRDYint=32768};

/* send a character out the serial port when it is ready */
static void put(BYTE x)
{
    while((U1SRint & TRDYint)==0)
        ;   /* wait until register available */
    U1TX.DATA =x; /* send the data out */
}


/* read in and echo a character through the serial port */
BYTE getchar(void)
{
    BYTE a;

    while((U1RXint & CHARRDYint)==0)
        ;   /* wait till character is ready */
    a=U1RX.DATA;
    putchar(a);
    return a;
}

/* read in a character with no echo */
BYTE getch(void)
{
    BYTE a;

    while((U1RXint & CHARRDYint)==0)
        ;   /* wait till character is ready */
    a=U1RX.DATA;
    return a;
}
```

```
/* read in a character from the serial port.  Do not
   check for the character ready. This routine should
   be used with kbhit(). */
BYTE getce(void)
{
    return U1RX.DATA;
}


/* return TRUE when a key has been hit and FALSE otherwise */
int kbhit(void)
{
    return U1RX.CHARRDY;
}
```

Listing 8-7: Direct I/O Functions

The last two functions shown in Listing 8-7 are useful when you need to exit a function if there is an asynchronous keyboard hit. The function kbhit() returns a logical TRUE or FALSE. Its return should be used as the argument to an if() or while() test. Whenever the keyboard is touched, kbhit() returns a TRUE. That means that a character has been entered into the keyboard. If you need to read and use the value entered, the test to determine if a character is ready is not necessary. Therefore, the function getce() was written to read in the value contained in the data register without involving a test to show that there is a character ready to be read. These two functions, kbhit() and getce() should be used together.

The next four functions shown below also access the serial input and output, but they all use the functions above for the direct access and have no computer specific code. The function putchar() checks to determine if the parameter x is a '\n'. If it is, the function put() is called twice with a '\n' and a '\r' argument. Otherwise, the character passed to the function is sent to put() where it is sent to the serial port.

```
/* Send a character to the serial port when the port is
   available. If a '\n' is received send a '\n' followed by
   a '\r' sequence. */

void putchar(BYTE x)
{
    if(x=='\n')
```

```
    {
        put('\n');
        put('\r');
    }
    else
        put(x);
}


/* send a string to the serial port */
void puts(BYTE *a)
{
    while(*a!='\0')
        putchar(*a++);
}


/* This function reads a string into the buffer a. The
   length of the buffer is max. If the string is less than
   max long, the function returns the number of characters
   entered. If the string is longer than the buffer, the
   buffer is filled and a -1 is returned. The input string
   is terminated by either a '\n' or a '\r'. */

int gets(BYTE *a,int max)  /* no echo */
{
    int i=0,c;

    do  /* read in data a byte at a time */
    {
        *a++=c=getch();
    }while(c!='\n'&&c!='\r'&&++i<max-1);
    *a='\0';    /* make it a string */
    return (i>=max)?-1:i;
}


/* same as gets() but data entered are echoed */
int getse(BYTE *a,int max)  /* with echo */
{
    int i=0,c;

    do
    {
        *a++=c=getchar();
    }while(c!='\n'&&c!='\r'&&++i<max-1);
    *a='\0';
    return (i>=max)?-1:i;
}
```

Listing 8-8: General Serial Input/Output Functions

The next three functions can be used to send and receive strings through the serial port. To use `puts()`, you use a string argument. This function will send characters from the string until it finds a zero value in the string. The get string functions perform similar to the standard `fgets()` function. The difference between `gets()` and `getse()` is that `getse()` echoes the characters read in to the serial output. Otherwise these two functions are identical. You pass `gets()` two parameters. The first is a pointer to a character array and the second is the dimension of this array. As characters are read into the program, they are stored in the character array. The string input is terminated by either a `'\n'` or a `'\r'` character. The input will also be terminated when the input character string is one less than the size of the array size. When termination is detected, the input is terminated with a character zero, making the data a string. The return to the calling program will be −1 in the event that the character array was completely filled. Otherwise, the return is the number of characters entered into the character array.

The function `gets()` reads data from the serial port with the `getch()` function. This input does not echo the data entered. `getse()` uses the `getchar()` to read the data in. `getchar()` always echoes the input data to the serial output.

The above routines were separated into a series of individual functions, which were combined into an archive library file. This file is found on the CDROM under the name `libserio.a`. This file can be linked during the linking operation like any other library file.

Handling Interrupts

All of the onboard peripherals found on the MMC2001 that need access to interrupts are set up to use the core processor Auto Vector capability. Shown in Table 8-1 is a copy of the interrupt vector table. The vector table is placed at a location called the VBA, Vector Base Address, when the chip is initialized. You will note that the table is 0x200, two hundred hex bytes, long. The vector numbers are consecutive. The vector addresses move in steps of 4 and are usually dealt with as hexadecimal values. This table must be filled in some way when the program is initialized. Each vector will contain the address of the function executed when the corresponding exception occurs. Most of these vectors are self-explanatory. Of import here is the Fast Interrupt Autovector location. Note that the offset of this vector is 0x2c.

The use of this table is similar to that shown with the MC68HC16 family. Here, though, we will concentrate on the use of the autovector. The autovector is accessed when an interrupt is called with the autovector line to the core processor asserted. When this type of interrupt is executed, control of the processor is automatically transferred to the function addressed contained in the autovector vector. Note in the table that there are two autovector locations, the normal autovector and the fast autovector. The DIAB compiler automatically encodes interrupts to use the fast autovector, so the offset vector location is 0x2c from the Vector Base Address.

The MMC2001 has an internal peripheral called the Interrupt Controller. This device handles all interrupt sources from the on-board peripherals. There are several 32-bit registers in the Interrupt Controller. You will notice in the MMC2001 Reference Manual, Section 10, that these registers are each just collections of 32 single bits. These registers control interrupts. The first register is called the Interrupt Source Register, INTSCR. Whenever an interrupt is requested by this controller, it is assigned a level from 0 to 32 and the corresponding bit in the INTSCR is set. There are two interrupt enable registers: Normal Interrupt Enable, NIER and Fast Interrupt Enable, FIER. The program must set the corresponding bit in one of these registers to enable an interrupt to be detected.

When an interrupt is requested, the corresponding bit in the INTSCR is set and this register is automatically ORed with the Interrupt Enable registers. The results of these operations are stored in the proper Interrupt Pending Register. These registers, NIPND and FIPND, then contain a bit pattern that show all of the pending interrupts for the system.

There is one very convenient instruction available in the MCORE instruction set. This instruction, FF1, indicates "find the first 1 set" in a memory location. This instruction uses the system barrel shifter and requires only one clock cycle. The priority of the several interrupting sources is established by their individual bit locations in the various registers in the controller. In the FIPND registers, bit 31 contains the status of the highest priority pending interrupt, etc. When the FF1 instruction is executed on the FIPND, the result identifies the bit number of the highest priority pending fast interrupt. Table 8-2 shows the interrupt assignments of all on-board peripherals on this chip.

The FF1 instruction returns a 0 if bit 31 is set and 31 if bit 0 is set. It also returns a 32 if no bit is set in the designated memory location. Therefore, if the PIT is the highest requested interrupt, the contents of the FIPND register are stored in R3, and the instruction

```
FF1 R3
```

is executed, R3 will contain the numeric value 24. We can use this sequence to vector to the correct interrupt service routine. Consider the code shown in Listing 8-9. This function is a general-purpose interrupt handler to control the autovector access. The program starts with the normal file inclusions. In this case, the interrupt controller is being used so the header `intctl.h` is included. The code that follows must access the contents of the fast pending interrupt register FPIND. The address of this register is placed into the location `fIpnd1` for convenient access by the assembly program that is generated later.

The next entry defines a struct Table, which contains an array of 32 pointers to functions that require no parameters and return the type void. An instance of this structure is created and named `table`. The 32 entries in the array are each filled with a pointer to the function `unused_vector()`. It is difficult to decide what to fill unused vectors with. The choice of zero is not realistic. When an interrupt occurs that needs one of these vectors, control is passed to the address contained in the vector. If the vector contains zero, the computer goes to zero and starts executing instructions found there. The program will surely run amuck until who knows when if it is told to execute the code found at zero! I usually create a simple interrupt service routine that does nothing to help here. The function in the listing below is called `unused_vector()`. In this case, if an uninitialized interrupt occurs, it will be ignored and control will be passed back to the executing program. This particular approach has the drawback that you are never aware of the occurrence of uninitialized interrupts unless you put a break of some sort in the `unused_vector()` routine.

In practice, you will need to place the names of interrupt service routines in the proper vector locations shown below. We will do this in later code to show you how.

```
#include "mmc2001.h"
#include "intctl.h"

UWORD const fIpnd1=INTCTL_+0X10;
```

```
void handler(void);
void unused_vector(void);

static struct Table {
   void (*vector[32]) (void);
};

struct Table table ={
   unused_vector, /* 31 unused */
   unused_vector, /* 30 unused */
   unused_vector, /* 29 unused */
   unused_vector, /* 28 INT7 */
   unused_vector, /* 27 INT6 */
   unused_vector, /* 26 INT5 */
   unused_vector, /* 25 INT4 */
   unused_vector, /* 24 INT3 */
   unused_vector, /* 23 INT2 */
   unused_vector, /* 22 INT1 */
   unused_vector, /* 21 INT0 */
   unused_vector, /* 20 ISPI */
   unused_vector, /* 19 UART1 receive */
   unused_vector, /* 18 UART0 receive */
   unused_vector, /* 17 UART1 transmit */
   unused_vector, /* 16 UART0 transmit */
   unused_vector, /* 15 PWM5 */
   unused_vector, /* 14 PWM4 */
   unused_vector, /* 13 PWM3 */
   unused_vector, /* 12 PWM2 */
   unused_vector, /* 11 PWM1 */
   unused_vector, /* 10 PWM0 */
   unused_vector, /* 9 unused */
   unsued_vector, /* 8 PIT */
   unused_vector, /* 7 Time-of-day alarm */
   unused_vector, /* 6 KPP control */
   unused_vector, /* 5 UART0 RTS_DELTA */
   unused_vector, /* 4 unused */
   unused_vector, /* 3 unused */
   unused_vector, /* 2 software3 */
   unused_vector, /* 1 software2 */
   unused_vector  /* 0 software1 */
};

#define Do_Interrupt() asm(" subi R0,32\n subi R0,28\n  \
                 stm R1-R15,(R0)\n lrw R2,table\n \
                 lrw R3,fIpnd1\n ldw R3,(R3,0)\n \
                 ldw R3,(R3,0)\n FF1 R3\n lsli R3,2\n \
```

```
        addu R2,R3\n ldw R2,(R2,0)\n \
        jsr R2\n ldm R1-R15,(R0)\n \
        addi R0,32\n addi R0,28\n rfi\n")

void unused_vector(void)
{}

void handler(void)
{

    Do_Interrupt();

}
```

Listing 8-9: Autovector Interrupt Handler

Following the table initialization, there is an assembly language sequence. This sequence is #defined as the function Do_Interrupt(), which is called in the interrupt service routine handler(). The first five assembly instructions clear 120 bytes on the stack, enough to save the contents of registers 1 through 15, and

Table 8-1 Exception Vector Assignments

Vector Number(s)	Vector Offset (Hex)	Assignment
0	000	Reset
1	004	Misaligned access
2	008	Access error
3	00C	Divide by zero
4	010	Illegal instruction
5	014	Privilege violation
6	018	Trace exception
7	01C	Breakpoint exception
8	020	Unrecoverable error
9	024	Soft reset
10	028	INT autovector
11	02C	FINT autovector
12	030	Hardware accelerator
13	034	(Reserved)
14	038	
15	03C	
16-19	040-04C	Trap #0-3 Instruction Vectors
20-31	050-07C	Reserved
32-127	080-1FC	Reserved for vectored interrupt controller use

save these registers on the stack. The following two assembly instructions load the address of `table` into R2 and the address of `fIpnd1` into R3. We want the contents of the value found in the FPIND register. Therefore, we must dereference the value in R3 twice, once to get the address of FPIND and then to get the contents of FPIND. This value has a bit set for every pending fast interrupt. When this value is operated on by the FF1 instruction, the register R3 will contain the 32 minus the bit number of the highest priority pending interrupt. If you look at how `table()` is constructed, you will find that the highest priority interrupts are listed first down to the lowest priority. Each entry in the table requires four bytes to hold a function pointer with this system. Therefore, we must multiply the value found in R3 by 4, shift it left 2, and then add the result to the value in R2 to find the address of the desired interrupt service routine.

All of that is exactly what is done with the assembly language insert shown above. The last instruction `jsr R2` passes control of the computer to the specified interrupt service routine. If you look at the C code generated by `handler()` you will find first that the necessary computer status is saved and then the assembly code in the macro `Do_Interrupt()` is executed. The last instruction here is the `jsr R2` instruction mentioned above. After the code needed for the interrupt service routine is completed, control is returned to `handler()` where the machine status is restored and control is returned to the interrupted program with an `rfi`, return from fast interrupt, instruction.

The compiler has a #pragma interrupt `isr_function`. This #pragma causes the function specified by the `isr_function` to be compiled as an interrupt service routine. These functions save a portion of the machine status, execute the code specified in the `isr_function`, and restore the machine status before returning to the interrupted program with an `rfi` instruction. When writing a general function such as the one above, it was found that the partial status save was not enough. No functions could be called from within the interrupt service routine. This limitation was too great for a general-purpose handler like that above, so in this routine, the entire status of the computer is saved and restored rather than the partial save and restore generated by the #pragma. It is safe to execute functions within the interrupt service routines in the handler above.

Table 8-2 Interrupt Controller Assignments

Bit Number	Use
0, 1, 2	Software
3	unused
4	unused
5	UART0 RTS_DELTA
6	KPP control
7	Time-of-day alarm
8	PIT
9	unused
10	PWM0
11	PWM1
12	PWM2
13	PWM3
14	PWM4
15	PWM5
16	UART0 transmit
17	UART1 transmit
18	UART0 receive
19	UART1 receive
20	ISPI
21	INT0
22	INT1
23	INT2
24	INT3
25	INT4
26	INT5
27	INT6
28	INT7
29	unused
30	unused
31	unused

A Clock Program

A clock program is an excellent program for demonstrating the use of interrupts on a microcontroller. The MMC2001 has a rather extensive timer subsystem. It is broken into three parts. There is a time of day clock, TOD, with a built-in alarm, a watchdog timer and a programmable interval timer, PIT. We will use the PIT in this example. Briefly, the timers are driven by a separate clock. The clock here is a 32768-Hz clock controlled by a watch crystal. The TOD clock is driven by the low-frequency oscillator frequency divided by 128 or 256 Hz. The watchdog and the PIT are both driven by the low-frequency oscillator frequency divided by 4 or 8196 Hz. The TOD keeps track of the time as the number of seconds since a specified time, perhaps midnight. There is no provision for automatic roll-over of the registers at the end of the day, so if you want to work from midnight each day, you must reset the TOD clock to zero each midnight. This clock also has an 8-bit fractional second register in addition to the regular 32-bit second count register.

The alarm system also contains an 8-bit fractional second register and a 32-bit second register. When the time in the alarm registers matches the time in the time of day register, a flag is set that can be polled or an interrupt can be requested.

The PIT system contains a register that is counted down. When this register underflows, it can be reloaded automatically, set a flag to be polled, or request an interrupt. Loading this register is done through a register called the ITDR, the Interval Timer Data Register. If a bit is set, data written to the ITDR will automatically transfer to the ITADR, the register that counts down. When this register underflows, the contents of the ITDR is automatically transferred to the ITADR. The contents of the ITDR can be changed at any time, and it can be changed without altering the contents of the ITADR until the next underflow.

The ITADR is clocked 8192 times per second, approximately 122 microseconds. Since it is clocked an exact number of times per second, it is possible to count in exact second intervals. It makes no difference how you break up the time intervals, the clock will decrement exactly 8192 times in one second. Therefore, if you want a basic interval time for other uses in the system, you can cause interrupts to occur at this faster rate and then count the number of interrupts until you achieve the magic number of 8192 ticks each second. For example, many systems need a 1 or 2 millisecond time interval. The system can be set to interrupt about every 2 milliseconds and then in the ISR, a counter incremented. This counter counts to about 500 each second and can be easily used in a clock controller. Of course, "about" is not allowed. The interrupt time must be 0.001953125 seconds and there will be exactly 512 interrupts per second.

The program starts out with the creation of a function that we will call keep_time(). This function is executed repeatedly inside of a loop. There are four external variables, hours, seconds, minutes, and count accessed by `keep_time()`. The variable `count` is incremented in an interrupt service routine that is executed each 1.953125 milliseconds. The constant `TIME_COUNT` will have a value of 511. When `count` becomes greater than `TIME_COUNT`, `count` is reset to zero and the parameter seconds are incremented. When `seconds` exceeds the value 59, `seconds` is reset to zero and `minutes` is incremented. When minutes exceeds 59, it is reset to zero and `hours` is incremented. Finally, when `hours` becomes greater than 12, it is reset to 1 which corresponds to one o'clock. Once each second, the function `output_time()` is executed.

```
#include "mmc2001.h"
#include "timer.h"

#define TIME_COUNT    511
#define MAX_SECONDS   59
#define MAX_MINUTES   MAX_SECONDS
#define MAX_HOURS     12
#define MIN_HOURS     1

/* function prototypes */
void output_time(void);
void keep_time(void);

/* external-global variables */
WORD seconds,minutes,hours,count;

/* the main applications program. count is incremented in
   the isr 512 times each second. Therefore, this routine
   must be executed at least once every two milliseconds. */

void keep_time(void)
{
    if(count>TIME_COUNT)
    {
        count=0;
        if(++seconds>MAX_SECONDS)
        {
            seconds=0;
            if(++minutes>MAX_MINUTES)
            {
                minutes=0;
                if(++hours>MAX_HOURS)
                    hours=MIN_HOURS;
```

```
        }
      }
    output_time();
      }
  }
```

Here we are starting to write a program. Therefore, it is smart to do all things correctly. For example, we will attempt to avoid magic numbers by defining mnemonics for the numbers used in the program. Also, whenever a new function is written, its function prototype will be immediately inserted into a function prototype list at the beginning of the program. A series of external variables is used. It is usually better to use local variables when working with parameters in several different functions. In that case, the parameters can be passed as arguments to the functions when they are called. This approach avoids debug problems where it becomes difficult to determine where variables are changed in different functions. In the case here, the variables hours, minutes and seconds are changed in only one function, keep_time(), and the variable count is changed only in the interrupt service routine or the reset time function. Therefore, there is no uncertainty as to where the variables are changed.

Note the structure of the if() statements in the function keep_time(). The nesting of these statements

```
if()
{
        .
    if()
    {
            .
        if()
        {
                .
        }
    }
}
```

causes the first test to be executed every time the function is executed. The argument of the first test must be TRUE when the second test is executed, and the argument of the second test must be TRUE when the third test is executed, and so forth. Most of the time when the function is executed, the total effort required by the above nested if sequence is the outside test only. Such an arrangement will require a minimum amount of computer time each time the function is executed. Many programmers will not nest the above tests so that each test is executed each time the function is called. It works, but it requires more computer resources than the nesting shown above.

In the function `keep_time()` above, note that the arguments of the several `if()` statements all involve a "greater than" test. This particular approach is more robust than any test that involves an "is equal to" test. In all cases but the hours, the tested parameter is reset to zero. Therefore, the zero is counted in the sequence and the maximum value is one less than the number that might be expected. For example, the range of seconds is 0..59 not 1..60, the range of count is 0.. 511 not 1..512. When the respective count exceeds the specified maximum value, the parameter is reset to its minimum value. If lightning should strike the chip and the value of seconds be set at 100, the "greater than" test would fix the error the next time that second were tested. In the event that an "is equal to" test were used, it would be a long time before the seconds would count through a wrap-around and be equal to the MAX_SECONDS value again.

Also note the `if` argument

```
if(++seconds>MAX_SECONDS)
{
```

The seconds are first incremented and then the test is completed. This sequence could be completed in two statements,

```
seconds=seconds+1;
if(seconds>MAX_SECONDS)
{
```

Some people prefer the latter approach, but the two approaches accomplish exactly the same thing. The important item is that seconds must be incremented before the test is completed rather than after.

At the bottom of the `seconds` loop, a call to `output_time()` is executed. This function will send the current time to the output device. Here again, the call to `output_time()` could be placed anywhere in the loop, but the bottom, as is shown above, is best because the output takes place after all of the parameters are updated and the time displayed will be now rather than before all of the updates are completed.

Let us now look at output_time(). For this program, we will send the time to the serial port to be displayed on a terminal. All of the parameters hours, minutes, and seconds each have a range of 0 or 1 to some maximum two-digit value. Therefore, these numbers will contain only tens and units. There will be no hundreds or thousands or fractions, or minus signs for that matter. To convert these values to characters to be sent to a function like putchar(), we have to do two things. First count the number of tens, convert this number to an ASCII digit and use it as an argument to putchar(). Then calculate the number of units in the number, convert it to an ASCII digit and send it to putchar(). These two operations are relatively easy to program. In fact, these little functions can be written as function-like macros easily, as follows:

```
#define hi(x) ((x)/10+'0')
#define lo(x) ((x)%10+'0')
```

Remember, the value passed to these two macro functions must lie between 0 and 99. The first function hi(x) determines the number of tens in the number and converts the result to an ASCII digit by adding the character zero to the result. The second calculates the number of units in the number by calculating the number modulo 10. This value is converted to an ASCII digit when the character zero is added.

With these two little macros in hand, the output_time() function is easy to write:

```
void output_time(void)
{
    putchar('\r');
    putchar(hi(hours));
    putchar(lo(hours));
    putchar(':');
    putchar(hi(minutes));
    putchar(lo(minutes));
    putchar(':');
    putchar(hi(seconds));
    putchar(lo(seconds));
}
```

The output_time() function consists of a series of putchar() function calls. The first has an argument '\r'. This command causes the cursor on the screen to be returned to its leftmost position. The next two calls send the `hours` to the screen. Then a colon is printed. This sequence is repeated for the `minutes` and the `seconds` except no colon is sent out after the `seconds`.

Our next problem is to set up the PIT. The PIT is to interrupt once each 1.953125 milliseconds. The clock that drives the PIT runs at 8192 Hz and we need a clock rate of 512 Hz. Therefore, we need to count 16 clock ticks between interrupts. I will call this time `TWO_MS` in the following code. The function below is an initialization function to set up the PIT to operate as needed. Recall, data written to the ITDR is transferred to the ITADR when the ITADR underflows. Data written to the ITDR is automatically transferred to the ITADR if the OVW bit is set when the ITDR is written. This bit will be set first and then the value `TWO_MS` is written to the ITDR. Next, the reload mode is enabled by setting the RLD bit in the ITCSR register. In this mode, the value in the ITDR is automatically reloaded into the ITADR when the ITADR underflows. This operation assures a periodic interrupt at the period set by the value stored in the ITDR.

The interrupt handler shown in the section "Handling Interrupts" will be used here. Recall that function contains a general-purpose interrupt service routine that can be used for any of the autovector interrupts in the MMC2001. For our purposes here, the interrupt vector number 8 in Listing 8-9 will have to be changed from `unused_vector` to the name of the PIT interrupt service routine. We will call that routine `pit_isr` and it will be written in the next section. The macro function `vector()` in the `mmc2001.h` header file will be used to put the address of the header into the Fast Autovector vector location. Also in the Interrupt Handler section you will find Table 8-1. This table shows the allocation of the interrupt vectors in the base vector table. This table is placed in a specified location by the linker program. With the set-up used for our programs in this book, I placed the system RAM at the address 0x30000000. The first 0x200 entries in this table are the vector table whose outline is shown in Table 8-1 and whose values are established by the interrupt handler routine in Listing 8-9. There you will note that the Fast Interrupt Autovector is offset 0x2c into the vector table. Therefore, the address of the `FAST_AUTOVECTOR` in our program must be 0x3000002c. The next instruction in the code below will place the address of the `pit_isr` interrupt service routine in this address.

```
#include "intctl.h"

#define TWO_MS       15

/* initialize the PIT */
void pit_init(void)
{
    ITCSR.OVW=ON; /* set-up ITDR to ITADR overload */
    ITDR=TWO_MS;/* interrupt each two milliseconds */
    ITCSR.RLD=ON; /* enable reload mode */
    ITCSR.ITIF=ON;/* clear the interrupt flag */
    ITCSR.ITIE=ON;/* enable the pit interrupt */
    FIER.EF8=ON;  /* enable the AVEC interrupt */
    ITCSR.EN=ON;  /* enable the pit */
}
```

The final four instructions in the `pit_init()` function get the chip ready to receive interrupts from the PIT. The first

```
ITCSR.ITIF=ON; /* clear the interrupt flag */
```

clears the interrupt flag ITIF. It is a quirk of the chip that the interrupt flag is cleared by writing a 1 to the chip rather than a 0. This flag should be cleared prior to enabling the interrupts for both the autovector and the PIT. If this bit is set when the interrupts are enabled, the system will immediately respond to the interrupt and that action is not usually what is wanted. The next instruction turns on the interrupt enable flag ITIE for the PIT. The fast interrupt bit 8 is set in the FIER register of the interrupt controller and finally, the PIT is turned on when the bit EN is set in the ITCSR. The FIER register is identified in the `intctl.h` header file. After this code is executed, the PIT is set up and ready to interrupt every 16 ticks of the 8192-Hz clock.

Of course, we will need an interrupt service routine. When the Interrupt Handler is used, the interrupt service routine need not be concerned with saving the system status. That is handled by the interrupt handler. All that is needed in this case is to turn off the interrupt flag in the ITCSR and increment the value of count. The entire interrupt service routine is

```
/*   in this isr, all that must be done is to turn off the
     pit interrupt request bit and increment the global
     parameter count */

void pit_isr(void)
{
    ITCSR.ITIF=ON;  /* ON clears the interrupt bit in these
                       registers */
    count++;
}
```

The interrupt handler takes care of restoring the status prior to returning to the interrupted program.

The next logical function that we will need is the function `main()`. Main is broken into two parts, the initialization portion of the program and the applications portion of the program. The first three instructions are the initialization portion of the program. Here, the serial I/O system is initialized and the baud rate is set to the value BAUD_RATE. The `pit_init()` routine written above is executed and then the system fast interrupts are enabled with the assembly language macro function defined in the header file `mmc2001.h`.

```
#define BAUD_RATE    38400
#define FAST_AUTOVECTOR 0x3000002c

main()
{
    inituart(BAUD_RATE);  /* initialize the UART */
    pit_init();      /* initialize the PIT */
    vector(handler,FAST_AUTOVECTOR); /* put handler address
                                  in FAST_AUTOVECTOR. */
    Enable_Fast_Interrupts();
    FOREVER
    {
        keep_time();  /* keep track of the time */
        if(kbhit())
            reset_time();
    }
}
```

The application portion of the program is quite simple. It is simply a FOREVER loop that repeatedly executes `keep_time()`. Also in the loop, a test is made to determine if a key has been touched on the terminal attached to the serial port. If it has, the function `reset_time()` is executed where the clock time can be set.

The function `reset_time()` follows. Recall from earlier discussions that the function `getch()` was written specifically to read in a character from the serial port after the function `kbhit()` indicates that the keyboard has received an input. This function, `getch()`, is used here. The character read in is saved in the location c and then a switch/case statement is set up to determine what to do with the input. It is assumed that the letters m, M, h, and H are the valid inputs. When either an m or an M is received, the minutes will be incremented. Detection of either an h or an H will cause the hours to be incremented. Any other input will merely be discarded, so there is no need for a default statement in the switch/case sequence. If an m, either upper or lower case, is detected, the minutes will be incremented. If the incremented value exceeds `MAX_MINUTES`, minutes will be reset to 0. Similarly if an h is detected, the hours will be incremented and if the incremented value exceeds `MAX_HOURS`, hours is reset to `MIN_HOURS`. After these transactions are completed, the function `output_time()` is executed. Without this final function call, the setting of the time will be very jumpy, definitely not smooth.

```c
void reset_time(void)
{
    int c;

    c=getch();
    switch(c)
    {
        case 'm':
        case 'M':   if(++minutes>MAX_MINUTES)
                        minutes=0;
                    break;
        case 'h':
        case 'H':   if(++hours>MAX_HOURS)
                        hours=MIN_HOURS;
                    break;
    }
    output_time(); /* send out a new time after each change */
}
```

That is the end of the program. The program seems quite simple, and it has been tested thoroughly. Shown below is the whole program pulled together into a single listing. Several items have been moved around in this listing to make the whole program easier to follow.

```c
#include "mmc2001.h"
#include "timer.h"
#include "intctl.h"
#include "serial.h"

#define TIME_COUNT        511
#define MAX_SECONDS       59
#define MAX_MINUTES       MAX_SECONDS
#define MAX_HOURS         12
#define MIN_HOURS         1
#define FAST_AUTOVECTOR   0x3000002c
#define TWO_MS            15
#define BAUD_RATE         38400

#define hi(x)  ((x)/10+'0')
#define lo(x)  ((x)%10+'0')

/* function prototypes */
void output_time(void);
void keep_time(void);
void reset_time(void);
void pit_init(void);
void pit_isr(void);
void handler(void);

/* external-global variables */
WORD seconds,minutes,hours,count;

/* the main applications program. count is incremented in
   the isr every two milliseconds. Therefore, this routine
   must be executed at least once every two milliseconds to
   keep the outputs to the serial port smooth. */
void keep_time(void)
{
    if(count>TIME_COUNT)
    {
        count=0;
        if(++seconds>MAX_SECONDS)
        {
            seconds=0;
            if(++minutes>MAX_MINUTES)
            {
```

```
                minutes=0;
                if(++hours>MAX_HOURS)
                    hours=MIN_HOURS;
            }
        }
    output_time();
    }
}

void output_time(void)
{
    putchar('\r');
    putchar(hi(hours));
    putchar(lo(hours));
    putchar(':');
    putchar(hi(minutes));
    putchar(lo(minutes));
    putchar(':');
    putchar(hi(seconds));
    putchar(lo(seconds));
}

/* initialize the PIT */
void pit_init(void)
{
    ITCSR.OVW=ON;    /* set-up ITDR to ITADR overload */
    ITDR=TWO_MS;     /* interrupt each two milliseconds */
    ITCSR.RLD=ON;    /* enable reload mode */
    ITCSR.ITIF=ON;   /* clear the interrupt flag */
    ITCSR.ITIE=ON;   /* enable the pit interrupt */
    FIER.EF8=ON;     /* enable the AVEC interrupt */
    ITCSR.EN=ON;     /* enable the pit */
}

/* in this isr, all that must be done is to turn off the pit
interrupt request bit and increment the global parameter
count */

void pit_isr(void)
{
    ITCSR.ITIF=ON;  /* ON clears the interrupt bit in these
                       registers */
    count++;
}

main()
```

```
{
   inituart(BAUD_RATE); /* initialize the UART */
pit_init();              /* initialize the PIT */
vector(handler,FAST_AUTOVECTOR);
   /* put handler address in FAST_AUTOVECTOR. */
   Enable_Fast_Interrupts();
   FOREVER
   {
      keep_time();   /* keep track of the time */
      if(kbhit())
         reset_time();
   }
}

void reset_time(void)
{
   int c;

   c=getch();
   switch(c)
   {
      case 'm':
      case 'M':    if(++minutes>MAX_MINUTES)
                      minutes=0;
                   break;
      case 'h':
      case 'H':    if(++hours>MAX_HOURS)
                      hours=MIN_HOURS;
                   break;
   }
   output_time(); /* send out a new time after each change */
}
```

Listing 8-10: Complete Clock Routine

Note that in this program the only functions that have direct access to the memory of the computer are those that involve the PIT. The pit_init() routine is truly chip specific as is the pit_isr(). All of the other routines could be used on just about any computer. It is always good to place the chip-specific routines in functions by themselves. Then when you need to change to another chip, all of the chip-specific code is lumped together and the general-purpose code is easy to lift and move to the new program.

All #define object-like and function-like macros are placed at the head of the program. Also, a list of function prototypes is included.

Every function in the program with the exception of main() has a function prototype at the beginning of the program. In this case, the functions are all void functions that require no arguments. It is important, however, that every programmer get into the habit of placing these prototypes at the beginning of their programs, either in the form of a header file or a direct listing in the program.

Keyboard

The MMC2001 chip has a built-in keyboard interface. This interface can be either synchronously polled by the program, or it can be asynchronously controlled through the use of an interrupt. We will examine the asynchronous control here. This program requires three parts. The first is the initialization routine. This routine sets up the operation of the keyboard to cause an interrupt whenever a key is depressed. The interrupt service detects which key is depressed and saves it in a buffer. The program also saves a semaphore passed to the initialization routine. This semaphore is released when the end of a line is detected in the input. Then the main program prints the buffer to the terminal screen.

The initialization routine is shown below:

```
#include "mmc2001.h"
#include "keypad.h"
#include "intctl.h"

static int semaph;
static BYTE *data,*savedata;
static int leng;

/* key board initialization routine */
void kpinit(int sem, BYTE *s, int length)
{

    KPCRN.LOROWS=0xf;      /* enable keypad rows */
    KPDRB.COLUMNS=0x0;     /* write 0 to KPDR[15:8] */
    KPCRN.LOCOLS=0xf;      /* make cols open drain */
    KDDRN.LOCOLS=0xf;      /* make cols outputs */
    KDDRN.LOROWS=0x0;       /* and rows inputs */
    KPSR.KPKD=ON;      /* clear KPKD status flag */
    KPSR.KDSC=ON;      /* clear key depress synchronizer */
    KPSR.KDIE=ON;      /* set KDIE bit to enable depress
                       Interrupt */
```

```
KPSR.KRIE=OFF;      /* disable release interrupt */
FIER.EF6=ON;        /* enable Fast Interrupt number 6 */
semaph=sem;         /* save the semaphore */
data=s;
savedata=s;         /* and array information for */
leng=length;        /* use in the isr */
}
```

It is expected that the calling program will attach a semaphore and provide a buffer into which the input data will be stored. The semaphore is passed to the initialization routine as sem, and the array by a pointer to its first element and its length. We plan to use a 4 by 4 matrix keyboard. This keyboard will be connected to the least significant four rows and columns of the keyboard controller interface. There are several registers that attach to these interfaces. The KPCR is a control register, the KPDR is the data register, and the KDDR is the data direction register. In creating the header file for the keyboard controller, I added an extra letter, N or B, to these pseudovariable names to indicate that they have been broken into groups of bits rather than a register of single bits. A register name that has the letter N appended is split into two 8-bit fields named COLUMNS and ROWS. Those with a letter B appended are split into four fields named HICOLS, LOCOLS, HIROWS and LOROWS. Therefore, the command

```
KPCRN.LOROWS=0xf;
```

will cause the least significant bits of the ROWS field in KPCR to be set to 1.

The comments above explain the various actions that take place in the initialization routine. The Key Pad Data Register, KPDR, is a register that connects to the external keyboard interface. The most significant 8 bits are to connect to columns and the least significant bits connect to the rows. The plan is to detect cross connections between a column and a row. The columns serve as outputs and the rows inputs. Therefore, the columns will be set as open drain and as outputs through the data direction register. In the KBSR, Key Board Status Register, the Key Pad Key Depressed bit, KPKR, is set whenever the system detects a depressed key. This bit is reset by writing a 1 to it. On both key depress and key release, there is a synchronizer that aims to hide any key bounce that might occur. The key depress synchronizer is cleared by writing a 1 to KPSR.KDSC.

Next the key depress interrupt is enabled and the key release interrupt is disabled. The parameters passed to the initialization routine are now saved in static external variables. These parameters are used in the interrupt service routine.

The column outputs on the four least significant bits are all set to zero so that any key depression will cause the normally high inputs sensed by the key inputs to be pulled low. This action will initiate the key depress synchronizer operation. Approximately four milliseconds later, if the input key has the same state, a key depressed interrupt will be requested. The key release operates similarly. As long as a key remains depressed, nothing happens. When the key is released, the key release synchronizer initiates its action. If the key has the same state after the synchronizer time-out, a key release interrupt can be requested. For these operations to work, the column outputs must all be held at zero.

It is most convenient to use both the key depress and the key release interrupts for the task at hand. Each time a key is depressed, it is detected and the position of the key in the keypad is saved in a buffer. Conditions to cause a key release interrupt are then set up. Then control is returned to the interrupted program. When the key is released, a key release interrupt is requested, and in the interrupt service routine, a key depress interrupt is set up so that the next key press will be detected. Each time a key position is entered into the buffer, the next position in the buffer is marked with a 0xff.

I speak of the key position. In the interrupt service routine below, it is first tested to determine if the interrupt was caused by a key depress or a key release. If it is a key depress interrupt, control is immediately passed to a `for` loop. Within this loop, the columns are first loaded with a 0xe, which will have the three highest bits in the column high and the least significant bit low. A test checks to determine if all of the column bits are high. If not the value of column is impressed on KPDRB.COLUMNS. The next test checks the value found on KPDRB.ROWS. If there is no button depressed that connects column 1 with any of the rows, the result will be 0xff. Therefore, control will go to the next column. If the value found on rows is not 0xff, the least significant four bits will indicate the row in which the switch is depressed. Then, the row and column data is combined into a single value that is stored in the data array. The column contains 4 bits, as does the row. These bits are packed into a single 8-bit value by the instruction

```
c=*data++=(~(KPDRB.COLUMNS<<4 | (KPDRB.ROWS&0xf))) & 0xff;
```

The column value is shifted left by four bits and the result is bit wise ANDed with the row value. These values are the ones-complement of the values we want, so the ones-complement is taken. The final result is ANDed with a mask of 0xff to delete any bits outside of the eight bits needed for the final result. This result is stored in the array back in the calling program. Also at this time, the result is stored in a local variable for later use. The count of the number of characters entered into the array are incremented. If this value exceeds the length of the array, there is an error and the input is completely terminated when the program is restarted at its very beginning.

```
#define COL_0 0x0e

/* key board interrupt service routine */
void kb_isr(void)
{
    UHWORD column,input;
    static BYTE c;

    if(KPSR.KPKD)   /* if a key has been depressed */
    {
        for(column=COL_0;column!=0xf;column=(column<<1|0x1) & 0xf)
        {
            KPDRB.COLUMNS=column;
            /* kill some time */
            if(KPDRB.ROWS!=0xff)
            {
                c=*data++=(~(KPDRB.COLUMNS<<4 |
                    (KPDRB.ROWD&oxf))) & 0xff;
                if(++n>=leng)       /* data exceed the array */
                {
                    puts("input data overflow\n");
                    _start();  /* bad error, restart the
                            program */
                }
            }
        }
    }
    *data=0xff; /* needed to terminate decode */
    if(c==0x11) /* '=' reached the end of the input */
```

```
    {
        decode(savedata);
        release_semaphore(semaph);
    }
    KPDRB.COLUMNS=0X0; /* write 0 to KPDR[15:8] */
    KPSR.KDIE=OFF;/* disable key depress interrupt */
    KPSR.KRIE=ON;  /* enable key release interrupt */
}
else  /* key release was detected */
{
    KPSR.KPKR=ON;
    KPSR.KPKD=ON;  /* clear key release and depress */
    KPSR.KRSS=ON;  /* set key release synchronizer */
    KPSR.KDSC=ON;  /* clear key depress synchronizer */
    KPDRB.COLUMNS=0X0;/* write 0 to KPDR[15:8] */
    KPSR.KDIE=ON;  /* enable key depress interrupt */
    KPSR.KRIE=OFF;/* disable key release interrupt */
}
}
```

Notice in the code above that after the columns value is set and before the row values are tested, there is a comment to kill some time. It turns out that some time is required to allow propagation from the command to the electrical connection on the external switch. The delay() routine with an argument of 1, a 1-ms delay, worked well, but this routine also uses the PIT. It is intended to use this routine with the clock shown in the earlier section, so it is very inconvenient to use the clock. Therefore, a series of instructions around the keyboard controller were put together that would use up the necessary time. These instructions would have to be executed to set up correct conditions prior to the return from interrupt, so it is not too much a waste of computer resources. To get enough delay, I used some of these instructions more than one time. The code for this time delay is shown below. The last six lines of code are required to set up for the exit from the isr, and the first four are duplicates needed to round out the time.

```
KPSR.KRSS=ON;
KPSR.KPKD=ON;    /* flip and flop some innocuous bits */
KPSR.KDIE=OFF;   /* just to kill time to synchronize */
KPSR.KRIE=OFF;   /* the command with data arrival at */
KPSR.KPKR=ON;    /* the chip's edge */
KPSR.KPKD=ON;
KPSR.KRSS=ON;
```

```
KPSR.KDSC=ON;
KPSR.KDIE=OFF;
KPSR.KRIE=OFF;
```

After the key position is detected, and its value stored in the array, the value 0xff is placed in the next location in the array. This value is used in the decode routine that changes the key positions to ASCII characters for display. A termination character is needed to show the end of the input data has been reached. There is an '=', equal sign on the keyboard, and it was decided to use it as the termination character. We will see later that the position value for the '=' symbol is 0x11. Therefore, if the value c is 0x11 at this point in the code flow, it is time to decode the buffer contents and proceed further. After the data are decoded, the semaphore is released and control is returned to the interrupted program. The COLUMNS bits are all set to zero, the key depress interrupt is disabled and the key release interrupt is enabled prior to the return.

It turns out that it is next to impossible to get your finger off a key soon enough to allow this code to work without detecting the key release. The key release interrupt is set up in the key depress interrupt service routine. The main operation that takes place in the key release isr is to disable the key release interrupt and set up for a key depress interrupt. Without this approach, each time the key is depressed, it will get through the isr several times before the key is released and the buffer will be filled with duplicate values each time the isr is executed. The last seven lines of code in the isr above are executed in the key release isr.

The decode function converts a value that represents the position of the depressed key on the keyboard to an ASCII representation of the character. The position data is an 8-bit unit with the upper 4 bits representing the column and the lower 4 representing the row. Only one column or row bit should be enabled at a time. More than one row or column is enabled if multiple keys have been depressed. The acceptable row and column values are 1, 2, 4, and 8. The circuitry fixes the row value of 1 corresponding to the bottom row with the value 8 as the top row. Also, the right-most column corresponds to column 1 and the left-most column corresponds to column 8. The inner rows and columns follow the pattern established above. A look at the keyboard has the top row, read from left to right, containing

'+', '7', '8', and '9'. The next row contains '-', '4', '5', and '6'. The third row from the top contains '*', '1', '2', and '3'. The bottom row '/', '0', '.', and '='. Remember that the right-most column is column 1 in the position and the bottom-most row is row 1. Therefore, one can build a matrix as shown below that converts the position of the key to its ASCII value. Now the positions must be indices into a two-dimensional array rather than the position values. The decode routine must convert the position values to indices.

The decode routine works on an array that contains several key positions. This data array is terminated with a character 0xff. The string is passed as a pointer to the beginning of the string. In the function, the value of the pointer is copied to another pointer to a type BYTE that is used in the program. A `while` loop is entered which will walk the length of the array that contains all of the key position data. This loop is terminated when the data 0xff is found in the input array. Each piece of input data is now split into two parts, the rows and the columns, and a switch/case statement is used to convert the 1, 2, 4, 8 values that the input data contains into the 0, 1, 2, 3 values used for indices into the two-dimensional array. In the event of a multiple key hit, a default value of 4 is passed to either the row or column. These values are tested for, prior to access to the `keys[][]` array. If a 4 is found in either case, the decode loop is skipped to the next character. This approach does not prevent errors in decoding, but it does keep control and allows the procedure to be restarted.

```c
static const BYTE keys[][4]=      {{'=','.','0','/'},
                                   {'3','2','1','*'},
                                   {'6','5','4','-'},
                                   {'9','8','7','+'}};
/* the key position decode routine */
static void decode(BYTE *s)
{
    int r,c;
    BYTE *p=s;

    while(*p !=0xff)
    {
        switch(*p & 0xf)
        {
            case 8:  r=0;
```

```
                    break;
        case 4:    r=1;
                    break;
        case 2:    r=2;
                    break;
        case 1:    r=3;
                    break;
        default:  r=4;
    }
    switch(*p & 0xf0)
    {
        case 0x10:   c=0;
                        break;
        case 0x20:   c=1;
                        break;
        case 0x40:   c=2;
                        break;
        case 0x80:   c=3;
                        break;
        default:     c=4;
    }
    if(r==4||c==4)
        break;  /* got a multiple key hit, skip it*/
    *p++=keys[r][c];
}
*p='\0';
data=savedata;
}
```

Lastly, after the character is read from the array and stored into the array, the loop is continued. When the loop is terminated, a null character is written to the next entry in the array. This character is a string terminator and the data can be used as such with any of the input/output functions. In the initialization routine, two values of pointers to the array a[] were saved. The value saved in data has been corrupted by now and it is necessary to restore it to its original value by assigning savedata to data.

Next, we have to test the code above. The following program tests the keyboard initialization and isr routines. To use these routines, you must first adjust the code in the program handler() and recompile it. This code must be linked to the following program. The address of the routine kb_isr() must be placed in the specified location in the vector table found with the handler() function.

This location is entry number 6 in the vector table. The `main()` function is broken into two sections, the initialization, and the applications section. In initialization, `inituart()` establishes connection with the serial I/O functions. The function `attach_semaphore('a')` connects a semaphore to this process. This semaphore is passed to the `kpinit()` function along with a pointer to the beginning of the array `a[]` and also its length. The address of the interrupt handler is placed into the autovector location by the `vector()` function call, and finally the fast interrupts are enabled when `Enable_Fast_Interrupts()` is executed.

In the `main()` application loop, the program waits for the release of the semaphore `sem` that was attached earlier. At this time, the keyboard handler has collected a buffer of data and has detected the input of an '='. Also, the decode has been completed so that it is possible to execute `puts(a)` to send the contents of the buffer to the serial port. After the data are written to the output, to continue operation, it is necessary to reattach the original semaphore if it is available and proceed. This particular program will remain in the inner loop forever reading inputs from the keyboard and outputting them to the serial port whenever an '=' button is depressed.

Usually at this point, a single listing of the whole program is included in the text. It is a duplicate of the various smaller segments shown above. Here I will refer to the program listing that is contained on the CD-ROM. It can be found under the directory `chapter8` and has the name `kbinit.c`.

Integrating Keyboard and Clock

The next piece of code will integrate the keyboard with the clock developed earlier. The purpose of this program is to show the ease of integrating these two programs and the use of more than one interrupt simultaneously with the interrupt handler developed earlier. The main change in the clock program will be to alter the `reset_time()` function. This new function will receive inputs from the keypad. The code developed in the previous section can be used, but it will have to be altered significantly to be of use in this program. Here, we will want to set up the keyboard input to interrupt when any key is depressed and send a flag to the executing program when one or two appropriate keys have been depressed. Let us increase the minutes when the '1' button is depressed and the hours increase when the '2' button is depressed. The program will need the interrupt service routine, but the recording of the data into a buffer and all of the attendant decoding is unnecessary here. We will use a single global variable to set when a correct key is depressed. That is all required for this application.

Let's start with the initialization routine. This initialization routine needs to do even less than was needed earlier. Consider the code:

```
/* key board initialization routine */
void kpinit(void)
{

    KPCRN.LOROWS=0xf;     /* enable keypad rows */
    KPDRB.COLUMNS=0x0;    /* write 0 to KPDR[15:8] */
    KPCRN.LOCOLS=0xf;     /* make cols open drain */
    KDDRN.LOCOLS=0xf;     /* make cols outputs */
    KDDRN.LOROWS=0x0;     /* and rows inputs */
    KPSR.KPKD=ON;    /* clear KPKD status flag */
    KPSR.KDSC=ON;    /* clear key depress synchronizer */
    KPSR.KDIE=ON;    /* set KDIE bit to enable depress
                        Interrupt */
    KPSR.KRIE=OFF;   /* disable release interrupt */
    FIER.EF6=ON;     /* enable Fast Interrupt number 6*/

}
```

The main change observed is that no parameters are passed to the initialization function and none of the variables saved earlier are needed here. Otherwise, this function is identical to the earlier kpinit() function.

The changes made in kb_isr() are instructive. There is no substantive change in the function until after the depressed key has been detected. For this program, we are interested only in the detection of a '1' or a '2'. These digits correspond to the position values 0x42 and 0x22 respectively. After the key has been detected, a simple case statement converts these specific values to the proper digits. If the position numbers are not one of the expected values, *data is assigned a value of zero. This value is used by the main() program to determine when a key has been depressed.

```
/* key board interrupt service routine */
void kb_isr(void)
{
    UHWORD column,input;
```

```
static BYTE c;

if(KPSR.KPKD)   /* if a key has been depressed */
{
    for(column=COL_0;column!=0xf;column=(column<<1|0x1) & 0xf)
    {
        KPDRB.COLUMNS=column;
        /* kill some time here */
        if(KPDRB.ROWS!=0xff)
        {
            c=(~(KPDRB.COLUMNS<<4 |
                (KPDRB.ROWS&0xf))) & 0xff;
        }
    }
    if(c==0x22)
        *data='2';
    else if(c==0x42)
        *data='1';
    else
        *data=0;
    c=0;
    KPDRB.COLUMNS=0X0; /* write 0 to KPDR[15:8] */
    KPSR.KDIE=OFF;  /* disable key depress interrupt */
    KPSR.KRIE=ON;   /* enable key release interrupt */
}
else                    /* key release was detected */
{
    KPSR.KPKR=ON;
    KPSR.KPKD=ON;   /* clear key release and depress */
    KPSR.KRSS=ON;   /* set key release synchronizer */
    KPSR.KDSC=ON;   /* clear key depress synchronizer */
    KPDRB.COLUMNS=0X0; /* write 0 to KPDR[15:8] */
    KPSR.KDIE=ON;   /* enable key depress interrupt */
    KPSR.KRIE=OFF;  /* disable key release interrupt */
}
}
```

The code executed by the isr when a key is released is unchanged from earlier.

Adding a Display

Access to the LCD port is through a memory-mapped register. This register contains two byte-wide fields, one called COMMAND

and the other DATA. This register is at address 0x2c3ffff0. The following `typedef` and `#define` makes this location available to the program.

```
typedef struct {
    BYTE COMMAND      :8;
    BYTE DATA         :8;
} Lcddrv;

#define LCDDRV (*(Lcddrv *)(0X2C3FFFF0))
```

Access to this memory area is through Chip Select number 3. This Chip Select must be set up and enabled. The following initialization function allows access to the above memory address. The chip select control registers are not set to a specified value at reset. It is best to put a specified value into this register, 0, and then set the necessary bits. The first two lines of code in the following function put a zero into the address `CS3CR`. The fields in the chip select control register allows insertion of wait states from 0 to 15. Fifteen was used here because the memory location is rarely accessed; when accessed, it deals with the outside world, and the long wait state will not appreciably degrade the computer performance. An extra dead cycle is added when there is a write to this address. The access is to a 16-bit port. The last instruction enables the chip select number 3.

```
/* set up memory to access the LCD */
void initperip(void)
{
    UWORD *CS3CRX=(UWORD *)&CS3CR;

    *CS3CRX=0;          /* zero the whole register */
    CS3CR.WSC=3;        /* 3 wait states */
    CS3CR.EDC=ON;       /* extra dead cycle on write */
    CS3CR.EBC=ON;       /* enable byte write access only */
    CS3CR.DSZ=2;        /* 16 bit port */
    CS3CR.CSEN=ON;      /* chip select enabled */
}
```

There are two output functions that are useful. These functions are:

```
/* Routines for the LCD Display */
/* send a command to the LCD */
void LCDCommand(BYTE command)
```

```
{
    LCDDRV.COMMAND=command;
    delay(25);
}

/* send data to the LCD */
void LCDData(BYTE data)
{
    if (data=='\r')
        LCDDRV.DATA=0X01;
    else
        LCDDRV.DATA=data;
    delay(25);
}
```

These functions do essentially the same thing. The main difference is that the Command function writes to the upper byte of the specified address and the Data function writes to the lower byte of this address. One other special operation has been put into the Data function. If the data string contains a '\r' character, it is intended that the cursor should be returned to the beginning of the line. The 0x01 command does just that. Otherwise, the data received is sent to the screen for display. It is recommended that a 25-ms delay be executed after each write to the LCD system. This delay is included in both of the above functions.

A sequence of events is needed to bring the LCD display into operation. The code below first executes the initialization routine above so that the LCD display is connected to the computer. There immediately follows a repeated execution of the LCD command 0x03 three times. These instructions prepare the LCD display for operation. There follows a series of commands: set the display for 2 by 40 character display, turn the display off, clear the display and move the cursor to the home position, automatically increment cursor position after each write, turn the display and cursor on, and finally shift the cursor right after each input. These commands are all executed by the following code.

```
/* initialize the LCD */
InitLCD()
{
    int i;
    initperip();         /* initialize memory block */
    for(i=0;i<3;i++)
```

```
LCDCommand(0x03); /* get things turned on */
LCDCommand(0x3c); /* 2x40 display */
LCDCommand(0x08); /* display off */
LCDCommand(0x01); /* clear display and home cursor */
LCDCommand(0x06); /* increment cursor position */
LCDCommand(0x0f); /* Display and cursor on */
LCDCommand(0x14); /* Shift cursor right */
}
```

The function `LCDData()` shown above performs the essential operation of the `putchar()` that we have used so often above. In our program above, we used `putchar()` everywhere. It seems that we should have a separate command for the LCD display, but at the same time, it would be desirable that we could write to the LCD with a `putchar()` and not have to worry about changing all of the occurrences of `putchar()` in the earlier code. A way around this problem is to use a simple macro

```
#define putchar(a) LCDData(a)
```

and now we can use `putchar()` in place of `LCDData()`. Of course, it is important that we do not link `serial1.o` to the program, and the inclusion of `serial.h` is no longer needed.

One item that must be considered here is the use of the function `delay()`. Recall that `delay()` uses the pit to clock the delay time. The function in Listing 8-1 turns the pit on at the beginning of the operation and off at the end. We cannot allow this function to turn the pit off because the pit is being used by the basic clock function. Therefore, the two lines of code

```
ITCSR.EN=ON;
.
.
.
ITCSR.EN=OFF;
```

must be removed from the delay routine. The function now will work correctly, and it will not interfere with the operation of the basic pit operation used by the clock. This modified function is `delay2.c` on the CD-ROM.

Usually, a complete listing of the program is included in earlier chapters of this text. In this case, the code is approximately 250 lines long, and it is not so important to see this amount of code. It is all shown above in bits and pieces. The complete listing is found in `newclock.c` found on the CD-ROM in the Chapter 8 directory.

Summary

The developments in this chapter show an incremental approach to producing a program. The whole project was understood at the beginning of the chapter, but rather than try to develop the whole project, a series of small tasks were developed and tested. The complexity of each of these individual tasks was minimal. Often, the code required for each portion was but a dozen or so lines long. Such an approach is often the best approach to the development of complex code. This approach is not mine—it has been around as long as I can remember.

With any new project, you should analyze the whole project and start the development by breaking the project into tasks. Then, break into subtasks and repeat the process until the tasks are so simple that the code to implement them is trivial. Then step back a level and repeat the process, making certain that the code implementation is as simple as possible at all times. As this code is integrated, each module should be compiled and tested. Therefore, your program will be built of tested modules. Never attempt to write a big block of code without first testing the various small blocks that comprise the whole program.

Index

!=, not equal operator, 17, 26
#, preprocessor command identifier, 2
#asm, 152
#define, 16, 431
#endasm, 152
#pragma, 115, 349, 418
%, modulo operator, 24
%d, integer format, 5, 7
&&, and operator, 26
&, address-of operator, 65
*, multiplication operator, 24, 66
-, subtraction operator, 24
/, division operator, 24
@far, 295
@port, 295
\?, escape sequence for question mark, 15
\\, escape sequence for backslash, 15
\', escape sequence for single quote, 15
\", escape sequence for double quote, 15
\a, escape sequence for bell character,15
\b, escape sequence for backspace, 15
\f, escape sequence for form feed, 15
\n, escape sequence for new line, 2, 15, 409, 411
\ooo, escape sequence for octal number, 15
\r, escape sequence for carriage return, 15, 409, 411
\t, escape sequence for horizontal tab, 6, 15
\v, escape sequence for vertical tab, 15
\xxx, escape sequence for hexadecimal number, 15
{, block or compound statement beginning indicator, 2
| |, or operator, 26
}, terminator of compound statement, 3
" ", file name delimiter, 2
+, addition operator, 24
<, less than operator, 26

<=, less than or equal to operator, 26
= =, equality operator, 26
>, greater than operator, 26
>=, greater than or equal to operator, 26
15-bit timer, 130, 166
16-bit timer, 130, 166, 178-181,
programming, 186-195

A

abs(x), *56*
accessing files, 110
acos(x), *117*
ADC, 132-133
analog-to-digital converter (ADC), 132-133, 195-201, 288
AND, *26*
ANSI standard, 294
argument, function, 51
arithmetic logic unit (ALU), 124
arithmetic operators, 24-25
arithmetic shift, 29
array boundary checking, 19, 74
array index, 19
array initialization, 80
array, 18-19
array, multidimensional, 80-87
ASCII characters, 352
asin(x), *117*
assembly codes callable by C6805, 150
assembly language for MC68HC12, 388
assembly language for MC68HC16, 328-332, 338, 345
assembly language, 152
assignment operators, 32
association, 7

associativity of operators, 34-35
asynchronous time service, 251
asynchronous, 126
atan(x), *117*
atan2(x,y), *117*
auto, *9*
automatic variables, 12-13
autovector interrupt handler, 416-417
autovector, 414
Axiom demonstration board, 394

B

background debug mode, 147
BAUD register, 276
BCD encoding, 207-209, 352
binary operator, 26
binary tree sort, 95
binary tree, 238
bit field, 107, 108-109
bit manipulations, 108
bitwise AND, 28
bitwise operator, 28
boot FLASH memory, 348
boundary checks, 74
break keyword, 9, 48
bubble sort, 74

C

C6805 compiler, 150, 295
calloc, *115*
case *keyword, 9*
cast operator, 28, 114
char, *9, 15*
character constants, 15-18
character tests, 117
chip-specific routines, 431
circular convolution program, 341
clear interrupt flag routine, 426
CLI *instruction, 174*
clock program for MMC2001, 419, 429-431
clock, 297
coding tips, 137-148
comments, 17

compilers, 52, 63, 150, 221-230, 305-318, 386, 393, 414
compound statement, 3
computer operating properly (COP), 131, 167, 219
concatenation, 74
conditional expression, 33
const *keyword, 9-10*
constant, 8-11
continue *keyword, 9, 48*
controlling DC motor, 254
conversion commands, 112
convolution, 332, 334, 341
cos(x), *117*
Cosmic compiler, 221-230, 285, 293, 295, 305-318, 335, 349, 386
counter register, 184
CPU16 core processor, 288, 289-296

D

data compression, 237-245
data storage memory, 159
date function, 81, 119
DC motor control, 255-275
debouncing, 255, 275
debugging programs with user-specified inter-rupts, 294
declaration statement, 4
decrement operators, 30-32
default *keyword, 9*
deference operator, 66
definition statement, 4
delay routine for MMC2001, 395-397, 401, 436, 446
development boards, 136
development environment for microcontroller programming, 134-137
Diab compiler, 393, 414
diagnostics, 119
digital input/output, 131
digital signal processor, 287
digital-to-analog converter (DAC), 209
Dirac Delta function, 335
display, LCD, 443-446
do *keyword, 9*

do/while *construct, 39-42*
Do_interrupt( ), *417-418*
dot product, 332
double *keyword, 9, 28*
DSP operations with microcontrollers, 326-345
　　M68HC16 and, 326-345
DSP register model, 327
dynamic debounce, 275
dynamic memory allocation, 101

E

EBDI, 394
EEPROM, 128, 153, 155, 159
　　Programming, 159
EK register, 290
else *keyword, 9*
enum, *9, 20-22*
erasable programmable read-only memory
　　(EPROM), 127, 153-154
escape sequence, 15
exception vector table, 290-291
exceptions, 125
exclusive OR, *28*
exit, *60*
expanded bus, 155
expression, 24-34
Extended Background Debug Interface, 394
extern *keyword, 9, 15*
external static variable, 13-14, 58

F

factorial, 62
fast interrupt enable register (FIER), 414
FFI *instruction for MMC2001, 414*
fgets, *111, 413*
Fibonacci number, 62, 77
FIER register, 426
filter, 332
FIPND interrupt pending register, 414
flags, 113
FLASH memory, 128, 153
float *keyword, 9*
floating point variable type, 4, 11

for *keyword, 9*
for *loop, 38-39*
FOREVER *loop, 174, 427*
formatting, 113
Fourier transform, 333
fp, *110*
fputs, *111*
fractions, handling, 25
free( ), *115*
function argument, 51
function prototype, 52, 432
functions, 8, 51-61

G

general purpose timer module (GPT), 288, 314
general purpose timer, 129
generic pointer, 69
get string *functions, 413*
getce( ) *function, 411*
getch *function, 413*
getchar, *17, 382-383, 409, 413*
gets *function, 72, 413*
getse *function, 413*
goto *keyword, 9, 48*
"greater than" test vs. "is equal to" test, 423

H

handling interrupts on MMC2001, 413-419
Harvard architecture, 124
HC11E9.H header file, 258
hc16.h header file, 290
header files, 116, 171, 211-221
hexadecimal numbers, 11
Hoare, C.A.R., 231
Huffman code, 237-244, 350, 356, 361-368

I

IARB field, 302
if *keyword, 9*
if/else, *42-44*
if-else if *sequence, 44-48*

include, *2*
inclusive OR, *28*
increment operators, 30-32
index registers, 290
index, of array, 19
initialization section of program, 188
inituart( ) *function, 409*
input capture, 130, 185
    MC68HC11 family, 253
input/output, 110, 129-134, 382-386
    memory-mapped, 129
int, *2, 9, 11*
integer variable type, 3
integrating keyboard and clock, 440-443
inter-modual bus (IMB), 288
interrupt controller, 414, 419
interrupt flag clear routine, 426
interrupt handler routine for MMC2001, 415-417, 425
interrupt pending register, 414
interrupt request, 125
interrupt service routine, 125, 252, 294, 415, 425-426
interrupt source register (INTSCR), 414
interrupt vectors, 290
interrupts, processing, 220, 301-302
    for MMC2001, 413-419
IRQ, 125
isalnum(c), *117*
isalpha(c), *117*
iscntrl(c), *117*
isdigit(c), *117*
isgraph(c), *117*
islower(c), *117*
isprint(c), *117*
ispunct(c), *117*
ISR, 125, 220
isr_function, *418*
isspace(c), *117*
isupper(c), *117*
isxdigit(c), *117*
ITADR, 395, 402, 420, 425
ITDR, 395, 402, 420, 425
ITIE flag, 426
ITIF flag, 426

J

jsr R2 *instruction, 418*

K

kbhit( ) *function, 411*
KBSR register, 432
KDDR register, 432
keep, *14*
key depress synchronizer, 434
key release interrupt, 434
keyboard interface for MMC2001 chip, 432-440
keywords in C, 9
KPCR register, 432
KPDR register, 432
KPKR register, 432
Kroniker Delta function, 333

L

label, 48
last in, first out (LIFO), 59
LCD display routines, 445
left shift operator, 29
letter analysis program, 359
library functions, 80, 116, 223
libserio.a *archive library file, 413*
local static variables, 13
log(x), *118*
log10(x), *118*
logic analyzer, 137
logical AND, 26
logical operator, 26
logical OR, 26
logical shift, 29
long, *9-10, 28*
longjmp *function, 49*
look-up table with slopes, 320-322
looping construct, 6
    eliminating, 325
lvalue, *69, 85*

M

M68HC08, 149
MAC multiplier input registers, 326
macro definition, 56, 60, 102
main( ), *1, 2, 3, 51, 427*
malloc, *12, 101-102, 114*
masked ROM, 127, 153
math functions, 117
MC68300, 288
MC68HC05 microcontroller, 142, 149
MC68HC05EVM, 152
MC68HC05EVS, 142
MC68HC11/12, 149, 211-286, 347-391
MC68HC16, 149, 287-296
MC68HC16EVB, 146
MCORE architecture, 131, 393-446
memory models, 336
memory allocation, 4, 10
memory management, 114-116, 156
memory types, 153
memory-mapped I/O, 129
microcomputer, 123
microprocessor, 123
MIX compiler, 386
MMC2001 microcontroller, 393
        macros for, 400
mmc2001.h header, 400, 407
mnemonics, 140, 422
modular arithmetic, 334
modular program development, 347-349, 391
monitor routine, 370-376
motor control routines, 255-275
multidimensional arrays, 82-83

N

names, 8
nested functions, 52
nested if statements, 422
Newton loop, 38
NIPND interrupt pending register, 414
noise spikes, 219
noisy switches, 255
normal interrupt enable register (NIER), 414
null pointer, 76, 98

null, *19, 60*
Number-to-character conversion, 424
numeric encoding/decoding, 352-356

O

one time programmable, 128
one's complement, 28
operators, 24-34
optimizing code, 325
OPTION register, 159
OR, 26
OTP chip, 128, 154
output compare, 130-131, 185, 246-253
output_time *function, 428*

P

P&E Microcomputer Systems, Inc., 146
page zero, 158
parameters, copies of, 52
periodic interrupt, 309-315
phone book program, 360
PIT (see also programmable interval timer), 314-
        315, 395, 402, 419-420, 425
pointers, 65-121
        and function arguments, 70
        array name as, 69
        assigning, 69
        comparing, 68
        incrementing/decrementing, 69
        null, 98
        subtracting, 69
        to functions, 84
        type void as applied to, 69
PORTA, *85, 109, 132*
portable code, 296, 404
pragma directives (C6805 compiler), 151
pragma, 151
precedence, 7, 30, 33, 34-36
preprocessor command, 2
print formatting, 113
printf, *2, 3, 5, 7, 18, 112*
printing routine, 378-380
program counter (PC), 289

program flow and control, 36-51
program memory, 159
programmable interval timer (PIT), 314-315, 395, 402, 419-420
programmable timer, 246-251
programming hierarchy, 347-348
prototype, function, 52, 432
PSR register, 400
pull( ), *60*
pulse width modulation program for MC68HC11, 245-251
pulse width modulation program for MC68HC16, 297-302
pulse width modulator (PWM), 201-207, 247-253, 255, 268
push( ), *60*
put( ) *function, 409, 411*
putchar, *111, 382-383, 411*
puts *function, 382, 384, 413*

Q

queued serial peripheral interface module (QSPI), 288
queued serial module (QSM), 305-308
quick sort, 231

R

RAM, 153
random access memory, 153
reading data from the keyboard, 371
read-only memory, 127
real time interrupt (RTI), 168, 176
realloc, *115*
recursion, 61-63, 96, 157
recursive code, 324
reed switch, 268
re-entrant function, 61
register, *9, 13*
relational operator, 26-27
reset function, 381-382
reset signal, 125, 169
reset time function, 428
return from interrupt, 126

return, *9*
rfi *instruction, 418*
right shift operator, 29
RISC microcontrollers, 324, 393-446
ROM, 127, 153
RTI, 168, 176
RTS, 176
rvalue, *69*

S

saving data to EEPROM, 376-378
SCCR1/SCCR2, 276
semaphore, 397-400, 403, 433
semicolon, use of in C, 53
sequence point, 11
serial communications control registers, 276
serial communications data register, 276
serial communications interface (SCI), 133
   MC68HC11 family, 275-285
   MC68HC16 family, 308
serial communications status register (SCSR), 276
serial I/O, 133
   with MMC2001, 404-413
serial peripheral interface (SPI), 133
serial port, 382
SET INCLUDE, *2*
setjmp *function, 49, 119*
Shell, D.L., 230
short, *9-10*
signals, 119
signed, *9-10*
sin(x), *117*
sinh(x), *117*
sizeof, *9, 34, 81*
Software Development Systems (SDS), 393
software watchdog, 301
sort, 74-76, 230-237
   bubble, 74
   entry, 230
   quick, 75, 230-237
   Shell, 75, 230-231, 234, 236, 356
square, *56*
SRAM, 288

stack pointer (SP), 289
stack, 59
standard I/O, 119
static variables, 13-15
static, *9, 58*
stdio.h *header, 2, 3, 17, 52, 116*
stdlib.h *header, 101*
STOP *instruction, 169*
storage classes, 12-15
strcat(s,t), *117*
strchr(s,c), *117*
strcmp(s,t), *117*
strcpy(s,t), *117*
string operations, 117
string, 2, 19
strlen, *72*
strncat(s,t,n), *117*
strncmp(s,t,n), *117*
strncpy(s,t,n), *117*
strrchr(s,c), *117*
struct FILE, *110*
struct, *9, 20, 23-24*
structure tag, 88
structures, 87-106
    pointers to, 88
    self-referential, 95
    types, 92
switch bounce, 255-258, 268
switch, *9, 20, 49-51*
synchronous, 126
SYNCR register, 299
system integration module (SIM), 288, 296-297
    chip selects, 298
    external bus interface, 297
    general purpose I/O, 298
    interrupts, 298
    reset and initialization, 298
    system clock, 297
    system configuration and protection module, 297

T

table look-up, 318-325
tag, 88

talloc, *97, 102*
tan(x), *11*
TCNT, *246, 254*
TCR, *168-169*
telephone book function, 348
terminal emulator, 145
testing philosophy, 348
testing using evaluation boards, 142
TFLG1/2 *registers, 304*
time of day (TOD) clock, 419
time-of-day (TOD) clock, 131
timer control register (TCR), 181
timer counter register, 246
timer processor unit (TPU), 130
timer status register (TSR), 182
timer subsystems, 129-131, 166-173, 245-288, 419
TOF, *168*
tolower(c), *117*
toupper(c), *117*
TPU, 130
trace buffer, 136
type conversions, 27-28
type declaration, 9-12
type, *9, 52*
typedef, *9, 93*

U

UART, 276
unary operators, 25
uninitialized interrupts, 415
union, *9, 20, 22-23, 107-108*
universal asynchronous receiver transmitter (UART), 276
unsigned, *9*
utility functions, 118

V

variable types, 4
variable, 8-10
vector assignment, 303
vector initialization routine, 293
vector table, 125, 387-388

void, *3, 9, 53, 86*
volatile, *9, 11, 227*
Von Neumann architecture, 124, 129

W

wait routine, 409
WAIT, *169*
watchdog timer, 131, 301, 419
weighting functions for digital filters, 333
while, *9, 17, 36-38, 72*